THE LOGIC OF SCIENTIFIC REVOLUTIONS

THE LOGIC OF SCIENTIFIC REVOLUTIONS

Chris Glynn

CheckPoint
Press

The Logic of Scientific Revolutions
ISBN-13: 978-1-906628-34-5
Published by Checkpoint Press, Ireland

 CheckPoint Press
Books With Something to Say..

CHECKPOINT PRESS, DOOAGH, ACHILL ISLAND, CO. MAYO,
REPUBLIC OF IRELAND
TEL: 098 43779
EMAIL: EDITOR@CHECKPOINTPRESS.COM

WEBSITE: WWW.CHECKPOINTPRESS.COM

***To Sheba and Bully**
*whose devotion and adoration
have been a constant source
of inspiration*

Preface

It has often been said that the writings of William Shakespeare constitute the greatest works of literature in history and that the dialogues of Plato are the foundation of all modern philosophy and enjoy a similar accolade. Isaac Newton's *Principia* is frequently said to be the greatest scientific work ever published.

The great works of philosophy, from the seminal expositions of Descartes, Kant, Spinoza, Schopenhauer, Locke, Berkeley, through to, Hume, Hegel and the logical investigations of Frege and Russell, have all contributed significantly to what is sometimes called the 'Great Conversation'; the philosophical and scientific investigation of the nature of reality, knowledge and being.

Now that all these worthy tomes have been superseded, I should say something about the motivation for the following pages. When new ideas present themselves, or old ideas are seen in a new light, there is an overwhelming desire to commit such revelations to print, as quickly as possible, lest the train of thought be lost or disturbed in the course of time. Thus, many writers develop and expand their original thoughts through the process of articulation that serious writing demands.

Few academic authors write for profit or financial gain. It is sufficient that their works are made available to their peers for scrutiny and review. If the work is successful and monetary gain thereby ensues, then that is merely a fortunate bonus. Financial reward rarely warrants consideration in relation to the far nobler objective of increasing human knowledge or understanding.

Yet, in writing this book I have found that I have been able to depart from this primary motivation. If these pages are successful and many sales follow, then I shall not be disheartened. Financial gain, indeed great wealth, would be sufficient reward for my labours. In this respect, there is no reason why you should not purchase another copy of this book, or indeed several copies. Neither, of course, should you forget your friends and family, for this book will make an excellent Christmas, birthday or anniversary present and in the modern visual age of computers and television, it is important that your children adopt the habit of serious reading. Please do not neglect your children.

Thank you for your purchase.

Acknowledgements

For stimulating discussions and useful advice, I would like to thank myself. I am greatly indebted to myself for many of the original ideas contained in these pages. This work has, of course, taken some time to complete. For making that time available, I would like to thank myself. For the arduous task of meticulously preparing and typesetting a difficult manuscript, frequently revised by a temperamental and demanding author and for showing much patience during such adversity, I am deeply indebted to myself. Finally, for financial support during the preparation of this work, as well as continued encouragement, I shall be forever grateful to myself.

Notation and Conventions

We assume that we have a sufficiently rich symbolic language, such as the Predicate Calculus, with which to describe theories in whatever detail required. This will include constant symbols, relation symbols, function symbols, sentential and logical connectives and indeed whatever else is needed. We shall generally denote scientific theories by an upper case roman S, with or without a subscript; S_n. We shall denote axioms by lower case Greek letters, generally towards the end of the Greek alphabet, again with or without subscripts: φ_i, ψ_j, χ, etc. Variations on this convention will be made clear by the context. We also introduce special symbols and terminology where needed.

Contents

Chapter 1

Introduction

Scientific Theories may, at least in the fundamental sciences, appear to be sophisticated mathematical constructs that attempt to capture, to a greater or lesser extent, something of the workings of the Natural World. Whilst such theories may well be idealised, they have nevertheless enjoyed great success in the explication and prediction of natural phenomena. Thus, the evolution of science is greatly dependent on the development of theoretical models of various degrees of sophistication. The elucidation of the manner in which such models are constructed and how they themselves evolve can, therefore, throw considerable light on both the scientific endeavour and its methods and indeed, the role of human interaction in this grand epistemological enterprise.

We consider that it is the *construction* of scientific theories, whether based on observation or experiment, or indeed, even human ingenuity, that is the most significant factor in the evolution of science as a theoretical enterprise. This requires that any theory be sufficiently well-developed to admit of detailed analysis in the wholly pragmatic sense that clear foundations and detailed structure can be at once discerned without ambiguity. In particular, a theory must have *laws*; by which we mean precise formulations of the way in which the *variables* of the theory behave under certain restricted conditions. Such laws are, in the mature sciences, usually expressed as mathematical equations or relations, but this need not be so. It is quite possible to express the fundamental tenets of a theory in Natural Language and indeed this is often done. However, in many, or perhaps most cases, this would tend to be rather cumbersome and the benefit of mathematical expediency quickly becomes apparent. This becomes ever more so as theories acquire greater and greater sophistication and complexity. The laws of a theory, so described, are the fundamental premises of the theory and must, of course, be empirically verifiable in some sense or other.

These criteria immediately preclude certain 'ill-defined' disciplines that purport to be scientific in the analytical sense. We do not mean to say that such areas of enquiry are 'unscientific', 'underdeveloped', or even perhaps 'insufficiently mature' to warrant investigation. Rather, we would say that these disciplines are 'non-analytical' in the sense that no clear, well-defined laws are available, either for sustained empirical verification or proper mathematical scrutiny.

The traditional 'pure' sciences: Mathematics, Physics, Chemistry and Astronomy, are generally, to a greater or lesser degree, considered to be the most 'mature', or most developed of the scientific disciplines. We would now include Biology and perhaps Geology in this list and we might well exclude Mathematics. Of course, there are many sub-disciplines and interdisciplinary areas that would satisfy the somewhat restrictive criteria specified above: Biophysics, Biochemistry, Geophysics, Astrophysics, etc. The list is endless. All of these worthy areas of enquiry are clearly scientific and analytical. They enjoy laws, theories and models constructed within their respective domains of inquiry which are then empirically tested, or at the very least, are in principle, testable.

Few will doubt that of all these scientific disciplines, it is Physics that is the most fully developed; the most mature of the sciences and the most fundamental. With regard to this latter point, we may take a reductionist stance and argue that all; at least *physical* or *physico-biological* sciences are, ultimately, reducible to the principles of physics. It would then only be necessary to consider physics in our investigations into the processes by which scientific theories develop. The evolution of physical theories could then be used as a primary example in order to illustrate the evolution of scientific theories in general.

Thus, there is no loss of generality in adopting this attitude in relation to physics. Whilst we may seek universal principles that are specifically illustrated by physical theories merely in order to illuminate our understanding, by so doing, we are not excluding any other scientific discipline. On the contrary, such methodology may well clarify the nature of theories in those very sciences that we regard as being currently in a less mature state than theoretical physics. In fact, it is because physics is such a well-developed and mathematically rich science that we *must* regard it as the most appropriate model of theory development and the growth of scientific knowledge in general. Indeed, it is really the *only* area of scientific growth that has both a reasonably long and detailed period of historical development together with a sufficient degree of sophistication to enable us to discern the most important characteristics of the evolution of scientific theories.

Hence, by examining the structure of scientific theories with physics as our primary illustrative example, we hope to gain an insight into how such models are constructed in the first instance and subsequently develop and change with time, i.e. evolve. It is the *modification* of theories and, in particular, the *mechanism* of the modification process that can be expected to reveal the most significant insights into the nature of the evolutionary progress of scientific modelling of the empirical world.

It is this modification mechanism that is at once so obvious to the practitioner (of the Natural Sciences) and yet so elusive from the philosophical perspective. This is partly due to the fact that there would appear to be a psychological component to the very methodology of scientific investigation itself. However, this would be to make an incorrect assessment of the scientific method. Whilst it is certainly true that scientists are human, with all the psychological paraphernalia and baggage that may accompany that condition, it is not the *humanity* of the scientist that is at all relevant to the method of science *per se*. A non-human, but sentient and intelligent being, living on a planet in a distant galaxy, may well conduct scientific investigations into the natural world, in an attempt to discover the nature of physical laws and would, doubtless, arrive at the same universal laws that have been found by us hapless humans here on Earth. And all this in spite of such a creature being endowed with its own, very different psychology.

Alternatively, we may imagine a sophisticated robot diligently pursuing the scientific endeavour. It is blessed with powerful analytical tools and much curiosity. This is because these latter attributes were initially programmed into the device. Furthermore, this remarkable robot has the ability to learn from its findings and to adjust its program accordingly. The job of this artificial intelligence is nothing more than to extend scientific knowledge. In so doing, it follows the usual methods of science and with great success. Very soon, it reaches that state of advanced scientific knowledge that enables it to 'understand' the workings of the Universe. The robot has had no psychological baggage to hinder, or 'misdirect' its work; it has merely followed the scientific method and has been rewarded accordingly. It is solely because it is the *method* of science, rather than any psychological or subjective peculiarities of scientific practitioners, that our robot has been so productive.

Then, we cannot take seriously the notion that scientific research, conducted by the highly successful method of hypothesis, observation and empirical verification, is influenced by the arbitrary psychological precepts of the individuals involved in that pursuit of knowledge, as some have suggested. Rather, we take the view that the very methodology of science is governed by an underlying logic that compels and indeed forces objectivity. It is this hidden structure that, once illuminated, will demonstrate with the utmost clarity, the inevitability of the scientific method as the only way in which objective knowledge of the world may be obtained.

This business of understanding the way in which science progresses may at first seem to be somewhat paradoxical: How, we may ask, can humans not fully understand the nature of scientific growth when they themselves are the very instigators of that process? But, we have seen that the scientific method is not unique to humans. It is an independent epistemological process that may be employed by any intelligent being. Thus, there is no paradox. The underlying assumptions and principles by which theories develop may well be human and even unjustified, but they are guidelines only; they are not the empirical laws that ultimately dictate the form and content of the theory. Each sentient entity may, therefore, make use of its own guidelines and assumptions, but it is Nature herself that determines the outcome and the correct final form of a physical theory.

Hence, whatever philosophical stance one takes in relation to science hardly matters. The laws of Nature will persist regardless and will be inevitably descriptive of the true nature of the universe, whatever motivational assumptions are made to determine those laws. Therefore, one should take the empirical laws themselves as fundamental and regard the assumptions which guided the investigative process as more subjective.

This does not imply, however, that theoretical musings or philosophical speculation is of no value in the pursuit of scientific knowledge. Indeed, such intellectual reflection, if of a serious kind, has often in the history of science, led to profound insights into the workings of the Natural World on both the microscopic and macroscopic scales and has revealed complexities thereof that it may be thought, would have been otherwise unseen. Yet, it is undoubtedly also the case that historically, purely philosophical contemplations of the nature of the World - when devoid of empirical content - have in greater measure and in far greater abundance, ascribed to Nature notions and dictates that have subsequently been found to be far removed from the reality of the external world as revealed by the sensible observation of natural phenomena. Then, if by such speculation, the nature of the external world is made apparent in only a minority of cases, whilst the greatest successes and revelations as to the true structure of the World are achieved by empirical means, one is compelled to conclude that the former philosophical and introspective methodology is without scientific merit, whereas the latter entirely empirical and objective approach is at once both the more efficient and singular method of elucidation in relation to obtaining a more perspicuous and empirically verifiable understanding of the real structure of the World.

If this is so, then a detailed examination of the methods of science should reveal something of the reasons for the apparent success of the 'scientific method'. Such an analysis should also make it abundantly clear, in no uncertain terms, exactly why this relatively recent approach, specifically, the empirical approach to scientific investigation, is methodologically superior to any other alternative analytical or investigative procedure, at least in relation to the fundamental quest of ascertaining the apparent structure and fundamental nature of the external world. Yet, it may be argued, that no such explanation is necessary at all. Surely, one may contend, it is quite obviously the case that because the methods of science are such that observation of the World will evidently and most explicitly reveal the workings of the latter, then undoubtedly and in virtue of this fact, science *must* work - simply by default. However, such a response does not adequately address the question for two principal reasons. Firstly, it by no means follows that the empirical, observational method guarantees that the 'Laws of Nature' are correctly discerned by this technique, for clearly, there is the problem of perception. It is stressed, however, that the latter does not imply subjectivity, for that is a personal issue involving the interpretation of perceived phenomena in accordance with the psychological state of the individual observer. Rather, perception relates only to the physiological characteristics of the investigator who, for this purpose, can be regarded as just a piece of scientific instrumentation. Secondly, as the objection continues, the scientific method 'works', simply because it works and if it did not work, then, *ipso facto*, no discussion of this kind could take place, or need be entertained. Any other method *may* work equally well, but if it did, then that would instead be the primary method of our interest and thus the topic of the present discussion. However, that this is evidently *not* the case, demands explication. Such issues have been expatiated upon *as nauseam* in the philosophical literature over a significant historical period and we shall continue in this tradition in the sequel, hopefully, with a different and possibly entirely new perspective.

Then, it is not the investigative method *per se* that should be our primary concern in understanding the explanatory power of science, for the latter is merely the 'chosen method', which may or may not be due to the lack of the availability of a more superior approach, if such exists. Let us be clear, we do not say that the scientific method, so refined over the centuries and yet at once so simple, is not in any way invalid, rather, we merely emphasise that of all conceivable routes of inquiry, it remains the best and most efficient way for the acquisition of knowledge of the external world, irrespective of the nature of the organisms conducting such research and, moreover, although other methods may be available, any such sentient organism will eventually fall upon the scientific method just because of its outstanding success.

Accordingly, a clear and transparent exposition of the methodology and development of scientific theories within their various disciplines is of paramount and fundamental importance in understanding the progress of the Natural Sciences. Then, as already alluded to, the means or *mechanisms* by which such theories are constructed, by whatever agency and, in particular, the structure, formation and subsequent evolution of those scientific theoretical constructs so devised should immediately, in virtue of such a revelation, allow us to envisage a more perspicuous representation of the scientific method *per se*. Hence, it is apparent that consideration of those mechanisms of theory-change and evolution, thus isolated, will be of primary importance in the elucidation of the philosophical aspects of the scientific process. Yet, as we have mentioned, the philosophical nature of scientific laws, although involving to some extent the seemingly non-empirical assumptions upon which they are grounded, need not concern us in any great detail in our quest for enhanced comprehension of the methodology of theory construction and development, providing that we take empirically 'given' laws as not just fundamental, but as the sole basis or indeed, the initial foundation for a scientific theory. It is the adoption of this maxim that shall be our guiding principle in all our subsequent deliberations. Then, with such an empiricist stance, since any theory consists of a set of axioms, we are now to regard empirical laws as the essential *a priori* axioms for a scientific theory. Thus, for the purpose of analysing a given theory as a distinct entity, at least in a reasonably systematic fashion, we may disregard the philosophical assumptions upon which they are founded, for the latter are not part of Nature in this context and *a fortiori*, essentially irrelevant in relation to the practical pursuit of the scientific enterprise taken *en bloc* as the primary method of discovery of Natural Laws by empirical methods. However, this does not imply that such axioms and hence, the theories that are built from the latter, cannot be determined by other means, for indeed, this is sometimes the case; we are merely noting that if other methods than the scientific-empirical are utilised for such a purpose, then the axioms so obtained will nevertheless need to be empirically verified and thus, still subject to the scientific method and the dictates of Nature. But, this is not to say either, that the formulation of axioms from fundamental assumptions, or 'first principles', is without merit, for as noted, in the history of science, especially physical science, such a procedure has on more than a few occasions provided useful, quantifiable, insightful and indeed verifiable results. It is to say, however, that such assumptions, which have initially seemed a trifle vague and tenuously formulated, or even presumptuous, have at once also been essentially empirical in origin and therefore, only of a different and less evident ontological status than the full scientific laws or axioms to which they lead. Thus, again, empirical processes remain an essential prerequisite for the determination of the structure of the Natural World and once more, Nature is the final arbiter. This is necessarily so, whatever may be the biological or indeed mechanical characteristics of the investigative organism or entity so involved in the epistemological enterprise.

The implication here is that, since Nature determines the axioms of a theory - and one wonders how it could be otherwise - then any sentient being, in attempting to ascertain the gross structure of the World must, in that endeavour, adopt a procedure by which it is possible to discover the axioms by whatever methods are available. Speculation and philosophical contemplation may be useful, but ultimately, such musings must be testable, or verifiable in the empirical sense. The net result is that a set of axioms is obtained that reflect the behaviour of the Natural World. Then, these axioms or 'laws' constitute a scientific 'theory' that describes certain phenomena that have been observed and moreover, they are readily perceived to work via the criteria of testability, verifiability, etc.; essentially, application of the scientific method. Yet, it is clear that this is true only where such axioms can be reasonably applied, for unless the 'laws' thus posited or empirically observed, are so general as to embrace the whole of Nature, then they will be applicable to only a restricted area of the physical world. This is, of course, quite inevitable, for again, it is not to be expected that mere observation of the evidential properties of the universe by only the finite means available to any sentient, intelligent entity, will be so encompassing as to embrace all possible Natural Laws, or axioms that govern the entire domain of creation. Rather, a subset of limited and highly restricted extent of that vast domain is obtained in this way. Then, to each axiom-set and hence, to every theory so constructed, there will correspond a particular and specific subset of the far greater domain that pervades the entirety of the Natural World. The scientific quest is then to extend the domains of individual theories until they are of so great an extent as to contain axioms that will describe all possible natural phenomena: an all-embracing theory.

To describe or specify an axiom or 'natural law' such that the latter is to correlate with observed phenomena of the Natural World is at once seen to entail a multiplicity of concepts and apparently *ad hoc* assumptions of a methodological and epistemological kind of such disparate and complex ontology that even at the outset, must surely render any analysis of the processes involved insusceptible to precise delineation or categorisation. If the way of science is empirical, then any axiom or law so determined would seem to be dependent upon the methodology of that process and hence, such dependency would be transmitted to the principle participant in such scientific enquiry; namely, whatever sentient being or entity is investigating the natural phenomenon concerned, thereby introducing a participatory aspect to the process with consequent implications regarding the objectivity of the scientific endeavour itself. Yet, this cannot be so, for no specific attributes have been assigned to the observer other than mere sentience and minimal analytical and data-processing skills and these are qualities possessed by both organic entities and suitably sophisticated mechanisms alike, providing that motivation for the pursuit of knowledge of the external world is inherent in the former and programmed in the latter. Then, there can be no question of lack of impartiality or observer neutrality in this respect, for the gross facts of the universe are available to the investigator independently of his – or its - particular constitution or condition. Accordingly, it must be the case that what is observed is identical for all observers, for there can be no meaningful empirical science otherwise. The Laws of Nature are observer independent in their general form. The physical phenomena with which science deals are thus universal and hence, so also must be the observational and methodological techniques themselves by which they are elucidated, which is just the scientific method.

We are forced to conclude that whatever the nature of the physical or natural phenomena subjected to scientific and empirical investigation by observers of any kind whatsoever, they will ascertain the same fundamental attributes of such phenomena and identify the quantities and variables that constitute their axiomatic theories in a similar fashion. Hence, whilst the language employed by the one or the other may, and in general will differ considerably, the physical laws or axioms obtained by each will be of the same form and content, being expressed through essentially similar relations between the basic observationally identified variables. Then, we appear to be adopting the realist stance that there are gross facts of Nature that may be perceived and interpreted in a similar fashion by all observers. This does not, however, yet imply that there is an external world to observe at all, but merely that all observers reach the same conclusion in relation to their observational and experimental data and this suggests only that a mechanism exists that is either external or internal to such investigators.

If such a correspondence exists by which a sentient organic being or even an inorganic but technologically capable mechanism may perceive the substantive content of the physical and external world, then the source of such an interaction must reside within the very fabric of the universe itself and indeed, within the entities that so participate in the observational pursuits of that empirical exercise - which thus results from the dual participatory nature that defines the essential structure of the World and, *ipso facto*, necessarily incorporates the observer, for the latter are at once not divorced from the subject of investigation, namely, Nature Herself. But, such proximity or intimacy does not preclude objectivity or imply the non-reality of the external world.

On the contrary, it is entirely reasonable to assume that the physical phenomena observed by any entity of whatever nature may reveal the facts of the World simply because such information is available and susceptible to scrutiny by sentient beings who themselves necessarily accord with the structure of the Natural World merely in virtue of their own physicality or corporality. Then, the interpretation of the World will be little more than a simple matter since the interpreter is an integral part of that which is to be interpreted. In this way, any difficulties of a perceptual kind become superfluous and disappear within the context of an integrated whole, for now the usual distinction between the external world and the internal world of the observer no longer obtains. In effect, the essential structure of the universe is reflected in all its observable physical aspects, which must thus include that of any sentient investigator, but it is only in this sense that the observer is participatory. Hence, the great surprise would be that the Laws of Nature could *not* be so easily determined.

On closer examination, it is found that the observer interacts with the physical world through the empirical process of observation and experimentation which is conducted in a manner such that certain phenomenological qualities of Nature are thereby isolated in relation to the particular phenomenon under scrutiny. Specifically, it will appear to the observer that some attributes of Nature change in time whilst others do not. Hence, certain *variables* are identified as basic properties of the system under investigation and additionally, *constants* are observed as unchanging variables, i.e. variables that do not change in time. Then, two or more sentient beings observing the same physical phenomenon will be expected to identify the same physical variables that characterise that phenomenon. Thus, if we suppose that the physical process concerned is the uniform motion of a massive body, then we imagine all observers to identify a variable describing the *velocity* of the body in question as the (directed) rate of change of position and this interpretation will hold for all observers. If the process is the heating of a substance to the point of a phase change, let us say a liquid, then we anticipate that all observers will perceive a variable quantity which they will interpret as *temperature* that reaches a specific value when the liquid boils and also that there is another quantity indicating *pressure* that influences the temperature at which the phase change takes place, but that a *constant* will not exhibit such variability. This interpretation will hold for all observers. Similarly, for other, ever more complex physical processes that any number of sentient beings may observe, the same interpretations will hold.

It might be wondered how this is possible; that different entities in their interaction with the Natural World are able to ascertain similar descriptions of the physical processes taking place therein when they may be species so biologically diverse as to possess no other feature in common other than this mutual interpretation of the World. The answer, of course, has already been given. Such beings are *sentient* in that they have at their disposal a perceptual apparatus that allows them to discern the gross properties of Nature - of which they are themselves participants. Then, whatever names or additional attributes are ascribed to the natural phenomena, the essential variables will remain the same in virtue of the fact that it is not perception *per se* that is a property of any particular individual, but rather the *mechanism* of perception that may be special or peculiar to the entity concerned. Hence, all biological organisms or, indeed, robotic artificially constructed entities endowed with a sufficient degree of sentience will necessarily, via the empirical process, interpret similarly the well-defined quantities or variables describing the essential features of the universe in which they find themselves. This is a circumstance dictated by Nature Herself and we may refer to this interpretation of natural phenomena by all participating observers as the *natural interpretation.*

Thus, interaction with the external world, whether it be by machine or biological organism, requires that all perceptions of physical variables be interpreted in a natural way that accords with that empirical interaction through a perceptual process that is common to all participating observers because it would not, *ipso facto* be possible for it to be otherwise the case just because all such investigators form an integral part of the Natural World. Accordingly, the natural interpretation is, for practical purposes, essentially the empirical process itself.

It is this fact, common to all observers capable of sentient and intelligent or to some degree, purposeful interaction with the external world that must necessarily admit the formulation or construction of initial ideas by which perceived quantities so behave. Such a being will, by both observation and the application of the analytical faculties provided by its intelligence and diligence in the examination of the physical processes thus ascertained, doubtless construct a relationship of a concrete and formal kind from the quantities observed. Then, given that all such entities must similarly perceive the World, principally via the natural interpretation, it is also the case that the physical or *natural variables* so observed shall also be available to all inhabitants of a universe in which the natural interpretation obtains. Hence, in this way empirical relationships between such variables are ascertained. Axioms are therefore formed, which are universal in the sense that they are the subject of perception of all possible biological or indeed, non-biological entities that may exist in the Natural World with such a capability, for it is the uniform characteristic and universality of the natural interpretation that so demands the empirical determination of axiomatic relationships and their expression in terms of natural variables.

Then, given that intelligent observers may identify appropriate natural variables and by the empirical method formulate verifiable axioms between them, it becomes at once apparent that the empirical 'laws' thus formed are of necessity dependent upon the precision by which such variables can at any instance be measured. Hence, for any species of observers or scientific investigators, unless the available technology is such that perfect accuracy of experimental and observational measurement is possible, then the sophistication of the current scientific apparatus becomes a limiting factor in the determination of the axioms obtained by the empirical method and therefore, theories so elucidated may at a later time, as technology progresses, be subject to change. Thus, in the absence of infinite wisdom or perfect observational scrutiny, scientific theories must undergo modification or, perhaps, even total abandonment as knowledge of the external world is revealed in ever greater detail through the continued development of scientific instrumentation of a more sophisticated variety and capability than was hitherto extant. Evidently, the system of axioms by which a scientific theory is defined must evolve from an initial primitive state to one that in increasing degree may afford a more representative description of the Natural World - as the newly acquired empirical facts so demand.

Then, it is appropriate to inquire as to how such changes may in practice be effected and what means or mechanisms may be envisaged for transitions of this kind and also, if the nature of change or modification of scientific theories is at all quantifiable in any meaningful way in relation to the diverse character of the sentient beings concerned. Evidently, no modification or alteration of an empirically determined scientific theory can obtain in the absence of interaction with the external world, for the latter is the very foundation and driving force of the scientific method that thus operates exclusively under the natural interpretation. Hence, since a scientific theory is to be merely a set of axioms, modification must consist of the replacement or alteration of one or more of those axioms through the empirical process and is, perhaps contrary to expectation, therefore amenable to analysis. Hence, theory-change is essentially a discrete process in which axioms are altered either individually or *in toto*, together with the introduction of new natural variables.

We may therefore imagine a number of specific processes by which theory-change can be precisely described. Such processes may be conceived as a set of particular operations that act upon the axioms of a given scientific theory, thus changing or transforming the theory so that it accords more closely with the newly acquired empirical data obtained through the continued observation of the physical world and the phenomena therein encountered. In such a scenario, axioms may be removed, changed, replaced or added to a theory and new variables may be defined, always in every such instance respecting the empirical scientific method and the natural interpretation. Such operations are of a simple and discrete nature, for each individual process of this kind makes just one change to a scientific theory and a succession or repetition of similar operations may alter or modify a theory to whatever degree is required or demanded by the empirical dictates of Nature. Then, to a greater or lesser extent, theory-change becomes a purely mechanical process in which these particular operations or transformations play an 'atomistic' role in the sense that they each constitute a minimal possible alteration that a theory may undergo with respect to the mode and nature of the change characterised by the exact process involved, yet collectively may substantiate major alterations in the structure and content of the theory. The mechanistic nature of such operations admits of detailed analysis of the manner in which scientific theories evolve and reveals much concerning the empirical process in general. We may therefore refer to these discrete processes or operations as the *mechanisms of theory-change*, an expression that we shall frequently employ in the sequel.

Whilst scientific theories must change and evolve in order to encompass an ever greater understanding of the various physical processes that appear to operate in the external world, if such evolution is necessary – as presumably must be the case for finite beings, then theory-change would be expected to embrace ever more of those natural phenomena than was hitherto the case and the area within which a theory is applicable, unless it was incorrect in the first instance, would therefore be appropriately enlarged. Consequently, if a given species is to employ the empirical method of science, then at each stage of development the domain of the theory should be suitably increased and the very essence of theory-change can only be justified by the expansion of the *domain of applicability* of the latter so that the new or modified theory, thus obtained by the mechanisms of theory-change, may now describe a greater range of physical phenomena and perhaps, in a hitherto not envisaged manner. Therefore, as must be abundantly clear, the very purpose of the scientific enterprise is to increase knowledge of the Natural World, which is thus tantamount to a similar and proportionate increase in the domain of applicability of a scientific theory. Hence, the mechanisms of theory-change are of such a nature as to automatically accomplish that task and are also thereby dictated by the empirical requirements of Nature Herself under the natural interpretation. Then, these mechanisms, once described, together with the natural interpretation, will embody the scientific method itself and therefore illuminate in a more precise way the workings of the empirical process.

This continual process of theory modification by precisely defined mechanisms through interaction with the physical world is thus available to all sentient beings merely by their involuntary participation as extant entities within that realm - and no alternative mechanism could obtain other than the empirical and scientific one, for the natural interpretation forbids such a possibility because it requires the participatory nature of the intelligences concerned to be operative by definition, which can now be the case only in a Natural World in which such beings persist and who cannot therefore adopt a different perspective of the universe in which they find themselves. Essentially, the manner in which one may perceive the external world is determined by the universe itself and the fact that one exists within it. Then, the natural interpretation is nothing more than the empirical scientific method, which must therefore be construed as the only possible means by which knowledge of the external world may be legitimately obtained and no alternative investigatory method into the physical domain is either reliable or realistic.

Hence, the very constitution of the external world is intimately related to the observation process itself, which is then understood to be an inevitable consequence of the natural interpretation and which, by extension of that realisation but also by inclusion of the observable and phenomenological nature of the observation process by sentient beings with that capability, demands and constrains such entities to perceive that which is at once perceivable by such participation and involvement. Then again, by such criteria, the scientific method is the only feasible epistemological option in the pragmatic sense, for there is evidently no philosophical alternative with equivalent content.

The elucidation of the structure of the World over its vast domain cannot be all at once revealed for finite beings, but must be a gradual process of refinement in which scientific theories seek to explicate the phenomena of Nature at ever deeper levels, thus enlarging their application at each successive stage of development to a greater and more varied range of those natural phenomena. Thus, the primary goal of a scientific theory is to maximise its domain of applicability so as to encompass, in its final stage, the physical world in its entirety. And during the course of the evolution of a scientific theory, the mechanisms of theory-change play a principal role in the scientific method and the empirical determination of natural variables *via* the natural interpretation. But, such mechanisms operate at different rates that are dependent upon the specific investigators and their sociological and economic circumstances, and sometimes, little progress may be made because of constraints imposed by such societal factors, whereas at other times very rapid theory-change and development may occur. However, it is still the theory-change mechanisms that are operating in either case, yet when they act upon a scientific theory in quick succession the impression is given of revolutionary scientific development of varying degrees ranging from minor to major. Such upheavals in the scientific enterprise appear as scientific *revolutions*.

Accordingly, scientific revolutions may be categorised in accordance with the actions of the theory-change mechanisms which operate upon a given theory and generally there will be many ways that this can occur. It is this variation and the combinatorial nature of theory-change that admits of revolution categories and types and further admits of their precise description in terms of the mechanisms of theory-change in conjunction with the empirical process and the natural interpretation. Hence, scientific revolutions and the discrete operations by which scientific theories evolve are equivalent processes with apparent differences measured only by the magnitude of the outcome that is determined, not by the theories themselves, but by the sociological structure and resources of the investigating culture. Thus, scientific revolutions are a matter of perception, but not in any absolute sense that would afford special status to such seemingly spontaneous revelations of the structure of the Natural

World; but rather of the continuous time variability and yet discrete process of successive application of the mechanisms of theory-change.

In this way it is understood that scientific revolutions and theory-change may be described in the same way, for by equivalence they are identical processes. If the spontaneity of revolutions arises from societal factors then the absence of that influence should not only demonstrate this equivalence, but by so doing, eliminate any distinction of that kind. In such case, scientific revolutions and theory-change mechanisms would then be indistinguishable. And so it must be for those sentient beings with the available resources and negligible societal restrictions placed upon their scientific pursuits. Yet, this can never be the case for biological entities, for they evolve slowly and strictly in accordance with that which their immediate environment can provide and for the greater part of their evolutionary development, such resources are rarefied and only by co-operation may their needs be minimised. Thus, a societal structure together with the constraints that necessarily ensue therewith will inevitably be restrictive for such naturally evolved entities. Hence, the scientific investigators of this society will, as it progresses, consequently experience the phenomenon of scientific revolutions.

This will not be the case for an automaton, which is not so constrained. Such a non-biological entity has no conception of the societal limitations that result in an inability to pursue a given project to completion and nor would it care, for its sole objective is to follow the instructions of its basic programming in a sequential fashion. If it proves to be necessary for this device to halt its predetermined course, it will not be for reasons of economy or sociological temperament, but merely through lack of scientific data or suitable instrumentation to access any new data that would enable it to proceed and then terminate the current action as defined by its program before continuing to the next, probably iterative instruction. Hence, the time that passes between halting and subsequent resumption is of no consequence or relevance to this automaton, which is not endowed with temporal perceptions and should a time-difference or delay be of such magnitude as to necessitate, once the new data is obtained, a number of successive applications of the theory-change mechanisms, then that would not be perceived by such a machine as a scientific revolution, although to an external biological observer, it may well be so construed.

The ambitious goal of science is ultimately the complete explication of all natural phenomena that may possibly occur within the physical world at all times throughout the history of the universe and the question arises as to whether such a final objective can be achieved within a finite time, or indeed if it is achievable at all, even in principle. If such a Final Theory is attainable, then for biological entities there will be a series of scientific revolutions that eventually converges to that theory, but for the automaton as we have described it, there is nothing more than successive application of the mechanisms of theory-change. The difference in perception, as we have noted, is quite irrelevant, for the same process of theory-change is in reality adopted in both cases, because they are not intrinsically different. The question is, therefore, whether the sequence of operations of the theory-change mechanisms before a Final Theory is found is finite - and this is just a computational problem. But no such problem actually obtains, for a theory is no more than its axioms, which are indeed finite. Hence, the number of applications of the theory-change mechanisms before a Final Theory is realised is also finite as long as the axioms of the theory are not entirely replaced and the theory is not renewed on each such application.

Since the primary objective of the scientific enterprise is to extend the domain of applicability of scientific theories to eventually embrace all natural phenomena, then theory-replacement would be a possible, but inefficient method of attaining that goal. At each stage a theory is replaced and a new set of axioms obtained with the old theory being discarded and with every such renewal the domain of applicability increased, for constancy of the domain and certainly any decrease thereof would defeat the primary objective. The scientific automaton may well proceed in this manner, essentially searching in a random fashion for axioms with ever greater domains of applicability, for this machine has no temporal or sociological restrictions that may prevent it from doing adopting such a procedure other than those previously mentioned, which may be assumed to be ineffective in this hypothetical scenario. But, for sentient and intelligent biological entities with only a limited time-span available for scientific research, the mechanism of theory-change in the form of theory replacement is not a viable option and therefore, for such beings, alternative theory-change mechanisms are preferable. As before, the mechanism employed must be such that the domain of applicability increases.

Since each axiom of a theory is applicable within a certain area of scientific enquiry, it has its very own domain of applicability. It is thus an empirically determined Law of Nature that may be effectively and usefully employed only with respect to that domain and not outside of it. Newton's Law of Gravitation seems to be truly universal and would appear to encompass the entire universe, but if it is to be incorporated within a theory in which the remaining axioms have a lesser domain of applicability, then it is those axioms that will define the domain of the theory since, clearly, the domain of a theory cannot extend beyond the domain of the axiom with the least domain, for otherwise the theory would have a greater domain than at least one of its axioms, which is absurd if a theory consists only of its axioms.

Amongst the numerous mechanisms of theory-change, the one that is most productive in relation to domain increase is when two or more theories are *unified* into a single theory which then acquires the domain of applicability of all of the component theories that participated in the unification process. This unification mechanism is not as simple as other methods of theory-change because when two theories are combined into a single theory it is necessary that there be some appropriate manipulation of the axioms of both theories in order to ensure domain compatibility. However, when this is done, the domain of the resulting unified theory is suitably enlarged and this is the most efficacious technique as far as biological entities are concerned in the search for a Final Theory. It is therefore expected that a Final Theory will be a unified theory.

We shall, in the sequel, describe the unification process in much detail, but there is one point that needs to be clarified at the outset. Since each of the component theories that partake in theory unification have their own domains of applicability, then the domain of the resulting unified theory should be no greater that the domain of that component theory with the least domain. But a moments reflection reveals that this is not so, for although there are single axiom theories, theories are not axioms, thus when theories are combined or unified it will only be because none of them share any part of their respective domains and, indeed, there would be little point in attempting to unify theories *unless* their domains were disjoint, which is why the axioms of the component theories must undergo an additional transformation before unification becomes possible. Thus, unification requires the incorporation of the domains of disparate theories, including the domains of all the axioms of all the component theories, into but one, uniquely singular domain.

Yet, there is another important aspect regarding theory-change and the mechanisms by which that is accomplished. When a theory is altered, then at least one of the axioms of that theory is modified in some way and however minimal the change may be, the theory is no longer the same, even if none of the remaining axioms are altered - hence our speak of theory *change*. Thus, revision of any axiom of any theory may be regarded as the replacement of that theory by a new theory, or at least as modification of the old theory. Whilst such an interpretation is always possible it is not how we expect theory-change to be effected by the mechanisms that we postulate except in the case of complete renewal and the reasons for this are empirical. Essentially, if a theory or its axioms have been obtained via the natural interpretation and subsequently empirically verified, then it must be scientifically *valid* in some sense and therefore, to discard it would be to reject the Natural World itself, for it is the latter that determines the form and content of the theory, axioms, or the Laws of Nature to which all physical phenomena must of necessity conform. Thus, once again the concept of *theory evolution* arises - not by natural means - but rather from the limitations and restrictions of intelligent biological entities with a predilection for scientific discovery. The mechanical automaton, of whatever degree of sophistication, understands the mechanisms of theory-change and with great diligence employs them, probably to great effect, but it does not comprehend the reasons for those mechanisms or indeed, as we have noted, scientific revolutions. We shall consider variations of this theme in the sequel.

All of these notions need to be clearly defined, for otherwise one merely enters the realms of philosophical and metaphysical speculation, which ultimately compounds ambiguity and resolves nothing. The concepts that we have introduced in the present discussion require a degree of precision of definition that may be sufficient only in the pragmatic sense, but if that can be achieved, then it will also afford measure of clarity within a philosophical framework. This, then, is the program upon which we now embark. It is not thought that we will resolve any major philosophical or epistemological problems, such as they are, but it is anticipated that the adoption of a *definitional* approach to the scientific method and the general nature of that technique may offer a degree of precision that admits of some logical analysis or, at the very minimum, a more systematic study of those methods that is not so imbued with ill-defined philosophical concepts and metaphysical ideas that have hitherto and without good reason, hindered even ordinary linguistic analysis.

We shall briefly summarise the content of the book. The methods we employ in the following pages come largely from set theory, as will already have been anticipated by our adoption of a theory as consisting of a set of axioms. The important concept of the domain of applicability of a theory can then be defined as the intersection of the range of the axioms conceived of as physical laws. Then, the mechanisms of theory-change are introduced as operations upon the axioms of a theory that alter or modify those axioms in various ways. The procedure is simple and straightforward and allows us to consider scientific revolutions as a particular application of the theory-change mechanisms, as we have already mentioned. We then classify and discuss various types of revolutions according to the manner in which the axioms of a theory are modified or changed and identify three main categories of scientific revolutions. Examples of these are chosen from various scientific fields, but principally physics since the latter presents the most highly developed structures and we examine such theories in some detail in order to discover whether our definitions obtain. Along the way we treat various philosophical problems and, in particular, the problem of emergent properties and the reduction of scientific theories. In the last chapter, we discuss Final Theories and identify three types of these and consider their different natures and structure. It is probably this objective that has served as an important motivation for the present work and we come to some interesting conclusions. All of this means that we are compelled to make a number of definitions ranging over the whole of the subject matter and whilst we introduce each one of them at the appropriate point, we collect them in an appendix for convenience of reference.

The chapters in this work are meant to be read consecutively, for that is how this book has developed – gradually and accumulatively - like a scientific theory, so they are not therefore independent of one another. Dipping into a given section at random will not make sense, for each is crucially dependent upon the previous sections or chapters. There is some repetition and additional emphasis at various points and we often treat the same topic in a different way. We introduce many concepts and structures that as far as we know, are entirely new. It is hoped that readers of this work will be stimulated to develop these ideas further. If that proves to be the case, then this writer will be more than satisfied.

Chapter 2

Theories and Axioms

The notion of a scientific theory as a set of axioms is introduced. Physical laws are defined as the axioms of a theory and the relation to foundational issues is briefly addressed. The idea of the domain of applicability of a theory is developed and its relation to the axioms of a given theory, or several theories, using simple set-theoretical notions is discussed. The universal domain is posited as the domain of all possible scientific theories.

2.1 Axioms

By a *scientific theory* we shall simply mean a set $S = \{\varphi_1,......,\varphi_k\}$ of k elements. The φ_k are the *axioms* of the theory. Thus, the theory S is completely defined by its axioms. In fact, S is no more than just the set of axioms, φ_k. The axioms are, of course, a subset of all the well-formed formulas of the language. We identify the axioms with the *laws* of the theory. Thus, a theory is completely specified by its laws[*]. We require, of course, that a theory be consistent in the sense that $\forall \varphi_k \in S \; \neg(\varphi_k \wedge \neg \varphi_k)$. We discuss the logical properties of the language and axioms more fully in §5.2. For the present purposes, we simply accept theories as 'given' and do not need now to consider the model-theoretic construction.

It may well be the case that not all the axioms of a theory are independent; i.e., one or more of the axioms in a theory are derivable from the others. If this is not the case and all the axioms are independent, then we shall say that the theory is *minimal*. The number of axioms in a given theory S, or the *axiom number* of S, is just the cardinality of the set S, which we denote as usual by $|S|$ or $card(S)$. If a theory is minimal, then so is $|S|$. We shall generally require that our theories be minimal.

It is quite possible for a theory to have only one axiom. Suppose S has k axioms φ_k, then we may form the conjunction;

---

[*] What we have called here a 'theory' is usually termed a 'model'. What we have termed 'axioms' are then interpreted according to this model. We avoid interpretations for the moment and continue with this somewhat "non-standard" usage.

$$\psi = \varphi_1 \wedge \varphi_2 \wedge, \ldots\ldots\ldots, \wedge \varphi_{k-1} \wedge \varphi_k$$

thus there is now only one axiom. However, the new axiom ψ is not really new at all, for it is just the concatenation of the previous k axioms.

The axioms of a theory, in the sense in which we are using this term, are not the same as the *principles* of that theory. Newton's laws of motion make no mention of absolute space or time, or inertial frames of reference. Rather these notions are contained in the principles of Newtonian mechanics, of which there are two, (the existence of inertial frames in which motion is linear and invariance under translation, which is momentum conversation) as is well known. The formulation of the three laws of motion is a consequence of those principles. Here, our concern is only with the *laws* themselves, which we take as axioms for the theory, not the philosophical foundations upon which those axioms are based.

It might be thought that Newton's first law:

A body continues in its state of rest of uniform motion in a straight line unless it is compelled to change that state by impressed forces.

is just a special case of the second law,

The rate of change of momentum is proportional to the impressed force and is in the direction in which the force acts, i.e.

$$F = \frac{dp}{dt} = m\frac{dv}{dt}$$

because, by integrating the second law in the form,

$$\int_{v_0}^{v} dv = \int_{0}^{t} \frac{F}{m} dt$$

(where we have assumed constant mass) gives;

$$v = v_0 + \frac{F}{m}t$$

where v_0 is the velocity at $t = 0$. A second integration gives;

$$\int_{x_0}^{x} dx = \int_{0}^{t} \left(v_0 + \frac{F}{m} t \right) dt$$

thus;

$$x = x_0 + v_0 t + \tfrac{1}{2} \frac{F}{m} t^2$$

where x_0 is the position at $t = 0$. Thus, when $F = 0$, the motion is linear and of constant velocity, which is the statement of the first law.

Then, we would conclude that these two laws are not independent. However, such a deduction would clearly be incorrect because the statement of the first law actually embodies the fundamental principles, or assumptions of Newtonian mechanics. We must, therefore, always bear in mind this distinction between fundamental principles and the axioms (laws) of our theory.

Nevertheless, since our axioms are already *interpreted* as physical laws, in the model-theoretic sense, we can largely dispose of, or at least not be unduly perturbed by, the philosophical principles and assumptions that underlie the foundations of our theories. If it is the case that one or more of these principles is itself an axiom, then all the better (e.g. constancy of the speed of light *in vaccuo*, Boyle's law, etc). Usually, principles with this property are generally entirely empirical and not derivable from more fundamental assumptions within the theories in question. They are 'facts' of Nature and verisimilitude ensures that they must be incorporated into our theoretical models.

This is not to say that fundamental or foundational principles of a philosophical kind are irrelevant to our theories, merely that we need not (always) take them into account. Hence, we are relatively free from considerations of an ontological nature in relation to our axiomatic structures, as so defined. Thus, we take Newton's laws as the axioms of Newtonian mechanics, Maxwell's equations as the axioms of Electromagnetism, etc.

Therefore, we imagine that the axioms of a theory are expressed as physical laws in equations of the form $f(x_i, c_j) = 0$, in which there are i variables x_i and j constant symbols c_j, and clearly this is always possible. Then, the *variables* of our language are just the usual variables and constants of whatever discipline we are considering. In the case of physics, they would be such measurable quantities as mass, pressure, temperature, the speed of light, etc., often in fact, the *observables* of the theory. Thus, Newton's second law above would take the axiomatic form $f(v,t,m) = dp/dt - F = 0$. More specifically;

Definition 2.1.1
An *axiom* is an expression of the form $f(x_i, c_j) = 0$, in which there are i variables x_i and j constant symbols c_j.

The axiomatic structure of a theory may be altered in various ways, two of which might be to either remove an axiom from the theory, or to add a new axiom. However, if an axiom is added to a theory, then that axiom must serve a useful purpose; it must *do* something. Just adding an arbitrary axiom at random will serve no purpose. The axiom must be applicable to the theory. If an axiom is removed from a theory, then it should be because either it follows deductively from the other axioms and is, therefore, not essential in a fundamental sense, or it was never suitable in the first place. The former case would mean that the theory, prior to the removal of the axiom, was not minimal and the latter, that it was not applicable to the theory or is invalidated by observation or inconsistency.

2.2 Domain of applicability
There is an abundance of theories in physics, each with its own particular area of application, or specialisation; Classical Mechanics, Quantum Mechanics, Optics, Electrodynamics, Hydrodynamics, etc. and many sub-disciplines of these. Axioms in one theory may be quite inapplicable in another theory. On the other hand, one or more axioms of a theory may also be axioms of another theory. It may well be the case that two theories have no axioms in common; they are disjoint:

$$S_1 \cap S_2 = \varnothing$$

On the other hand they may have at least one axiom in common:

$$S_1 \cap S_2 = s \neq \varnothing$$

where $s \subset S_1$ and $s \subset S_2$ and s contains at least one axiom.
  Theories may also have *all* their axioms in common and are, therefore, identical:

$$S_1 = S_2$$

so that $S_1 \subset S_2$ and $S_2 \subset S_1$.

The question of *similarity* between theories therefore, arises; what is the relationship between theories with some axioms in common, or no axioms in common? Obviously, if two theories have identical axioms, then they are the same theory. If two theories have one axiom in common, then we may wish to know whether this common axiom allows us to assert that the theories are *similar* in some manner and in a way that is over and above the simple fact that each theory possesses the same axiom. We can continue to the case where theories have two common axioms; or three, etc., right up to the point where all the axioms are identical and the theories are the same.

Now, scientific theories are constructed in order to describe the Natural World; the phenomena that are observed in Nature. It is clear that if two theories contain common axioms, then each of them must, to a greater or lesser degree, describe similar phenomena. This means that these two theories are *applicable*, at least in part, to the same *domain* of the Natural World. It is this 'area of applicability' that constitutes, at least partially, the similarity between theories that we alluded to above.

Thus, theories have a *domain of applicability*, or just *domain*. We denote this by $\mathcal{D}(S)$ for a theory S, or if we are dealing with several theories, we add subscripts; $\mathcal{D}(S_n)$. There are three possible cases here. We shall say that theories are *compatible* if they have precisely the same domains, in this case, *all* of the axioms of one theory are applicable in the domain of the other theory and *vice versa*. Then there are no axioms of one theory that are not applicable in the other theory. If, however, *none* of the axioms of one theory are applicable in the domain of another theory, we shall say that the theories are *not compatible*, or *incompatible*. If *some* of the axioms of one theory are applicable in another theory, then we shall say that these theories are *partially compatible*.

Thus, two theories are incompatible if their respective domains are disjoint:

$$\mathcal{D}(S_1) \cap \mathcal{D}(S_2) = \emptyset$$

They are compatible if their domains are equal:

$$\mathcal{D}(S_1) = \mathcal{D}(S_2)$$

In this case, of course, $\mathcal{D}(S_1) \subset \mathcal{D}(S_2) \wedge \mathcal{D}(S_2) \subset \mathcal{D}(S_1)$. Note that $\mathcal{D}(S_1) = \mathcal{D}(S_2)$ does not mean that $S_1 = S_2$, or that these two theories have any axioms in common; domain compatibility is quite different from theory equality.

A theory S_1 is partially compatible with a theory S_2 if their domains are not disjoint:

$$\mathcal{D}(S_1) \cap \mathcal{D}(S_2) \neq \emptyset$$

Axioms too, have domains of applicability. Indeed, if we take a single axiom $\varphi \in S$ and form the set $S = \{\varphi\}$, then S is itself a theory and, therefore, has a domain. Now, given a theory S, each of the axioms φ of S must have at least the same domain as S, i.e. $\forall \varphi \in S \, [\mathcal{D}(\varphi) \supset \mathcal{D}(S))]$. This is because any φ that does not have the domain of S has no business being an axiom of S. The domain of a theory must be at least the domain of its axioms and one can hardly have an axiom that does not have the domain of its theory. Of course, an axiom can have a 'larger' domain than the theory that contains it, but that does not matter, for it still shares the domain of the theory. Hence, we may define the domain of a theory in terms of the domains of its axioms. The domain of a theory is the domain of the axiom with the smallest domain, so that if we are given any arbitrary theory $S = \{\varphi_1, \varphi_2, \ldots, \varphi_{k-1}, \varphi_k\}$, we shall have;

$$\mathcal{D}(S) = \mathcal{D}(\varphi_1) \cap \mathcal{D}(\varphi_2) \cap, \ldots\ldots\ldots, \cap \mathcal{D}(\varphi_{k-1}) \cap \mathcal{D}(\varphi_k)$$

If two theories have the same domain, then as we observed above, that does not mean that they are the same theory. Consider two theories $S_1 = \{\varphi_1, \varphi_2, \varphi_3\}$ and $S_2 = \{\psi_1, \psi_2\}$, if they are partially compatible then:

$$\mathcal{D}(S_1) \cup \mathcal{D}(S_2) = [\mathcal{D}(\varphi_1) \cap \mathcal{D}(\varphi_2) \cap \mathcal{D}(\varphi_2)] \cap [\mathcal{D}(\psi_2) \cap \mathcal{D}(\psi_2)] \neq \emptyset$$

which means that for some $\varphi \in S_1$ and some $\psi \in S_2$, we must have that $\mathcal{D}(\varphi) \cap \mathcal{D}(\psi) \neq \emptyset$. Hence, if two theories are partially compatible, then there is at least one axiom from each theory with intersecting domains. Clearly, this generalises to any number of theories.

Suppose now, that S_1 and S_2 have an axiom in common, say ψ_1, so that $S_1 = \{\varphi_1, \varphi_2, \psi_1\}$ and $S_2 = \{\psi_1, \psi_2\}$, then;

$$\mathcal{D}(S_1) \cup \mathcal{D}(S_2) = [\mathcal{D}(\varphi_1) \cap \mathcal{D}(\varphi_2) \cap \mathcal{D}(\psi_1)] \cap [\mathcal{D}(\psi_1) \cap \mathcal{D}(\psi_2)] = \psi_1 \neq \emptyset$$

and S_1 and S_2 are partially compatible. Thus, if two theories share an axiom, then they are partially compatible. Clearly, this generalises to any number of theories. Of course, if all the axioms of a theory are identical then the theories are the same and their domains are equal, but again we emphasise that the equality of their domains does not imply that they are the same theory.

We consider four special cases of partial compatibility that are significant as far as theory similarity is concerned. To this end, let $S_1 = \{\varphi_1, \varphi_2, \varphi_3\}$ and $S_2 = \{\psi_1, \psi_2\}$. Firstly, the axioms of S_1 may be applicable to only a subset of the domain of S_2. In this case the domain of S_1 is entirely contained within the domain of S_2. Here, the axioms of both theories are applicable in the domain of S_1 and nowhere else. This is illustrated in Figure 1 below:

Figure 1
The domain of S_1 is a proper subset of the domain of S_2. The axioms of both S_1 and S_2 are applicable in $\mathcal{D}(S_1)$.
$$\mathcal{D}(S_1) \cap \mathcal{D}(S_2) = \mathcal{D}(S_1)$$

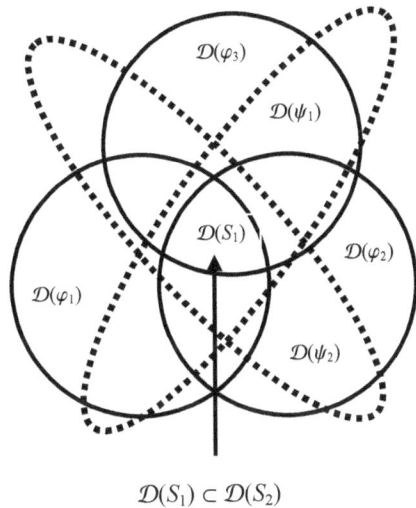

$$\mathcal{D}(S_1) \subset \mathcal{D}(S_2)$$

Secondly, the opposite situation may obtain. The axioms of S_2 are applicable only within a subset of the domain of S_1. Then, the axioms of both theories are applicable only in the domain of S_2 and nowhere else. This time, the domain of S_2 is entirely contained within the domain of S_1. This situation is illustrated in Figure 2.

Figure 2
The domain of S_2 is a proper subset of the domain of S_1. The axioms of both S_1 and S_2 are applicable in $\mathcal{D}(S_2)$.

$\mathcal{D}(S_1) \cap \mathcal{D}(S_2) = \mathcal{D}(S_2)$

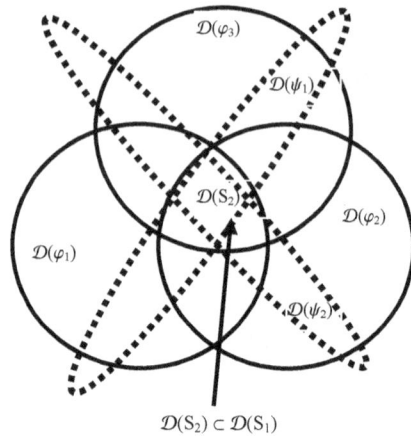

$\mathcal{D}(S_2) \subset \mathcal{D}(S_1)$

Thirdly, as a special case of the above, there is the important situation of domain equality. Here, S_1 and S_2 share the same domain. All the axioms of each theory are applicable within the domain of either theory. Figure 3 represents this.

Figure 3
The domain of S_1 is equal to the domain of S_2. The axioms of S_1 are applicable in all of $\mathcal{D}(S_2)$ and the axioms of S_2 are applicable in all of $\mathcal{D}(S_1)$, i.e.

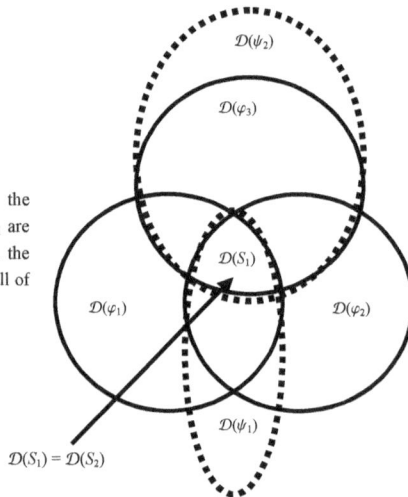

$\mathcal{D}(S_1) = \mathcal{D}(S_2)$

Finally, it may be the case that the axioms of both theories are applicable only within a subset of the domain of each theory. Thus, there is a subsets, as shown in Figure 4, where the axioms of both theories are applicable.

Figure 4
The axioms of S_1 and S_2 are both applicable only to a subset $ of both $\mathcal{D}(S_1)$ and $\mathcal{D}(S_2)$.
$\mathcal{D}(S_1) \cap \mathcal{D}(S_2) = $

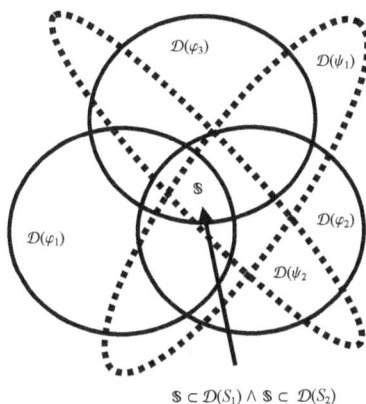

$ \subset \mathcal{D}(S_1) \wedge \subset \mathcal{D}(S_2)$

It may be the case that an axiom of S_1 is applicable to S_2, but that axiom is not a member of the set S_2. We must be careful to make the distinction between the domain of applicability of our axioms and the *presence* of those axioms in our theories. Observe that if a collection of theories have all their axioms in common, then their domains are equal. However, theories having equal domains are not necessarily equal theories. Also, theories can have no common axioms, but have equal domains.

Domains are *additive* in the set-theoretical sense. Given a set of n theories $S_1, S_2, \ldots\ldots, S_n$, the union of their domains is domain of their union;

$$\mathcal{D}(S_1 \cup S_2 \cup, \ldots\ldots\ldots, \cup S_n)$$
$$= \mathcal{D}(S_1) \cup \mathcal{D}(S_2) \cup, \ldots\ldots\ldots, \cup \mathcal{D}(S_{n-1}) \cup \mathcal{D}(S_n)$$

as one would expect if domains of applicability are to make any sense for multiple theories. We shall find later, however, that there are inherent difficulties in forming simple set-theoretical unions of the domains of theories in this way.

Of course, theories have sub-theories. Given a theory $S = \{\varphi_1, \varphi_2, \varphi_3\}$, any subset of S will be a theory, because a theory is completely determined by its axioms. The sub-theories of S are the following eight theories:

$$S, \{\varphi_1\}, \{\varphi_2\}, \{\varphi_3\}, \{\varphi_1, \varphi_2\}, \{\varphi_1, \varphi_3\}, \{\varphi_2, \varphi_3\}, \emptyset$$

We do not normally need to consider the empty theory, or S itself, but we include them here for completeness. Obviously, this is just the power set of S. Thus, given n axioms of a theory, there are 2^n sub-theories, including the null theory and the theory itself. Since each axiom has a domain, each sub-theory will also have a domain.

For example, the domain of the sub-theory $\{\varphi_1, \varphi_3\}$ of S above is simply $\mathcal{D}(\varphi_1) \cap \mathcal{D}(\varphi_3)$, in accordance with the foregoing. There may be many axioms and, therefore, many sub-theories of a given theory. We may, then, define a theory in terms of the domains of its axioms and in particular, the intersections of the latter, which define the domain of the sub-theories. Obviously, if $s \subset S$, then $\mathcal{D}(s) \subseteq \mathcal{D}(S)$.

Additionally, we may build a theory from axioms taken at random, provided the intersection of the domains of those axioms is not the null set. Thus, let us take six axioms: φ_1, φ_2, φ_3, φ_4, φ_5, φ_6 and consider the domains of these axioms. If we take the first three of these six axioms, we may form a theory $S_3 = \{\varphi_1, \varphi_2, \varphi_3\}$ and the domain of S_3 is;

$$\mathcal{D}(S_3) = \mathcal{D}(\varphi_1) \cap \mathcal{D}(\varphi_2) \cap \mathcal{D}(\varphi_3)$$

with similar expressions for other possible sub-theories. This is illustrated in arbitrary way in Figure 5 below for a number of possible intersections of the domains of the axioms.

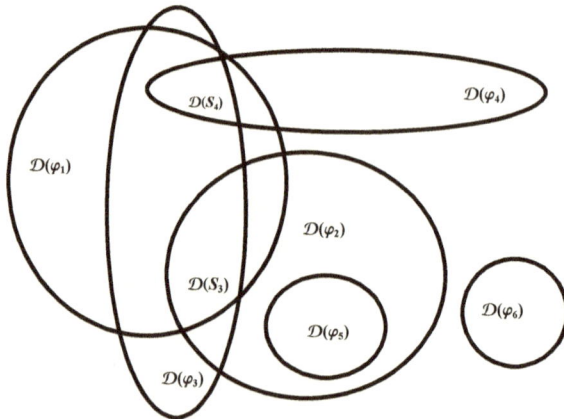

Figure 5

Note that one of the six axioms, namely φ_6, does not share its domain with any of the other axioms and is thus the domain of the theory consisting of only this axiom. We have also indicated in the figure the domain of the theory $S_4 = \{\varphi_1, \varphi_3, \varphi_4\}$ and the domain of the theory $S_4 = \{\varphi_1, \varphi_2, \varphi_3\}$. Note that compatibility applies only to the *domain* of the theory, not to theories themselves, as we have observed above. Compatible theories are not identical; they may be completely different.

An example of two compatible theories is the Ptolemaic description of planetary motion by epicycles and the alternative Keplerian-Copernican model. Both theories have the same domain of applicability, but differ widely in their formulations. Neither do they have any axioms in common.An example of two partially compatible theories is Newtonian mechanics and classical electrodynamics. Newtonian mechanics is only partially compatible with (classical) electrodynamics because Newton's third law is inapplicable to electromagnetic forces, or indeed, to any forces which propagate with finite velocities.

The domain of applicability of a scientific theory is the set of all those (natural) phenomena which fall under the embrace of the axioms of all the theory; i.e. for which the axioms are applicable. If, given a theory S, a (natural) phenomenon is found that cannot be described by the theory, then that phenomenon is not within the domain of the theory S. The domain of a theory is exactly those phenomena that can be described by the theory; the phenomena to which the theory is applicable. The detailed nature of domains of applicability will be seen to be fundamental in the sequel, wherein further developments will ensue.

We may summarise much of the above more succinctly as follows. We consider a 'universal set', N, the domain of which, $\mathcal{D}(N)$, is the 'universal domain', which is the domain of all possible valid scientific theories[*] and hence in particular, all axioms of such theories. The domain of any valid scientific theory, or Law of Nature, must be contained within the universal domain. The universal domain contains the domains of all possible valid scientific theories, so if $S \subset N$, then, of course, we must also have $\mathcal{D}(S) \subset \mathcal{D}(N)$. Thus, we may give a clear definition of domains as follows:

Definition 2.2.1
Given any axiom $\varphi = f(x_i, c_j)$, the domain of φ, $\mathcal{D}(\varphi)$, is precisely that subset of the universal domain $\mathcal{D}(N)$ for which the expression $f(x_i, c_j) = 0$ holds.

---

[*] *Not*, of course, the set of all sets, which does not exist.

Thus, for any axiom φ, we always have $\mathcal{D}(\varphi) \subseteq \mathcal{D}(N)$. Then for theories we have:

Definition 2.2.2
Given k axioms, φ_k, of a theory S, the domain of S, $\mathcal{D}(S)$, is the intersection of the domains of each of the axioms φ_k;

$$\mathcal{D}(S) = \mathcal{D}(\varphi_1) \cap \mathcal{D}(\varphi_2) \cap, \ldots\ldots\ldots, \cap \mathcal{D}(\varphi_{k-1}) \cap \mathcal{D}(\varphi_k)$$

And this is, therefore, evidently the smallest subset of the universal domain for which the axiom system;

$$\varphi_1 = 0$$
$$\varphi_2 = 0$$
$$\because$$
$$\varphi_{k-1} = 0$$
$$\varphi_k = 0$$

holds, as we have noted above. Observe that if $S = \{S_1, \ldots\ldots, S_n\}$ is the set of *all* current scientific theories, then the complement of the domain of N with respect to the domain of S, i.e. $\mathcal{D}(N) - \mathcal{D}(S)$, represents those areas of scientific investigation for which there are no corresponding theories.

Definition 2.2.3
By the *axiom of least domain* in a theory S we shall mean that axiom with domain such that its inclusion in the theory minimises the domain of the theory to its current domain and the removal of which from the theory results in an increase in the domain of the latter.

We shall need to discuss further the domain of applicability and those of the axioms of a theory in the next chapter.

Chapter 3

Theory Change and Adaptation

The manner in which scientific theories change over time is discussed. Mechanisms for this process are postulated and considered in detail. Specific examples of the processes involved are drawn from the sciences. The important concepts of 'axiom combination' and 'theory unification' are introduced and elaborated upon in connection with earlier ideas.

3.1 Theory change

We do not yet possess a Final Theory, a Theory of Everything, in the physical sciences. Whilst that Holy Grail of physics seems to be close at the beginning of the twenty-first century, it has nevertheless proved elusive. Hence, until such time when this goal is achieved, scientific theories are never static because they are never complete. Theories are altered, changed, added to, unified and even abandoned. New theories are constructed and old theories modified. Empirical evidence, being the dictate of Nature, is always the arbiter in this respect. All this, of course, is part of the scientific endeavour. But precisely *how* this is done, the mechanism of theory-change, is of considerable interest and in principle, should greatly illuminate the epistemological nature of the scientific method. We address this question now. Clearly, a scientific theory may evolve by many mechanisms. New axioms may be added to a theory, with the effect of extending its domain of applicability. Axioms may also be removed from a theory because they have been 'falsified' empirically, or they are already consequences of the other axioms of the theory. Some axioms of a theory may be replaced by more apposite physical laws, that better accord with the observational or experimental data, or they may be simply 'modified' in some way to fit with that data. In the extreme case, all of the axioms of a theory may be so altered, resulting in a modification of the entire theory. In this latter case, it is still the same theory, but in a modified form. Theories may also be *replaced*. Here, none of the original axioms survive in either their old form, or in a recognisable modified form. The old theory has been abandoned as untenable and a new theory with completely new axioms constructed. This process will occur when empirical evidence contradicts the theory directly, or internal logical inconsistencies are found; the theory contradicts itself. The axiom number of the new theory will not necessarily be the same as that of the old theory in such a case.

We may also wish to consider the common axioms of several theories as a 'foundation' for a new theory. This may happen when several theories, in their entirety, encounter difficulty with empirical verification and the common axioms have already been substantiated, in the sense of empirical verification. This is rather like axiom removal as mentioned above, but on a somewhat grander scale. It may also happen that a new formulation of a theory may be found. The axioms of a theory may be cast into a new, but equivalent mathematical form, but are not *changed* thereby, merely simplified and retain the same domain of applicability. This is very common in physics. The theory is re-formulated in a new but equivalent way. It is still the same theory, but in a different guise.

Theories may be 'added together', or combined to form a new theory. Thus, 'unification' of this sort is also common in the natural sciences, especially physics. This is not a straightforward process and requires certain modification of the axioms of the 'component' theories before it can be accomplished. We shall have much to say about this important process in the sequel.

3.2 Mechanisms of theory modification
From the foregoing discussion we can discern several mechanisms by which theories may be altered or modified. We list below those that we consider to be the most relevant to theory re-construction:

Axiom Addition
Theories may be modified by the addition of new axioms:
$$S_{new} = S \cup \varphi_k$$
where k new axioms φ_k are adjoined to the theory S.

Axiom Subtraction
Theories may be modified by the removal of an axiom:
$$S_{new} = \{\varphi_1, \ldots\ldots\ldots, \varphi_{i-1}, \varphi_{i+1}, \ldots\ldots\ldots, \varphi_k\}$$
where the axiom φ_i has been removed from the original theory S to produce the new theory S_{new}.

Theory Modification

Theories may be *changed* by modification of the axioms themselves:

$$S_{new} \cap S_{old} = \{\varphi_1, \ldots\ldots, \varphi_i, \psi_{i+1}, \ldots, \psi_{j-1}, \varphi_j, \ldots\ldots, \varphi_k\}$$

where the new axioms $\psi_{i+1}, \ldots\ldots, \psi_{j-1}$ have replaced exactly j of the φ_k of the theory S_{old} to produce the modified theory S_{new} and the total number of axioms is unchanged.

Theory Replacement

Old theories may be replaced by entirely new theories:

$$S_{new} \cap S_{old} = \varnothing$$

$$S_{new} \neq S_{old} \Leftrightarrow \forall k (\varphi_k \in S_{new} \rightarrow \varphi_k \notin S_{old})$$

with or without the equivalent number of axioms and no axioms are common to both theories.

Theory Reduction

A common set of axioms from various theories may be combined to form a *core* theory:

$$S = S_1 \cap S_2 \cap, \ldots\ldots\ldots, S_{n-1} \cap S_n = \bigcap S_n$$

where we assume that each of the S_i are non-empty and not disjoint.

Theory Re-formulation

A theory may be cast into a new mathematical form, with new axioms, but be equivalent to the old form:

$$S_{new}(\varphi_1, \ldots\ldots, \varphi_m) \equiv S_{old}(\psi_1, \ldots\ldots, \psi_n)$$

where the axioms of the theory have been replaced by new, but equivalent axioms, resulting in equivalent theories. Note that the number of axioms in the reformulated theory does not necessarily equal the number of axioms in the old theory.

Theory Unification

Theories may unified into a single, more embracing theory:

$$S = S_1 \sqcup S_2 \sqcup, \ldots\ldots\ldots, \sqcup S_{n-1} \sqcup S_n$$

where the meaning of the "square union' will be made clear in the sequel.

We will also need an important principle that will be vital in the sequel:

Axiom Combination

Given two or more axioms, either from different theories or the same theory, we may combine those axioms into new axioms:

$$\psi = \varphi_1 \square \varphi_2 \square, \ldots\ldots\ldots, \square \varphi_{k-1} \square \varphi_k$$

This being not just simply axiom concatenation.

Theory-change results from the application of one or more of these mechanisms and often several at once. They are the mechanisms of *theory-change*. Each process will be significant in the modification of scientific theories to a greater or lesser extent depending on the particular theory concerned and the mechanism in question. Their objective is to effect a specific change of a given theory in a well-determined way.

We consider the above list to be the *primary* mechanisms of theory-change, but they are not entirely independent and there is some degree of redundancy. Additionally, the above may also not be exhaustive and other mechanisms may be identified, but they should certainly be sufficient for our immediate purposes. We merely wish to isolate the basic processes involved and do not preclude other possibilities or generalisations of these concepts.

However, before we can see how theory-change occurs, we must first examine the principle of *axiom combination* given above. We will then be in a position to discuss each of these mechanisms in far greater detail.

3.3 Axiom combination

When new axioms are added to a theory, several things may happen. Firstly, an inconsistency may result. The addition of the new axiom to the existing axioms of the theory may result in a contradiction. Then, from the theory S with some new axiom ψ, we have $S \cup \psi \vdash \varphi \wedge \neg\varphi$. Clearly, this is unacceptable and we must always insist (as mentioned in §2.1) on consistency, namely that $\forall \varphi_k \in S \neg(\varphi_k \wedge \neg\varphi_k)$. Secondly, the domain of applicability of ψ may not be that of S. In this case, the addition of ψ to S achieves nothing at all, other than increase the axiom number $|S|$ by unity. The new axiom is completely superfluous. For example, adding Ampere's or Coulomb's laws to Newton's does not result in electrodynamics.

We also noted earlier (§2.1) that simply forming the conjunction of two axioms does not produce anything new in a theory S. Thus, there is no advantage, from the point of view of theory-change, of replacing $\varphi_1, \varphi_2 \in S$ with the axiom $\psi = \varphi_1 \wedge \varphi_2 \in S$, for this is mere concatenation. The axiom has not changed. However, there is now no problem with domain of applicability since both φ_1 and φ_2 already enjoyed the same domain. Nevertheless, this kind of process is often done in physics and can, furthermore, be extremely convenient. A well-known example from Electrodynamics illustrates this point. If one takes as φ_1 and φ_2, the (source-free) Maxwell equations:

$$\nabla.\mathbf{B} = 0$$

$$\nabla \times \mathbf{E} + \frac{1}{c}\frac{\partial \mathbf{B}}{\partial t} = 0$$

then one may express them in tensorial form:

$$\partial_\lambda F_{\mu\nu} + \partial_\mu F_{\nu\lambda} + \partial_\nu F_{\lambda\mu} = 0$$

and now taking φ_1 and φ_2 as the second pair of Maxwell equations (with sources):

$$\nabla.\mathbf{E} = 4\pi\rho$$

$$\nabla \times \mathbf{B} + \frac{1}{c}\frac{\partial \mathbf{E}}{\partial t} = \frac{4\pi}{c}\mathbf{j}$$

we may similarly write:

$$F^{\mu\nu}{}_{,\nu} = \frac{4\pi}{c}J^\mu$$

There has been no change to the theory here and the axioms remain as they always were, as indeed does the domain of applicability. Furthermore, the tensorial form of Maxwell's equations *appears* to have reduced the axiom number. Of course it hasn't; the equations are simply expressed in a more compact form through the use of four-vectors and the brevity of tensor notation.

Whilst this procedure is common in the physical sciences and may be very useful in gaining insights into the structure of particular scientific theories, it does not, as we have mentioned, produce new theories because the latter require new axioms. Thus, processes of the sort described above, although part of genuine scientific research, are more akin to 'theory-manipulation', rather than theory-modification or theory-alteration. Then, we need to describe a procedure that will take the axioms of a theory and 'transform' them in some way into new axioms and hence a different or modified theory. We shall call such a procedure *axiom combination* and to prevent confusion with logical conjunction, denote it by '□'. Thus, given two axioms, φ_1 and φ_2, the new axiom ψ formed from axiom combination is $\psi = \varphi_1 \square \varphi_2$.

Axiom combination is a mathematical process. It is a possible algorithm and a series of operations. From two or more axioms, a new axiom is obtained. The new axiom is not derivable from either of the old axioms separately, but is a result of the combination process. Axiom combination may not always be possible, or may be very difficult, requiring much insight and 'intuition', but it is not mysterious and we shall shortly give real examples of this process from physics. The domain of applicability of the new axiom ψ is the union of the domains of the 'component' axioms $\varphi_1, \varphi_2, \ldots, \varphi_n$;

$$\mathcal{D}(\psi) = \mathcal{D}(\varphi_1) \cup \mathcal{D}(\varphi_2) \cup, \ldots, \cup \mathcal{D}(\varphi_{n-1}) \cup \mathcal{D}(\varphi_n)$$

and, of course, if all the φ_k belong to the same theory, then ψ will already have the appropriate domain. We shall find that axiom combination is especially important in the process of theory unification and will also play a significant role in the consideration of a Final Theory in Chapter 11. As a mechanism of theory-change (or modification) it is unique in this respect, as we shall be at pains to point out in the sequel. Then, with the notion of axiom combination now available to us, we are in a position to consider in greater detail the various mechanisms of theory-change.

3.4 Axiom addition

If an axiom is added to a theory, we expect it to satisfy certain conditions. Firstly, it must not cause inconsistency and secondly, its domain must be at least that of the theory to which it is being appended. Randomly adding axioms to a theory will be of little use as far as theory-change is concerned.

If an axiom ψ is added to a theory $S = \{\varphi_1, \ldots, \varphi_k\}$ containing k axioms, then three possibilities arise. Firstly, $\mathcal{D}(S) \cap \mathcal{D}(\psi) = \varnothing$, in which case the axiom contributes nothing to S, it is not applicable and then, though it may be consistent with the theory because no contradiction can arise, it is nevertheless redundant. Secondly, $\mathcal{D}(S) \cap \mathcal{D}(\psi) \neq \varnothing$, in which case the new axiom is applicable within the domain of S. If $\mathcal{D}(\psi) \subset \mathcal{D}(S)$ then the domain of the theory, after adding ψ will be reduced, it will in fact be $\mathcal{D}(S) \cap \mathcal{D}(\psi)$. Thirdly, if $\mathcal{D}(\psi) = \mathcal{D}(S)$ or $\mathcal{D}(\psi) \supset \mathcal{D}(S)$ then the domain of S will be unchanged. These three cases are illustrated in Figures 6, 7 and 8 below, where the theory S consists of the three axioms $\{\varphi_1, \varphi_2, \varphi_3\}$ and the new axiom to be added (in the normal set-theoretic sense of the union) is ψ:

Figure 6
The domain of S is disjoint from the domain of ψ. The new axiom ψ is inapplicable to the theory S.
$\mathcal{D}(S) \cap \mathcal{D}(\psi) = \varnothing$. The domain of the theory is therefore unchanged by the addition of the new axiom.

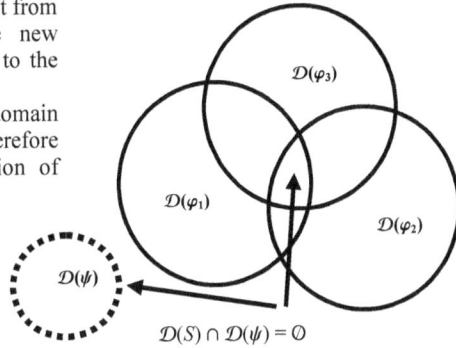

$\mathcal{D}(S) \cap \mathcal{D}(\psi) = \varnothing$

Figure 7
The domain of $S \cup \psi$ is a subset of the domain of S, but may be equal to $\mathcal{D}(S)$.
$\mathcal{D}(S) \cap \mathcal{D}(\psi) \neq \varnothing$ and $\mathcal{D}(S \cup \psi)$ $\subseteq \mathcal{D}(S)$. The domain of the theory is therefore reduced by the addition of the new axiom.

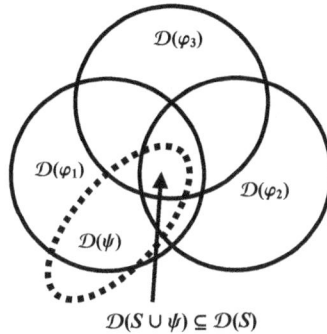

$\mathcal{D}(S \cup \psi) \subseteq \mathcal{D}(S)$

Figure 8
The domain of S is a subset of the domain of ψ, but may be equal to $\mathcal{D}(\psi)$. $\mathcal{D}(S) \cap \mathcal{D}(S_2) \neq \varnothing$ and $\mathcal{D}(S)$ $\cup \mathcal{D}(\psi) \subseteq \mathcal{D}(S)$. The domain of the theory is therefore unchanged by the addition of the new axiom.

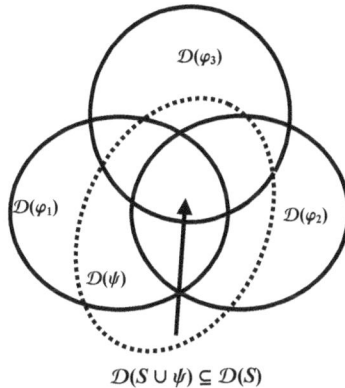

$\mathcal{D}(S \cup \psi) \subseteq \mathcal{D}(S)$

Hence, with axiom addition the domain of a theory is *never* increasing. We have noted earlier that an axiom is itself a theory, for it may constitute a theory of only one axiom. Hence, axiom addition in such case will consist of the set-theoretic union of the theory S and the singleton $\{\varphi\}$, namely, $S \cup \{\varphi\}$. Thus, generally, axiom addition, when repeated, is equivalent to set-theoretic union of several theories and without axiom combination, such a process must be subject to all the restrictions mentioned earlier in relation to domain compatibility and inconsistency.

3.5 Axiom subtraction

The removal of an axiom from a theory can have only two effects on the domain of the theory. Since the latter is the intersection of the domains of the axioms of the theory, only the removal of the axiom with the least domain (Definition 2.2.3) can result in a change in the domain of the theory. In all other cases, the domain of the theory must remain as it was. Thus, there are three possibilities. Firstly, the axiom with least domain is removed. Then, the domain of the theory is determined by the intersection of the remaining axioms and this is always greater than the original domain. Secondly, an axiom with a greater domain than that of the axiom with the least domain is removed, resulting in the same domain for the theory, because of the definition of domain as that of the intersections of the axiom domains. Thirdly, if all the axioms of a theory have equal domain of applicability, then the domain of the theory is unchanged by the removal of any one of them. Hence, with axiom removal (or subtraction), the domain of the theory is *never decreasing*. We can symbolise the procedure of axiom subtraction in the following diagram:

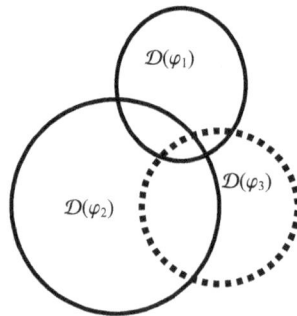

Figure 9
The domain of the theory $S = \{\varphi_1, \varphi_2\}$ with φ_3 removed is $\mathcal{D}(\varphi_1) \cap \mathcal{D}(\varphi_2)$

where we have removed the axiom φ_3 with domain $\mathcal{D}(\varphi_3)$ resulting in a theory with the enlarged domain $\mathcal{D}(S) = \mathcal{D}(\varphi_1) \cap \mathcal{D}(\varphi_2)$ as expected. In this case none of the domains of any of the three axioms are subsets of the domains of the others. Similarly for the other cases as depicted below in the next two Figures 10 and 11;

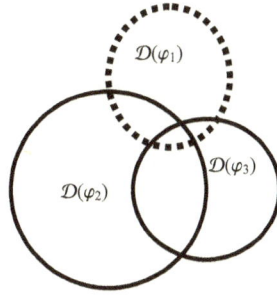

Figure 10
The domain of the theory $S = \{\varphi_2, \varphi_3\}$ with φ_1 removed is $\mathcal{D}(\varphi_2) \cap \mathcal{D}(\varphi_3)$

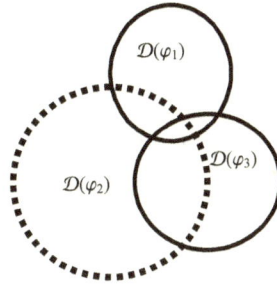

Figure 11
The domain of the theory $S = \{\varphi_1, \varphi_3\}$ with φ_2 removed is $\mathcal{D}(\varphi_1) \cap \mathcal{D}(\varphi_3)$

There are of course many possible variations on the above and one may wish to remove several axioms of a theory if empirical investigation or logical consistency so dictates. An interesting case arises when the domains of each of the axioms are subsets or supersets of the domain of the other axioms:

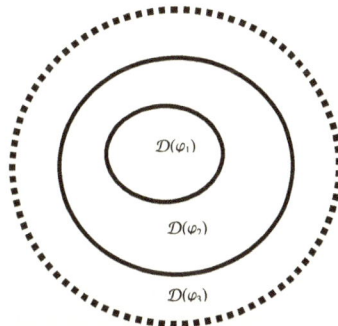

Figure 12
The domain of the theory $S = \{\varphi_1, \varphi_2\}$ with φ_3 removed is $\mathcal{D}(\varphi_1) \cap \mathcal{D}(\varphi_2) = \mathcal{D}(\varphi_1)$

Figure 13
The domain of the theory $S = \{\varphi_1, \varphi_3\}$ with φ_2 removed is $\mathcal{D}(\varphi_1) \cap \mathcal{D}(\varphi_3) = \mathcal{D}(\varphi_1)$

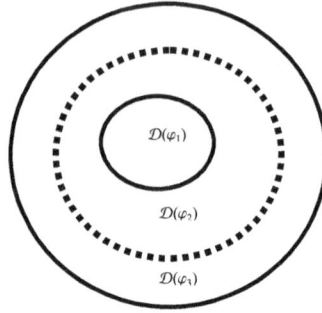

Figure 14
The domain of the theory $S = \{\varphi_1, \varphi_3\}$ with φ_1 removed is $\mathcal{D}(\varphi_2) \cap \mathcal{D}(\varphi_3) = \mathcal{D}(\varphi_2)$

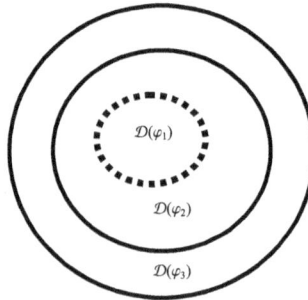

We see that in this case, the only instance in which the domain of the theory is increased is, as expected, when the axiom with least domain is removed. Axiom subtraction is somewhat rare in the physical sciences; it will most likely occur when an axiom has been added to a theory and found to be empirically untenable or logically inconsistent. The removal of an axiom from a theory may also be useful from the point of view of the study of the internal structure of the theory, where the effect of axiom removal will illustrate the role of that axiom in foundational studies.

3.6 Theory modification

Sometimes, an axiom may be found to be useful only in a very restricted domain, or may be empirically invalidated, or found to be logically inconsistent. In the first case, one may wish to extend the domain of the axiom by suitably modifying it. In the second and third cases, one may wish to remove the axiom and replace it with another, more appropriate one. In any event, one is essentially replacing an axiom, or several axioms of a theory with new ones. Essentially one has, therefore, a new theory.

Theory modification may be viewed as a partial function acting upon the set of axioms of a theory;

$$\mu : S_{old} \to S_{new}$$

that replaces one or more axioms with new axioms,

$$\mu : (\varphi_1, \ldots \ldots, \varphi_k) \to (\varphi_1, \ldots \ldots \varphi_i, \psi_{i+1}, \ldots \ldots, \psi_j, \varphi_{j+1}, \ldots \ldots, \varphi_k)$$

where j new axioms ψ_j have replaced j of the old axioms φ_k. When the modification function μ is total:

$$\mu : (\varphi_1, \ldots \ldots, \varphi_k) \to (\psi_1, \ldots \ldots, \psi_k)$$

then theory modification becomes theory replacement.

Theory modification is thus a process of adaptation, according to the restrictive criteria of either empirical verification, or logical consistency. It produces a new theory, but only in the sense that the old theory has been 'adapted' to new, previously unperceived empirical laws, or inconsistencies in the original formulation. The domain of the theory, after modification, is still the domain of the axiom of least domain, but may be greater than, less than or equal to the domain prior to modification.

A famous example of theory modification is in General Relativity and cosmology. As is well known, Einstein modified his field equations:

$$R_{\mu\nu} - \tfrac{1}{2} R g_{\mu\nu} = \frac{8\pi}{c^4} T_{\mu\nu}$$

by the introduction of the cosmological constant;

$$R_{\mu\nu} - \tfrac{1}{2} R g_{\mu\nu} - \Lambda g_{\mu\nu} = \frac{8\pi}{c^4} T_{\mu\nu}$$

in the belief that this would describe a static universe. Of course, this is not so, for the above system actually describes a universe that is unstable, either expanding or contracting. Subsequently, as is also well-known, the Λ term was dropped when Edwin Hubble found that the universe was expanding, by interpreting the red-shift in the spectra of external galaxies as a Doppler effect. Recently, however, it has been re-introduced as a 'vacuum energy', which actually acts as the driving force for the expansion of the universe; a further and startling example of theory modification.

Cosmology is full of such examples of theory modification. Particularly interesting is the case of the (now abandoned) Steady State Theory, proposed as an alternative to the very well-established evolutionary cosmological models, such as the 'Big Bang'. Here, the Einstein equations were modified by the introduction of the 'C' field for matter creation:

$$R_{\mu\nu} - \tfrac{1}{2} R g_{\mu\nu} + C_{\mu\nu} = \tfrac{8\pi}{c^4} T_{\mu\nu}$$

But this is not as simple as it seems. Recall that homogeneous relativistic cosmologies are based on the *Cosmological Principle*, which asserts that the universe is (on sufficiently large scales), both homogeneous and isotropic. In the Steady State theory, this is replaced by the *Perfect Cosmological Principle*, which extends the former to include homogeneity and isotropy throughout all time as well. Hence, we see that here not only have the axioms been modified, but so has a foundational (philosophical) assumption. This is hardly surprising. One would expect that theory modification would often result from the alteration of fundamental principles, but the converse may not necessarily be the case, although it was so in this example. It is clear, therefore, that in this and indeed, in all of the mechanisms of theory-change, one cannot divorce oneself from the fundamental philosophical basis of the theory in question.

3.7 Theory replacement

Theory replacement is, essentially, the abandonment of an existing theory. *All* the axioms are replaced, or at least modified – which amounts to the same thing – by new axioms. Thus, the axioms of the new theory may have very different domains than those of the old theory. Consequently, the new theory may also have a much modified domain of applicability. However, if a theory is entirely replaced in this fashion, then clearly, it might be thought that a minimal requirement is that the new theory should, at the very least, embrace the entire domain of the old theory, otherwise there is little point in formulating the new theory, which presumably has been done for the express purpose of replacing the old. Hence, the new theory should, therefore, in general, cover the domain of the old theory.

However, we cannot say that the new theory will *definitely* embrace the entire domain of the old theory; it may not. Empirical criteria may force new axioms, indeed, *in toto* in this respect, but there is no logical *sine qua non* that insists on domain equality, for the latter is determined by the nature of the laws and the axioms themselves. Nevertheless, as we observed above, it is most appropriate for the domain of the new theory to entirely encompass that of the old. Therefore, we shall *define* theory replacement as that process of axiom change or modification in which the domain of the new theory does indeed embrace that of the old theory; i.e., the domain of the old theory is a subset of the domain of the new theory. Hence, we envisage the situation as in Figures 1 and 3 of §2.2 to obtain. If each one of the axioms of the old theory is replaced by exactly one axiom of the new theory, then the axiom number of both theories will be equal. This will not, however, necessarily be the case and we may expect that the axiom number is not generally preserved. Theory replacement is the most extreme of the mechanisms of theory-change. It occurs quite rarely and represents a complete conceptual change that is invariably compelled by empirical considerations.

Then, the replacement of an entire theory by a completely new theory with different axioms, but with domain at least that of the old theory, consists in the radical step of theory abandonment. Such a profoundly reformist procedure could only result from empirical demands. The old theory has been found to be untenable; to conflict with observed facts of the Natural World and needs to be replaced, rather than merely revised. This, therefore, constitutes a revolution in science and may well involve a re-consideration of the fundamental principles and foundations upon which the original theory was based. Since this process is so radical and deeply related to the conceptual foundations, often of a philosophical nature on which the principles of the theory were originally based, it is an infrequent occurrence in science. For, indeed, if it were not, then it would be difficult to see how scientific theories could admit of reasonable stability for a sufficiently sustained period of time to allow scrutiny and substantiation. Therefore, a revolution of this kind could only result from the empirical falsification of the axioms, or consequences thereof of the theory in question.

As an extreme example of theory replacement, one may take the Biblical account of the origin of mankind as described in Genesis, as opposed to the position in evolutionary biology in the modern Darwinian synthesis. The Biblical creationist account may be thought not to be a scientific theory in the generally accepted understanding of that term, but it is nevertheless, still a theory. Empirical evidence in support of neo-

Darwinism, both from the fossil record and geological data, makes the creationist view untenable. Thus, the latter is replaced in its entirety, with no surviving axioms. As an additional example of this sort, but perhaps less extreme, we mention again the contrast between the Ptolemaic model of planetary motion in terms of epicycles and the heliocentric hypothesis of Copernicus. In this case, both theories can be said to be properly scientific and both are supported by empirical evidence. However, the Copernican model was simple and elegant from a geometrical and mathematical viewpoint and was supported to a greater degree by the observational evidence than the less accurate epicycles of Ptolemy, as astronomical observations became more precise.

3.8 Theory reduction

Given that the domain of a theory is determined by the intersections of the domains of the axioms of that theory, then obviously, if we form the intersection of two or more theories, the domain of the new theory so obtained, is the intersection of the domains of the axioms of all such 'component' theories. Thus, the domain of the new theory formed by such intersections, can never be greater than the domain of the axiom with the least domain. It is, as usual, necessary that the theories involved in this process have domains of applicability that are not disjoint, since arbitrarily intersecting theories at random can serve no useful purpose as far as purposeful theory development *per se* is concerned.

Since the domain of the resulting theory after forming such intersections is so restrictive, information concerning the philosophical assumptions upon which the 'component' theories are founded can be obtained in this way. Thus, one purpose of theory reduction may be just the perspicuity that results thereby. Philosophical insight may be the sole objective and when that is so, it is clearly useful in the determination of the limits of and the scope of a scientific theory; this being made evident simply because of the intersection process, which thus exhibits the mutual domain of all the axioms of each of the component theories.

One may ask if this process ever actually occurs in practice. Whether, in the serious scientific endeavour of theory construction, one employs this very set-theoretic method of ascertaining the domains of the axioms and hence, the fundamental assumptions and foundations of a number of (related) theories within a scientific discipline? Evidently, if the answer to this question was not in the affirmative, then major changes and deep foundational investigations into the nature of related scientific theories, or entire sets of sub-disciplines of related areas of research, would not be possible. However, the intellectual process by which such philosophical investigations are pursued in practice tends to be 'intuitive' rather than 'mechanical' or 'mathematical' in the above set-theoretic sense. Nevertheless, it is essentially the same process. When one considers the domain of applicability of a theory, one is forced to consider the domains of the axioms of that theory. Thus, examination of the intersections of the domains of the axioms of the theory becomes a pre-requisite, a *sine qua non*, for understanding the domain of the whole theory and hence its foundational principles. Similarly and by extrapolation, when such considerations are applied to a multiplicity of theories, possibly and most likely within the same discipline, then the domains of all the theories involved will be likewise considered, i.e. their intersections will determine the common foundational principles upon which each of the component theories are based. This, then, is just the intersection process of theory reduction, but is often accomplished within an intuitive and philosophical spirit. In some cases this process may proceed almost unconsciously, as one mentally sifts through the philosophical assumptions of the axioms and domains of the theories involved in the search for foundational principles. Even so, it is still a mechanistic process which admits of precise description. Once again, we observe the close connection between the axioms of a theory and the philosophical foundational principles upon which it is built.

3.9 Theory re-formulation

It is often the case, especially in the fundamental sciences, that a more 'eloquent' expression of the axiom content of a theory may be constructed by a suitable manipulation of the existing axioms of the theory. This is to say, that it is often the case, with appropriate insight, that a more perspicuous representation of the theory may be obtained by formulating the axioms of the theory in a different manner. This process is equivalent to re-writing a sentence in a natural language in order either to enhance the clarity of the original formulation, or to reveal new insights that were hitherto hidden, or at least not obviously evident in the earlier formulation. In the case of scientific theories, particularly in the physical sciences, this is generally a process of mathematical manipulation of the axioms. Nothing with respect to the content of the theory is gained and the domain of the theory is unchanged, but such re-formulation may often provide insightful recognition of the deeper nature of the theory involved which, although was always present in the original formulation, is now made more evident in the new and perhaps more elegant mathematical or linguistic depiction. Moreover, the increased transparency thus provided can often accelerate theory modification.

The process of theory re-formulation is extremely common in the Natural Sciences. The philosophical insight that is often gained by what, essentially, amounts to a re-examination of the axioms of a theory, can serve as a foundation for future developments. This is particularly well-illustrated in physics and we consider a familiar example from Classical Mechanics. Given the kinetic energy T, potential energy V and generalised co-ordinates q_j, with $\dot{q}_j = dq_j/dt$, then we can obtain from Newton's equations, the Lagrange equations:

$$\frac{d}{dt}\left(\frac{\partial L}{\partial \dot{q}_j}\right) - \frac{\partial L}{\partial q_j} = 0$$

where $L = T - V$ is, as usual, the Lagrangian of the system. These equations are entirely equivalent to the Newtonian formulation and no additional assumptions have been made. Furthermore, from the above we may also derive Hamilton's equations:

$$\dot{q}_j = \frac{\partial H}{\partial p_j}, \quad \dot{p}_j = -\frac{\partial H}{\partial q_j}, \quad \dot{H} = -\frac{\partial L}{\partial t}$$

where p_j are the generalised momenta, $H = T + V$ is the Hamiltonian and differentiations are with respect to time t. Again, this formulation is equivalent to the above. A particularly fashionable approach in modern physics arises because the Lagrange equations above are just the conditions for the action functional:

$$I = \int_{t_1}^{t_2} (T - V)\,dt = \int_{t_1}^{t_2} L\,dt$$

to have an extremum. Here, t_1 and t_2 are arbitrary instants of time. This is Hamilton's principle, which admits of yet another formulation of a mechanical system, this time with minimal axioms.

Thus, in this example, Newtonian mechanics has been re-formulated in a way that adds precisely nothing to the theoretical content of that model, but certainly affords insight into the structure of the theory itself. For now, we are dealing with a 'phase space' in terms of the generalised co-ordinates, which geometrically is a symplectic manifold. This was not immediately obvious before. Yet no new information is gained, only new perceptions, accomplished by an ingenious manipulation of the axioms. Although the domains of the theories are the same, the axioms are not. It might be thought, therefore, that for a mechanical system possessing a Lagrangian, Hamilton's principle could be used to replace Newton's laws in a fundamental sense. Of course, this is not so. The Hamiltonian formulation is derived from Newton's and carries the assumptions of Newtonian mechanics with it. The assumption that the co-ordinate system is fixed relative to an inertial frame of reference follows from Newton's first law, which defines an inertial frame. Additionally, the definition of mass is contained in Newton's third law, which we have assumed. However, the Hamiltonian formulation *can* serve as a fundamental starting point for the development of Quantum Mechanics, especially the action integral and historically, this was indeed the case. Thus, not only is insight gained into the theoretical structure of mechanics by this re-formulation, but also that re-formulation has provided a foundation for future developments, as we mentioned above. Another particularly pertinent example of insight gained by theory re-formulation is provided by the covariant form of Maxwell's equations that we discussed in §3.3. It was not realised for a long time that the electromagnetic equations were Lorentz invariant until they were expressed in this form, but when they were examined in this new formulation, it was apparent (at least to Einstein) that Maxwell's equations already contained the special theory of relativity and would require the four-dimensional space-time formalism of the latter to be consistent with Newtonian mechanics.

3.10 Theory unification

If one is given several theories, $S_1,........,S_n$, each with their own respective domains of applicability, $\mathcal{D}(S_1),........,\mathcal{D}(S_n)$, then simply forming a new theory S, by taking the set-theoretic union of each of the component theories S_i:

$$S = S_1 \cup S_2 \cup,..........,\cup S_{n-1} \cup S_n$$

results in little more than repeated axiom addition, which we discussed earlier (§3.4). The domain of the new theory $\mathcal{D}(S)$, so formed, can, of course, only be the domain of the axiom of least domain, of all the axioms of the component theories S_i. If the domains of the latter are disjoint, then little is gained by the formations of such a union. If the domains of the component theories have non-null intersection, then one has merely increased the axiom content of the original theories, with no additional structure. Thus, by ordinary set theoretic union, no significant results emerge with respect to the primary objective of 'theory-change' by this process.

Hence, it is clear that mere set-theoretic union, which is essentially axiom addition, can serve no fruitful purpose in this sense, given the conditions of axiom domain, unless further criteria are satisfied. If one wishes to create a new theory from a set of existing theories, then clearly, as noted previously, a new procedure is necessary. We have earlier identified just such a pre-requisite process and called it 'axiom combination'.

In forming the set-theoretic union, three possible cases may arise. Firstly, all of the domains of the component theories may be disjoint. In the terminology of §2.2 the theories are incompatible. Secondly, if some of the axioms of each component theory are applicable within the domain of the other theories, then again, in the terminology of §2.2, the theories are partially compatible. Thirdly, if all the component theories have equal domains, then the component theories are compatible. Thus, the set-theoretic union of various theories must take these situations into consideration.

Firstly, if the component theories are compatible, then their domains are equal and axiom combination cannot take place, simply because it would achieve nothing; the axioms of the component theories are already applicable within their respective domains. In this case, theory modification or theory re-formulation may be appropriate, but not theory unification.

Secondly, if the component theories are partially compatible, then there is no point in combining those axioms of the components that have mutual domains, for the same reason as above and anyway, such axioms can be subsequently adjoined to the theory by axiom addition, for they play no part in the unification process. However, the axioms with different domains may well be combined by the process of axiom combination. But that is no different from dealing with incompatible theories. To elaborate, if two theories are partially compatible so that they share at least one axiom, then the shared axioms do not participate in the unification process since there is no need for them to do so. It is only those axioms of the theories that have incompatible domains that may be unified and these may be sub-theories of the original two theories. It is the latter that will be unified, but the shared axioms will retain their domain within the unified theory because they are compatible. The essential point here is that any axiom common to the component theories will automatically be included in the unified theory. We therefore conclude that, for the purposes of theory unification, we need consider only incompatible theories, which may well be the incompatible subsets of the component theories. On reflection, this should be obvious, for if theories are to be unified into a single theory, then part of that process, exemplified by axiom combination, is to extend the domain of one theory into that of the other theory, for each of the component theories involved. Thus, axiom combination and hence theory unification, is the mathematical process by which incompatible theories are made to be compatible and thereby share the same domain. Hence, theory unification is a two stage process. Firstly, we form the set theoretic union, as above and secondly we subject the axioms to the operation of axiom combination.

When theories are unified, it might be thought that the total number of axioms in the resulting union is the sum of the axioms of each of the theories involved in the unification process. Certainly this would be the case for simple set-theoretic union. But if that were the case, then no new theory would result and no scientific progress made. All that would happen is that the theories would gain more and more axioms. What is required for the new unified theory to be different from the old theories is that new axioms appear; otherwise the whole exercise is pointless. Since no *new* axioms are added, axioms must instead be modified. Individual axioms are not modified, for that is theory modification, which we are not discussing here. Therefore, in the process of unification axioms must combine to form new axioms. Since no such combination took place in any of the pre-unification theories, each of the axioms in such a combination must come from different theories. This is why theory unification is a two stage process, first we have ordinary set-theoretic union, which is then followed by axiom combination. In order to avoid confusion between ordinary set-theoretic union and theory unification with axiom combination we shall denote the latter by the 'square union' symbol, '⊔'. Thus, theory unification consists of the operation of ordinary set-theoretic union '∪', followed by axiom combination '□'.

Consider two theories that are unified into a new third theory; $S_3 = S_1 \sqcup S_2$. Two axioms, φ_1 and φ_2 for example, with $\varphi_1 \in S_1$ and $\varphi_2 \in S_2$, combine to form the new axiom $\psi_2 = \varphi_1 \square \varphi_2$. This new axiom is of a different nature from each of the φ_1, φ_2; it is changed, but contains each. Thus, when unification takes place, axioms from different theories are combined into new axioms. This of course must be so, since otherwise the new theory would be inconsistent. In the combination of axioms during the unification process, axioms in one theory that would be inconsistent with axioms in another theory must combine in order to produce a consistent theory.

Consider again the two theories as above. Let us suppose that S_1 contains three axioms and that S_2 contains four axioms. How many axioms will be contained in S_3 after unification? Now, It may well be that all of the axioms of S_1 combine with three of the axioms of S_2, producing four axioms in the unification process (because each of the three axioms of S_1 combine with one of the axioms of S_2 and the remaining axiom of S_2 is unchanged). There are many other possible combinations also. We could have two of the axioms of S_1 combine with only one of the axioms of S_2 and the remaining axiom of S_1 combine with one of the remaining axioms of S_2, or even perhaps two, or even three of the remaining axioms of S_2. In fact, there are 3^4 possibilities.

Let us take the first of the above examples for illustration. Let $S_1 = \{\varphi_1, \varphi_2, \varphi_3\}$ and $S_2 = \{\psi_1, \psi_2, \psi_3, \psi_4\}$. Then we first form the set-theoretic union:

$$S = S_1 \cup S_2 = \{\varphi_1, \varphi_2, \varphi_3, \psi_1, \psi_2, \psi_3, \psi_4\}$$

and now we combine these seven axioms as follows;

$$S = S_1 \sqcup S_2 = \{\varphi_1 \square \psi_1, \varphi_2 \square \psi_2, \varphi_3 \square \psi_3, \psi_4\}$$

to produce a new theory with just four axioms. As mentioned, there are many other possibilities. What is important here, however, is that the set-theoretic union must *precede* axiom combination.

This makes it impossible to say precisely how many axioms there will be in S_3. However, an upper bound can be set. The nature of the combination process ensures that S_3 must always contain fewer axioms than the sum of the axioms in S_1 and S_2, which is seven. We can generalise this as follows. In the unification;

$$S = S_1 \sqcup S_2 \sqcup, \ldots \ldots \ldots, \sqcup S_{n-1} \sqcup S_n$$

we have, for the axiom number of S;

$$|S| < |S_1| + |S_2| +, \ldots\ldots\ldots, + |S_{n-1}| + |S_n|$$

We emphasise again that in theory unification, axiom combination occurs only between those axioms that do not share domains of applicability. This is because the unified theory will be inconsistent if it contains axioms with different domains of applicability. If two incompatible theories are unified purely set-theoretically, without axiom combination, then the resulting theory will be 'inconsistent'. This is to say that component theories, prior to unification, have different domains of applicability and that no one theory can be applied to the domain of another theory without inconsistency. It does not mean that axioms and thus their domains cannot be common to the component theories. It is important to distinguish between the domain of a theory and the domain of its axioms. The domain of a theory is one thing and the domain of an axiom within that theory quite another. The example above of unification with axiom combination carries the axiom ψ_4 into the unified theory but as we noted earlier, it is not itself involved in axiom combination and because the domain of ψ_4 belonged to the domain of one of the component theories prior to unification, its domain can still be included in the unified theory. This is why only incompatible theories need be considered for theory unification.

We re-iterate that in order for unification to work, the component theories must be incompatible; otherwise the unification process achieves exactly nothing, except enlarge the existing theories by increasing the number of axioms. The same applies to axiom addition. Axiom addition is not unification unless the new axiom is incompatible with the existing theory. The point here is that the union of the axioms of the component theories results in inconsistency because the theories are incompatible. By this, we mean that a theory contains axioms applicable to different domains. It is this inconsistency that 'forces' axiom combination and hence modification, in the unification process. Logical consistency, in respect of domain applicability, is the 'guiding' force behind unification. Then, theory unification (with axiom combination) enlarges the domain of applicability to that of each of the component theories and the resulting unified theory thus enjoys a domain of applicability that each of the component theories separately enjoyed.

In theory unification, some or all of the axioms of the component theories will combine, but it is never the case that *no* axioms combine. Thus, theory unification reduces the total number of axioms involved. This is why the resulting unified theory has fewer axioms than would be the case for pure set-theoretical union.

> ***In order for a set of theories to be unified, the domain of applicability of each of the component theories (or sub-theories) must be incompatible.***
>
> ***When theories are unified, then axioms from each theory are modified into new axioms in the resulting unified theory by the process of axiom combination.***
>
> ***When theories are unified, the axiom number of the resulting unified theory is always less than the sum of the axiom numbers of the component theories.***

The unification process has produced a new theory. Since the unified theory was built from separate theories, it is possible in principle to 'undo' the unification. This will also mean resolving any new axioms into their individual component axioms in the original theories. This can always be done. Then we have:

> ***If a unified theory is consistent, then it contains the component theories that contributed to the unification.***

Thus empirically verified theories are still valid within their respective domains of applicability.

It is helpful to consider a concrete example in order to solidify the above concepts. We choose a familiar illustration from physics, namely Magnetohydrodynamics or (MHD). This combines electrodynamics, fluid mechanics and, of course, already includes Newtonian mechanics in the hydrodynamical equations. We consider only the non-relativistic approximation and do not treat the full MHD equation set. We need both fluid mechanics and electromagnetic theory for this purpose and label the latter two theories as S_{HD} and S_{EM}. The unified theory will then be $S_{MHD} = S_{HD} \sqcup S_{EM}$. We select only axioms from each theory that we will need for illustrative purposes and label them φ, ψ, χ, with appropriate subscripts.

First, we have from hydrodynamics, the continuity equation expressing the conservation of mass;

$$\frac{\partial \rho}{\partial t} + \nabla \cdot (\rho \mathbf{u}) = 0 \qquad (\varphi_1)$$

where ρ is the density and $\mathbf{u}$ is the fluid velocity. We also have the momentum equation;

$$\frac{D\mathbf{u}}{Dt} = -\frac{1}{\rho}\nabla P - \nabla \Phi \qquad (\varphi_2)$$

with pressure P and gravitational potential Φ. This is Euler's equation, which is Newton's law for fluids. Here,

$$\frac{D}{Dt} = \frac{\partial}{\partial t} + \mathbf{u} \cdot \nabla$$

is the convective derivative. We have from electromagnetic theory the Maxwell equations:

$$\nabla \times \mathbf{B} = \frac{4\pi}{c}\mathbf{j} \qquad (\psi_1)$$

$$\nabla \cdot \mathbf{B} = 0 \qquad (\psi_2)$$

$$\nabla \times \mathbf{E} + \frac{1}{c}\frac{\partial \mathbf{B}}{\partial t} = 0 \qquad (\psi_3)$$

where we have, under this approximation, neglected the displacement current term. We also have Ohm's law in the form;

$$\mathbf{j} = \sigma\left[\mathbf{E} + \left(\frac{\mathbf{u}}{c} \times \mathbf{B}\right)\right] \qquad (\psi_4)$$

Writing (ψ_3) in the form $\partial \mathbf{B}/\partial t = -c(\nabla \times \mathbf{E})$ and using (ψ_4) to substitute for $\mathbf{E}$ gives;

$$\frac{\partial \mathbf{B}}{\partial t} = -c\nabla \times \left[\frac{\mathbf{j}}{\sigma} - \left(\frac{\mathbf{u}}{c} \times \mathbf{B}\right)\right]$$

and from (ψ_1) as $\mathbf{j} = (c/4\pi)(\nabla \times \mathbf{B})$ we obtain;

$$\frac{\partial \mathbf{B}}{\partial t} = \nabla \times (\mathbf{u} \times \mathbf{B}) - \nabla \times \left(\frac{c^2}{4\pi\sigma} \nabla \times \mathbf{B} \right)$$

$$= \nabla \times (\mathbf{u} \times \mathbf{B}) + \frac{c^2}{4\pi\sigma} \nabla^2 \mathbf{B} \qquad (\chi_1)$$

where we have used ψ_2. This is the Hydromagnetic equation. Now, in the approximation under consideration, the electric field is negligible in this frame, thus net force on a fluid element is $\mathbf{j} \times \mathbf{B}/c$, which, on using (ψ_1) again as $\mathbf{j} = (c/4\pi)(\nabla \times \mathbf{B})$, we see that the additional magnetic force is $[(\nabla \times \mathbf{B}) \times \mathbf{B}]/4\pi$. Hence, Euler's equation φ_2 becomes;

$$\rho \frac{d\mathbf{u}}{dt} = -\nabla P + \frac{1}{4\pi} (\nabla \times \mathbf{B}) \times \mathbf{B} \qquad (\chi_2)$$

Finally, the equation of continuity φ_1 remains unchanged. From these results can be derived many more interesting phenomenological results, such as expressions for magnetohydrodynamic waves, etc. Note that we have not included thermodynamic quantities in the above, since we only wish to illustrate the process of unification in the simplest possible terms.

Let us see what has happened here. The hydrodynamic theory is $S_{HD} = \{\varphi_1, \varphi_2\}$ and the electromagnetic theory is $S_{EM} = \{\psi_1, \psi_2, \psi_3, \psi_4\}$, while the unified theory is $S_{MHD} = \{\chi_1, \chi_2, \varphi_2\}$. The axiom χ_1 has been obtained from ψ_1 and ψ_3 purely from within the domain of S_{EM}, i.e. without recourse to S_{HD}. However, this is not the case for χ_2 because we have used Ohm's law to obtain the magnetic force acting on a fluid element, thus we have combined axioms (with a little manipulation) to form $\chi_2 = \varphi_2 \,\square\, \psi_4$. Hence, the former case is mere 'theory manipulation', whereas the latter case is axiom combination. Both are algebraic processes, but if it is the intention to produce a new theory, then axiom combination should take place between axioms of theories with incompatible domains. This is an important distinction, but since it will always be clear from the context whether we are combining axioms from the same, or different theories, we make no distinction and continue to refer to both processes as 'axiom combination'. Note two things here, firstly, not all of the axioms have combined; it was not necessary in this case and secondly, the axiom number of the final theory, as expected, is less than that of the component theories.

It is also clear from this example that something more than mere axiom combination has taken place. There has been some algebraic manipulation of the axioms, firstly, in the sense of 'theory manipulation', as mentioned above, but secondly, in the sense also of theory re-formulation as discussed in §3.9 earlier; for the algebraic manipulation of the axioms is essentially nothing more than theory re-formulation, albeit on a lesser scale than might be envisaged for that process. This, also, is an important point that is exemplified in the unification process. Sometimes, unification may not be possible without first performing other mechanisms of theory-change. This may often be necessary in order to find a suitable path for axiom combination, which would otherwise be hidden from view, or at least not immediately apparent from the existing formalism.

We could have considered the full MHD theory, with similar results. We could also have gone about the procedure of unification in a completely different way. There are many roads by which we could have arrived at the above equations; we simply chose one simple method. Thus, theory unification is not unique in its methods and sometimes a great deal of ingenuity may be required. It is, however, still a mechanical procedure, even though it may be necessary to make some underlying assumptions, as we needed to in the above illustration. Philosophical and foundational matters *do* play a part, but once the basic assumptions are specified, the process is entirely mechanical, at least in principle. Thus, the mathematical expressions of the two component theories have been manipulated in a way that produces a new expression, such that the content of the new expression implicitly contains the total content of the old. Essentially, the axioms have been combined in an 'algorithmic' sense, although they were formerly disjoint with respect to their domains of applicability. How is this possible? We emphasise again, that the domain of a theory bears no relation to the mathematical or grammatical structure of the theory. A theory may well seem to have an all-embracing domain, but this is not our usage here. Scientific Laws may appear to be universal, but they are restricted within the domain of the theory that contains them. Axioms are equally restricted. They are restricted not only by their own domains of applicability, but also by the domains of the other axioms of the theory of which they are a constituent part. This is so because of the intersection criterion of axiom and theory domain. It is therefore, of great importance to minimise the axiom number of a theory whenever possible; the best and most universal theory will have only one axiom.

It is not absolutely necessary that the theories undergoing the unification process be of a mathematical nature. To be sure, simplification results if they are, but mathematical representations of theories are by no means a prerequisite for unification. What is, however, a *sine qua non* for the latter, is domain incompatibility of the component theories. Mathematics is a language. It is particularly apposite for expressing complicated situations in a concise and readable form. Even so, anything that can be described mathematically can also be described in Natural Language, usually with the aid of scientific or technical terminology providing appropriate brevity of expression. The use of such terminology is equivalent to the employment of mathematical formula. Then, in principle, any set of theories that are sufficiently well-formulated in precise, but non-mathematical terms may also be unified if the appropriate criteria are satisfied. For this to happen, however, the axioms of each component theory must be formulated in a manner that is sufficiently clear so as to admit of analysis, without ambiguity and their respective domains of applicability must also be reasonably well-defined.

Therefore, as an *essentially* non-mathematical example, consider Darwinian evolution (circa 1859) and the Mendelian theory of heredity. Then, the unification;

$$S_{MS} = S_{DARWIN} \sqcup S_{MENDEL}$$

represents what has come to be known as the Modern Synthesis, or neo-Dawinism in evolutionary biology. Evidently, the axioms and the domains of each theory have been identified accurately enough to effect such unification. Furthermore, neo-Dawinism is indeed a new theory in the sense that it provides explanatory power with respect to the mechanism of Natural Selection that was hitherto absent. It is not, therefore, simple theory re-formulation, which could not achieve such new insights.

We may imagine many such areas of scientific enquiry where theory unification might seem desirable or appropriate. It is well to observe, however, that unification is only possible for sufficiently precisely defined disciplines and proper account must be taken of both the fundamental principles and the domain of applicability of the area under investigation. Unification is thus inappropriate for the less well-developed sciences.

Chapter 4

The Evolution of Scientific Theories

Theories are likened to biological organisms and the mechanisms of theory change are formulated as operators acting upon scientific theories. The question of the formulation of theories is considered in the context of theoretical and empirical criteria. It is shown that the mechanisms can be expressed in terms of just two simple operators and 'evolution' equations are given for the development of scientific theories.

4.1 Theory species

Scientific theories are living entities. They are born (formulated), they grow and adapt (theory-change) and they die (theory replacement). During its lifetime, a theory may undergo many changes and modifications, sometimes quickly and sometimes more slowly. The detailed mechanisms of theory-change, by which such modifications take place, were discussed in the last chapter. However, the lifetimes of individual theories are variable. One theory may live for a hundred, or even a thousand years, whilst another may last for only a few years, or perhaps even less. Thus, scientific theories are themselves very much like the Natural Phenomena that they purport to describe. They are, therefore, analogues of their own subject matter, behaving in a manner that is exactly analogous to the natural processes that govern the physical universe. We learn from the theory of stellar evolution that the more massive a star, the shorter is its lifespan. Thus, theories are like stars; the more 'massive' or complicated they are, the more quickly they will perish. Simple theories survive longer because they have undergone minimal adaptation and this is so because the simplicity of the theory results in fewer axioms that can be empirically refuted, just as low mass stars live longer because they do not need to pass through the many complicated (nuclear burning) processes that heavier stars need to in the later stages of stellar evolution. A similar situation persists amongst the many different organic life-forms (species) found on planet Earth and one can easily find analogous behaviour in many areas of science.

None of this, however, should be surprising. Whereas Natural Phenomena are governed by the Laws of Nature, which drives the evolution of those processes, so theory evolution is governed by the mechanisms of theory-change, which drives the development of theories in a precisely similar fashion. Thus, to continue our astrophysical analogy, as stars may evolve into new kinds of entities (white dwarfs, neutron stars, planetary nebulae, etc.) and as biological organisms may evolve into new species (speciation, adaptive radiation), so will theories evolve into new forms, new theories, or face extinction. The evolution of the former physical processes is determined by natural laws; which are described by the theories themselves and the evolution of the latter is determined by the mechanisms of theory-change. Hence, one may think of scientific theories as various species which evolve through the mechanisms of theory-change, with the latter now regarded as the 'Natural Laws' governing the entities or structures called 'scientific theories'. Then, in the context of the development of science, we may regard the mechanisms of theory-change as the 'Natural Laws' of scientific epistemology and the phenomena to which these laws apply, as the set of all possible scientific theories.

Thus, scientific theories are subject to an evolutionary process of 'Natural Selection' and very much in the Darwinian sense. Certainly, a theory cannot survive under the scrutiny of adverse or contradictory empirical evidence, but it may be modified in some appropriate way in order to accommodate such new criteria. Whilst the theory-change mechanisms we have described will determine such theory evolution, there may well come a point where further evolution, modification, or theory alteration is no longer viable. Thus, the abandonment of a theory may be necessary under the accumulated weight of experimental and observational data and theory replacement then becomes essential. This is the extreme case, but illustrates most clearly that it is Nature Herself that dictates the final fate and evolution of a scientific theory. Yet, there is much more to this than is apparent from so simple an approach and many questions remain unanswered. We shall address briefly some of these questions now, reserving a more complete discussion for the sequel.

Processes in the physical or natural world evolve strictly within the confines, restrictions and stipulations of the gross Laws of Nature, as directly empirically determined or inferentially adduced in a deductive fashion from either the former, or philosophical and foundational principles. We make take all such laws, from a reductionist stance, to be essentially the laws of physics. This position does not affect any of our arguments and no loss of generality ensues thereby, for whatever the nature of the laws governing the evolution of any tangible process within the domain of science, they are always subject to empirical verification or falsification. This is the point alluded to above; that the content of scientific theories are ultimately determined by the dictates of Nature and not by the individual investigator, current fashion or personal preferences.

Then, although theory-change is dictated by Nature, it is not natural laws in the scientific sense that determine such change, for it is the laws or axioms of a given theory themselves that must undergo one or more of the modification processes described by the mechanisms of theory-change and that happens either for empirical reasons, or for reasons of internal logical consistency. Evidently, this is how science progresses. However, the mechanisms of theory-change do not *merely* describe what is commonly referred to as the 'scientific method', for it is quite possible to apply those mechanisms, entirely at random, to any theory or set of theories, without consideration of observational or empirical data. Indeed, there are occasions when this is actually done in the more developed of the physical sciences. A recent example of this is Superstring Theory and M-Theory, where empirical evidence is distinctly lacking and may well continue to be so for the foreseeable future. The point here is that whilst theory-change is dictated by empirical or observational evidence, this need not necessarily be so. Empirical evidence is not required for theory-change *per se*. Hence, the mechanisms of theory-change are not empirically determined, as are the laws of physics, but exist independently and on a different level than the theories themselves. If one takes any given theory S and apply successively to S, in any order, various of the mechanisms of theory-change as described in §3.2, perhaps entirely at random, then one will arrive at a new theory S' that is very different from S, entirely without empirical motivation or justification. This, of course, is not the 'scientific method' as usually understood. However, it clearly *becomes* the scientific method when observational and empirical criteria are the sole motivation for the application of theory-change and theory modification mechanisms. We are, therefore, led to the conclusion that the mechanisms of theory-change allow us to construct all possible theories, but that only those theories which have been constructed with empirical criteria in mind will accord with the workings of the Natural World. Yet, in principle, successive application of the mechanisms of theory-change should produce all valid scientific theories, however sophisticated and ultimately, a so-called 'Theory of Everything'. These mechanisms are thus theory *generators*. The real question is: Can this be achieved in a finite number of steps? We shall return to these matters later. Now, we must take a closer look at theory evolution itself. We shall begin with the initial formulation of a theory.

4.2 Formulation of scientific theories

In the absence of previous knowledge, a new scientific theory can be formulated in only two ways; firstly, by the simple act of hypothesising and secondly, by empirical observation of natural processes. In either case, an axiom is produced and since, for us, an axiom constitutes a theory, then a theory is produced. But merely positing that something is the case, does not, of course, make it so. One can hypothesise indefinitely, but unless some method of verifying the hypothesis is available, such musings amount to nothing more than ill-informed speculation. Thus, the hypothesis, which we can regard as an axiom, should be empirically testable. This, however, is not always possible. For example, the hypothesis 'God made the World in six days', is not verifiable, for there is no clear way to demonstrate the falsehood of this assertion, nor to properly define the terms used. But the hypothesis, 'Under standard atmospheric pressure, water boils at 100° Celsius', can certainly be tested by observation and experiment, if the terminology is understood and the quantities involved are unambiguously measurable in some sense. Care must therefore be taken in formulating a hypothesis in such a way as to allow comparison with observable quantities, which in turn, must be suitably defined.

On the other hand, if an axiom is obtained purely from the observation of natural phenomena, then there is little room for speculation with respect to the apparent validity of the axiom, for it is provided by Nature Herself, but still, continued verification will be necessary, lest errors or misinterpretations have been made. Again, the definitions of the terms used must be amenable to proper measurement. Then, in the formulation of a scientific theory, the importance of clearly defined and directly (or at least indirectly) measurable quantities is paramount if the axioms are to mean anything at all. We therefore presume that given an axiom $f(x_i, c_j) = 0$ the variables x_i and the constant symbols c_j are well-defined and measurable observable quantities or can at the very least, be expressed in terms of such quantities in a way that allows testability in the sense of empirical verification. Then, for the variables and constant symbols of a theory to meet these criteria, some assumptions will have been necessary.

We have mentioned this several times and pointed out that the philosophical nature of such foundational assumptions, although instrumental in the formulation of a theory and are certainly not to be ignored from the epistemological viewpoint of theory construction, are not of primary importance *per se* once the axioms have been formulated.

What is of greater concern in the formulation of theories is that the axioms should be reasonably *general*, but not so general as to be vacuous, or devoid of verifiable content. Thus, our creationist hypothesis above is too general to be of use and is not even testable. On the other hand, the empirical law relating to the boiling point of water is eminently verifiable, but rather narrow in scope. One would not really wish to build a 'theory of water-boiling', even though it might be perfectly acceptable to do so. Rather, one would wish to generalise and consider changes of state of a large class of substances, which then includes H_2O as a special case. We see, then, that the degree of generality of a theory or axiom is intimately related to its domain of applicability and this in turn depends on the choice of the variables x_i and the constant symbols c_j. This is, of course, why we have defined the domain of an axiom as just that subset of the universal domain for which $f(x_i, c_j) = 0$ holds (Definition 2.2.1). The more general an axiom, then the greater is its domain and the less deep or fundamental that axiom will be. Thus, in framing or formulating a theory or axiom, one is also defining the domain of applicability.

It might be thought that the second method of obtaining axioms directly from observation is preferable to the first method of simply forming hypotheses. But, both methods are equally valid if the 'scientific method' of verification, falsifiability and testability are adhered to, for both will always be founded on some set of assumptions or other and no one method can be said to be 'superior' to the other, given such a relation. In this regard, the second method might, because it is in some sense 'given', appear to provide, by virtue of that fact, an already pre-determined domain of applicability for the axioms so obtained, whereas, with the first method, there is an apparent measure of arbitrariness with respect to the domain. But this is not so. We have observed that the domain of an axiom is very much dependent upon the variables employed and these are related to certain underlying assumptions and suppositions of an ontological and philosophical nature and this is clearly true for both of the methods discussed. Thus, whilst the empirical approach of the second method would seem to be the most appropriate for theory formulation, it in fact has no special value in this respect. However, once a theory is formulated and the axioms, variables and constants are well-defined, then the empirical development and evolution of the theory will be entirely determined by the dictates of Nature. There are cases, however, when it is not possible to formulate a theory by hypothesising from more fundamental notions. This is the case with the axiomatic development of (non-relativistic) quantum mechanics. The axioms are simply 'given' as fundamental properties of the Natural World and even if it is the case that the current state of physics does not admit of the derivation of these axioms by the first method, then it is nevertheless true that a valid scientific theory has been obtained empirically, in spite of the absence of a complete understanding of its foundations.

All this is very clear and obvious and is just little more than a reflection of the 'scientific method', but with some differences. Specifically, in the formulation process, we have seen that the domain of applicability of an axiom (or theory) cannot be determined without a precise understanding of the variables employed within the axiom(s), whether constructed by either of the two methods utilised for that purpose. It is this circumstance that constitutes the primary difficulty in the initial formulation of a theory and a judicious choice of the appropriate parameters may require considerable insight and ingenuity in virtue thereof, so as to admit of maximum perspicuity. Yet, especially in practice, this need not always be so. Suitable parameters may well present themselves to the observer through interaction with the physical system under consideration and in this way it may become clear which measurable quantities can best describe the gross properties of the system. However, the revelation of such parameters is achieved by the interaction of the observer with the natural world and this is essentially an empirical process. Thus, although the two methods are equivalent insofar as theory or axiom formulation is concerned and particularly in an ontological sense, there is a distinct advantage in the empirical approach. Nevertheless, the empirical method may well hide the underlying reasons for the truth of the axioms of a theory, because all that is presented is a scientific 'law'. This may be considered to be a disadvantage of this purely 'observational' method. On the other hand, formulating a theory from first principles will clearly display the foundational basis of that theory, with all the assumptions that have been deemed necessary for such a formulation, but it does not follow that the same theory cannot be constructed from alternative principles and in fact, a scientific theory may have many formulations that lead to the same axioms. This is certainly the case for some physical theories and evidently, it is what makes theory reformulation possible.

Let us illustrate these points with a somewhat whimsical example. It is known that the English hold two things to be of fundamental importance in their society; firstly, the beverage known throughout the civilised world as 'tea' and secondly, social status. The former is a solution to any and all of the myriad difficulties and trials that life may inflict upon the hapless mortal, whilst the latter, as exemplified by a rigid class division system, is apparently the necessary foundation for a stable and civilised society. Now, a 'nice cup of tea', as the English would have it, can be of variable density, it can be weak and it can be strong and it can be of any strength between some minimal value and some maximum value. Thus 'tea-strength', T, has a relative density (specific gravity). Suppose there is a diligent sociologist who meticulously measures the tea-strength of several groups of people of variable social status s. After examining his results, he formulates the following 'law', or axiom;

$$T = \frac{k}{s}$$

which says that tea-strength is inversely proportional to social status[*], with constant of proportionality k. Now, this law has been determined empirically. It is true simply because it has been observed to be the case. Continued verification will be necessary because there is no time-dependency in this simple equation. The parameters (variables) involved are T and s and we have assumed that they are well-defined (the constant symbol k, can be eliminated by a suitable choice of units). In actuality, these two quantities were defined prior to the formulation of the above axiom in quite different ways. Tea-strength is both observable and measurable, it is a quantity determined by the laws of chemistry and quite independent of the observer. Thus, physical laws determine the value of T. Social status, on the other hand, has no clear relation with the Natural World (unless one wishes to invoke some evolutionary consideration). The parameter s seems to be entirely arbitrary, perhaps even a consequence of antiquated societal prejudices. Thus, an empirically determined theory (the theory relating tea-strength to social status) is dependent upon the

---

[*] The writer formulated this deep and profound sociological law in his youth by observing his friends and peers. It was a source of much amusement.

definition of the parameters utilised in the formulation and more particularly, the domain of the theory also depends on those variables. In this case, the domain is the set of all English citizens.

Now, let us try to formulate the above in an entirely non-empirical manner. This will be the procedure of the first method of theory formulation described at the beginning of this section. We are, therefore, working from 'first principles' and hence, will need to make a number of initial assumptions before we can proceed. The assumptions usually made in theory formulation are of various types. Some are empirical, some are philosophical and others are of a purely mathematical kind. Amongst the latter, we stipulate that all our variables are continuous and in many instances, infinitely differentiable. We shall discuss some of these latter points shortly. For the present purpose of our seemingly trivial example, we need concern ourselves only with those characteristics of the model that may be reasonably assumed to be true. Then, we make the following observations relating to the properties of the model that may be considered to be 'self-evident', or at least acceptable in this context:

i) There are various types of tea. Initially, there is 'pure' tea, which is just plain tea and is rather strong. In fact it is the 'strongest' tea in the sense that the tea strength T attains a maximum value in this case. But then, there are more sophisticated varieties of tea. These are 'blended' teas and the blending process has the effect of reducing the tea strength. In fact, the more blended a tea, then the less strong that tea will be and we assume the following relationship holds;

$$T = \frac{\alpha}{B}$$

where B is some measure of the 'blending' and α is a proportionality constant.

ii) Pure tea, it is supposed, is relatively inexpensive. But the cost, C, of tea increases in proportion with the degree of blending, because of the greater costs in manufacture. Thus, we have the linear relationship;

$$B = \beta C$$

with proportionality constant β.

iii) When the English buy their tea there will clearly be cost considerations to be made in the selection of the type of tea purchased. Those in the lower income bracket will purchase less expensive tea, whilst those with greater financial resources will be able to purchase the more expensive varieties, depending on income. The cost C of the tea purchased is thus proportional to the income I, of the purchaser;

$$C = \rho I$$

with proportionality constant ρ.

iv) People of higher social status will certainly have greater incomes and again we assume direct proportionality;

$$I = \sigma s$$

where σ is again some constant of proportionality.

These are the principle assumptions that we consider it reasonable to make and it does not concern us here as to whether they are justified or not. Our purposes are illustrative only. Now from (i) and (ii) we have;

$$T = \frac{\alpha}{\beta C}$$

and from (iii) and (iv);

$$C = \rho \sigma s$$

hence;

$$T = \frac{\alpha}{\beta \rho \sigma} \frac{1}{S} = \frac{k}{s}$$

where the constants α, β, ρ and σ have been incorporated into the constant k *and* tea-strength is inversely proportional to social status

Although our example is trivial, very restrictive and indeed, may not even be true, it nevertheless illustrates many of the points made above with regard to theory formulation. We have certainly derived the 'tea' axiom from first principles and the question of the validity of those assumptions does not concern us. Yet, if it is thought that the assumptions and suppositions made here are unreasonable, then consider the assumptions discussed in §2.1 in connection with Newtonian mechanics. The 'reasonability' of the latter is certainly questionable and it is only a matter of degree whether or not our assumptions are similarly dubious. Newtonian theory is based on many other assumptions, such as the continuity of space and time, which are implicitly contained in the foundations of that theory. One could, then, formulate Newtonian mechanics on even more fundamental principles and this would regress, ultimately, into a detailed philosophical discussion that, in the final analysis, becomes merely linguistic or semantic. Then, assumptions are unavoidable and however counter-intuitive they may appear to be for humans, their justification must ultimately be empirical. Yet, we could easily have formulated our tea axiom with a different set of suppositions, or indeed many such schemes. Hence, it is not the assumptions that are of primary importance in theory formulation, for if there are several ways to construct a theory from first principles, then evidently, this merely demonstrates our ignorance of the deeper underlying and much more fundamental principles that have, thus far, escaped the embrace of science. Such basic principles will be of a unifying nature and hopefully reveal the foundation of Nature Herself. But this is an issue we need to discuss separately in the sequel. This brings us to the other class of suppositions subsumed under theoretical science.

Among the many additional assumptions that must necessarily be pre-supposed in order to construct any rational view of the world, perhaps the most important is the nature of the language and symbolic structure thereof. In the developed sciences, this is mathematics. Thus, the full symbolism of mathematics is used to express the axioms of our theories. Then, all the assumptions at the foundations of mathematics are simply held to be true. We accept the real and complex number systems, set theory and anything else that may be at our disposal. Additionally, the principles of logic and all rules of inference, are taken as given and this, whether we are using a symbolic or Natural Language. But, we have taken the axioms themselves as fundamental theories rather than constructing theories using the model-theoretic approach. This is because we are not concerned with foundational or ontological issues here. Our primary goal is to understand how scientific theories evolve and thereby their methodological status. Thus, we may assume that axioms are obtained in their entirety and consider foundational issues only when necessary. Thus, the structures we call 'scientific theories' are axiom schema together with an interpretation of the logical variables and constants. As theories, they are sets of axioms and logical consistency requires that they be complete theories. They are models in the conventional sense, if one accepts that they are interpreted theories from the outset. By taking this approach, we can more readily discuss the gross properties of scientific theories in the context of the development of science itself and indeed, independently of the agents of scientific discovery. We shall have many opportunities in the sequel to expand upon these important epistemological and ontological issues and in fact will return to them constantly.

4.3 Theory change operators

Once a scientific theory has been formulated, by either of the methods discussed in the last section, it will then be subject to a battery of empirical tests in order to substantiate or invalidate the theory, in the usual, familiar way. If the theory survives such scrutiny, then all is well, at least temporarily. In this case one may consider extending the theory, which may well mean extending its domain of applicability, or one may wish to combine the theory with another, or cast it into a different form. Thus, even if a theory is successful in its description of the physical processes within its domain, it will still not, in general, remain static, but will undergo changes of various sorts in the course of its lifetime. Evidently, such changes will be accomplished by means of one or more of the mechanisms of theory-change that we have postulated in the previous chapter. This may occur simultaneously with several scientific theories, with each one tending to increase its domain of applicability to the maximum possible subset of the universal domain that is attainable within the bounds of the given theory set by its axiomatic structure. If the domain of a theory is ever equal to the universal domain, then clearly no further modification, other than re-formulation, is possible. In this case one would, of course, have a Final Theory. Hence, we may think of theories as organisms whose primary evolutionary objective in life is to extend their domains to the limits of the universal domain and this is why even a satisfactory theory will continue to evolve.

On the other hand, if a theory is internally logically inconsistent or fails to satisfy empirical criteria, then obviously theory-change is inevitable and will generally proceed quite quickly. Note however, that if a theory is not consistent with another, by which we mean here that the axioms of the one are logically inconsistent with the other, then this only means that one of the two theories requires modification and precisely which one remains to be determined. In any case, once again, the mechanisms of theory-change will be brought into play and scientific progress will inexorably proceed toward the final goal.

Now, the progress of science is often thought of as a continuous process. Theories, knowledge and consequently the understanding and comprehension of the Natural World is a gradual process, that continues unabated as new generations of scientists refine and improve the achievements of a former generation. Only occasionally are really *radical* steps taken and an actual 'scientific revolution' occurs. The latter will usually be an entirely new theory, with new conceptual foundations and fundamental axioms. However, between such relatively rare and catastrophic events, science slowly progresses by a process of minor theory refinement and empirical and mathematical book-keeping. But this is not really so. The picture of scientific advancement painted here is far too simplistic. In the first instance, truly catastrophic events do not occur in science, for in retrospect it is seen that such apparently major upheavals are merely the culmination of many smaller steps that may have gone unnoticed by many practitioners of the period, giving the impression of great theoretical change, that is now seen to be quite inevitable with the benefit of hindsight. Secondly, scientific progress is not continuous. This is clear from the mechanisms of theory-change, which are specific, discrete events, far from continuity even over a significant interval of time.

There are many ways in which one can think of the mechanisms of theory-change as acting upon a theory. One particular view is to consider each of these mechanisms as an operator T that takes a theory S and produces a new theory S';

$$TS = S'$$

with a different operator T_i for each of the mechanisms introduced in §3.2. A theory is defined by its set of axioms and therefore, the operators T_i act on the axioms of the theory S and this axiom set is then the domain of the operator. We may list the mechanisms of theory-change by their respective operators:

Mechanism	T_i
Axiom Addition	T_+
Axiom Subtraction	T_-
Theory modification	$T_\exists$
Theory Replacement	$T_\lor$
Theory Reduction	$T_\cap$
Theory Reformulation	$T_\equiv$
Theory Unification	$T_\sqcup$
Axiom Combination	$T_\square$
Set-theoretic Union	$T_\cup$

where we have included axiom combination $T_\square$ and ordinary set-theoretic union $T_\cup$, although the latter is not strictly to be regarded as distinct from T_+. We shall write T^2 for an operator that acts twice in succession on a theory S, as in $T[T(S)] = TTS$ and in general T^n for the case of n such operations. Note that $T^0 = I$ is the unit or identity operator.

The operators of axiom addition and subtraction, T_+ and T_- simply alter by unity the axiom number of a theory, thus;

$$T_+ S = S'$$
$$T_- S' = S$$

such that $|S'| = |S| + 1$ in the first case and $|S| = |S'| - 1$ in the second. From the first of these equations we have, in accordance with §3.4;

$$\mathcal{D}(S') \le \mathcal{D}(S)$$

whilst in accordance with §3.5, the second equation gives;

$$\mathcal{D}(S') \ge \mathcal{D}(S)$$

from the above, we also have;
$$T_- T_+ S = T_- S' = S$$
$$T_+ T_- S' = T_+ S = S'$$
and one is led to conclude that;

$$T_+ T_- = I = T_- T_+$$

where I is the identity operator. But, these operators are only inverses of each other if it is the *same* axiom φ, that is being both added or removed. If this is the case, then the last equality holds and one should more properly indicate this fact by writing T_+ as $T_+(\varphi)$ and similarly for T_-. We continue with a brief discussion of each of these operators.

The theory modification operator T_3, takes a theory S and replaces some of the axioms $\varphi_j \in S$ with j new axioms ψ_j, to produce the modified theory S';

$$T_3 S = S'$$

but in this case, of course, the axiom number is constant, $|S'| = |S|$. The domain of S' is dependent on the domains of the new axioms in accordance with definition 2.2.2, and may be equal to, less, or greater than that of S. If one operates with $T_\exists$ a second time;

$$T_\exists^2 = T_\exists T_\exists S = T_\exists S' = S''$$

then we observe that if $S = S''$ then $T_\exists T_\exists = I$ and $T_\exists$ is its own inverse. But this can only occur when the j axioms ψ_j, which on the first application of $T_\exists$ replace the axioms φ_j are by the second application of $T_\exists$, themselves replaced by the original axioms φ_j and in this very special case we obviously have $\mathcal{D}(s) = \mathcal{D}(s')$. We see that theory modification can be expressed in terms of the operators for axiom addition and subtraction, T_+ and T_- since for a theory S with axiom number $|S| = n$;

$$T_\exists S = T_+^k T_-^k S = S'$$

where $k < n$.

The theory replacement operator $T_\forall$ takes all of the axioms of S and replaces them by new axioms;

$$T_\forall S = S'$$

and obviously, $|S'| = |S|$ only in the case where each one of the old axioms are replaced by exactly one new axiom. Theory replacement is therefore an extreme case of theory modification. Note that if each $\varphi_j \in S$ is replaced by the same axiom φ_j, then $S' = S$ and $T_\forall$ is just the identity operator I. If we are given a theory S with $|S| = n$, then clearly;

$$T_\forall S = T_+^k T_-^n S = S'$$

where all n axioms of S have been removed and replaced by k new axioms and we shall have a new theory S'. Then, theory replacement is just repeated application of axiom subtraction and axiom addition. This applies also to theory modification, but in the latter case only to a subset of the theory S. As mentioned earlier, theory replacement is essentially the complete reconstruction of a theory and therefore preserves the domain, but not necessarily the axiom number so that k may be different from n.

On the other hand, theory reformulation is a mathematical process. The new presentation of the theory S is obtained by manipulation of the axioms of S to obtain the theory S'. Thus theory reformulation is essentially axiom combination applied to a single theory. Then, this process can be written using the axiom combination operator $T_\square$;

$$S' = T_\equiv S = T_\square^k S$$

where the operator $T_\square$ has been applied k times. Of course, for the case of a single theory, we must have that $\mathcal{D}(S) = \mathcal{D}(S')$. We saw an example of this in §3.9 for the case of Hamiltonian mechanics. Clearly, if $|S| = n$, we require that $k \le n/2$.

The theory reduction operator $T_\cap$ is rarely used and is no more than set-theoretic intersection, it was discussed in §3.8 and operates over a number of theories S_n, we therefore write;

$$T_\cap S_n = \bigcap_{i=1}^{n} S_i = S$$

and similarly, set theoretic union is the operator;

$$T_\cup S_n = \bigcup_{i=1}^{n} S_i = S$$

This is no more than a change in notation, which we adopt for uniformity.

Then, with this latter notation, we recall from §3.10 that theory unification proceeds first with ordinary set-theoretic union, followed by axiom combination. Thus;

$$T_\cup S_n = T_\square^k \left(T_\cup S_n \right) = T_\square^k T_\cup S_n = T_\square^k S = S'$$

where, of course, $T_\cup$ is applied once over all the S_n, followed by k applications of $T_\square$.

It is apparent from the above that the mechanisms of theory-change fall into two distinct categories. There are those dealing with a particular *single* theory and those which operate with a multiplicity, or plurality of theories. Of the first category, we have the operators T_+, T_-, $T_\exists$ and $T_\equiv$. Since the latter three can be expressed, at least from an operational point of view, in terms of the first two, we must regard T_+ and T_- as more 'elementary' in some sort of fundamental sense. They can be regarded, in the context of single theory-change, as bearing a similar status to 'atoms' in relation to the gross structure of matter, or prime numbers with respect to the positive integers.

In the second category, we have the operators $T_\cap$, $T_\cup$ and $T_\sqcup$ which operate on several theories at once and possess no meaning otherwise. The first two of these are simply the familiar set-theoretic operations of union and intersection and require no further explanation. The last, however, theory unification, $T_\sqcup$ is of a different status because it is a dual and ordered operation. It involves two processes, the first of which is set-theoretic union and the second axiom combination.

Now, axiom combination has the property that it applies to both single and multiple theories. It is therefore more embracing in this respect than the other operations; it is more 'universal' and as such enjoys a different ontological status because, although it operates on axioms, it does so by a manipulative process that is essentially deductive. In fact, we may state the following:

> **When axiom combination is applied to a single theory, it is Theory Re-formulation and the domain is constant.**

> **When axiom combination is applied to the set-theoretic union of several theories, then it is Theory Unification and the domain increases.**

We saw an example of the latter assertion in §3.10 when we considered the Magnetohydrodynamic (MHD) approximation.

4.4 Theory evolution

When a scientific theory is first formulated, by whatever methods, it consists merely of axioms and nothing else other than definitions of the variables and constants that partake in the expression, be it mathematical or linguistic, of those axioms. The axioms are subject to the various processes described in the previous section, i.e. they are *operated* upon and hence *changed* by the actions of those operations. This is a necessary process, brought about by empirical demands that require an axiom to be substantiated, verified, or falsified. There is no subjectivity here, only the requirement of verisimilitude. Nor is a sentient being a necessary agent in this process. It is important to realise, from our viewpoint, that it is the theories themselves that must satisfy the empirical criteria of verification because they are representative of the Natural World and hence, a part of the latter. Consequently, they must accord with the dictates of that final authority. The intervention of sentient beings in this process results in nothing more than a series of observations of the natural world, which, as a result, provides a formulation of the principles or axioms contained therein. It must not, therefore, be thought that such sentient beings themselves construct theories, for we take the view, mentioned repeatedly here, that it is the axioms alone that constitute the theory, they *are* the theory and the whole of the theory and the theory is no more than its axioms. Such inquisitive creatures may well think and ponder a great deal *about* the axioms and even construct them from non-empirical methods, as we discussed in relation to theory formulation, but the axioms themselves are indifferent to the personal prejudices and musings of any particular intelligent and sentient agent, or whatever biological organism may be endowed with such qualities.

So it is with the theory-change mechanisms that we have described. The latter are universal in the sense that *any* sentient being must employ them in order to ascertain the correct Laws of Nature. Whilst these mechanisms can be expressed in many forms, they become most simply realised as the operators of the last section and it is in this form that we may most perspicuously observe their actions in the process of theory evolution and development.

The first thing to note is that if a single theory $S = \{\varphi\}$, containing just one axiom is acted upon repeatedly by the operators T_- and T_+ as follows;

$$S' = T_+^n T_-^n S$$

which continually removes and replaces the axiom φ with a new axiom, then eventually, as n increases, a more suitable and representative axiom will be found. Indeed, as $n \to \infty$, it is certain that a true Law of Nature will result. Clearly, this is a random process that is quite independent of the investigator. It is a simple example of the *evolution* of a scientific theory by a random process of 'natural selection' and evidently, we can generalise this to systems of ever greater complexity. Thus, we may discard the notion that the scientific investigator, by employing the scientific method, is in any way subjective, for the operators do not suffer from such a handicap and simply act in accordance with mathematical and probabilistic rules. But, this is exactly what the investigator does also, except, perhaps, with a little empirical guidance.

Since a theory consists only of its axioms, any theory with n axioms can be written as the union of n single-axiom theories. Thus;

$$S = \{\varphi_1, \varphi_2, \ldots, \varphi_n\} = S_1 \cup S_2 \cup, \ldots, \cup S_n = \bigcup_{i=1}^{n} S_i = T_\cup S_n$$

where the index i corresponds to each of the single-axiom theories S_i.

Then, the former expression for a theory with multiple axioms can be written as;

$$S' = \bigcup_{i=1}^{n} T_+^k T_-^k S_i = T_\cup T_+^k T_-^k S_n = T_\forall S$$

which replaces each axiom of the theory S with a new axiom to produce the theory S' and is therefore equivalent to the theory replacement operator $T_\forall$, now applied to a single theory with multiple axioms. We can see how this works if we write out this expression explicitly;

$$T_+^k T_-^k S_1 \cup T_+^k T_-^k S_2 \cup, \ldots, \cup T_+^k T_-^k S_n$$
$$= T_+^k T_-^k \{\varphi_1\} \cup T_+^k T_-^k \{\varphi_2\} \cup, \ldots, \cup T_+^k T_-^k \{\varphi_n\}$$

and each axiom is replaced k times. The new theory S' is then just the union of all the new axioms. Again, this is an entirely mechanical procedure. We may further generalise by allowing different operators to have different powers, i.e. each operator acts either k or l times on a given axiom φ_i, or theory S_i. We indicate the axiom that is operated upon by a subscript i on the powers k and l;

$$S' = \bigcup_{i=1}^{n} T_+^{k_i} T_-^{l_i} S_i = T_+^{k_i} T_-^{l_i} \bigcup_{i=1}^{n} S_i = T_+^{k_i} T_-^{l_i} T_\cup S_n = T_\exists S$$

which is theory modification. Theory replacement is then the case when $l = k$. Again, writing this out explicitly for the single axiom theories S_i;

$$S' = T_+^{k_1} T_-^{l_1} S_1 \cup T_+^{k_2} T_-^{l_2} S_2 \cup, \ldots, \cup T_+^{k_n} T_-^{l_n} S_n$$
$$= T_+^{k_1} T_-^{l_1} \{\varphi_1\} \cup T_+^{k_2} T_-^{l_2} \{\varphi_2\} \cup, \ldots, \cup T_+^{k_n} T_-^{l_n} \{\varphi_n\}$$

Let us make this more explicit with a specific example, let $S = \{\varphi_1, \varphi_2, \varphi_3\}$, then $S_1 = \{\varphi_1\}$, $S_2 = \{\varphi_2\}$ and $S_3 = \{\varphi_3\}$, $S = S_1 \cup S_2 \cup S_3$. Now, if we take $k_1 = 0$, $l_1 = 1$, $k_2 = 1$, $l_2 = 1$, $k_3 = 1$, $l_3 = 1$, then the above expression becomes;

$$S' = T_+^0 T_- \{\varphi_1\} \cup T_+ T_- \{\varphi_2\} \cup T_+ T_- \{\varphi_3\}$$
$$= \{\psi_2\} \cup \{\psi_3\} = \{\psi_2, \psi_3\}$$

Note that the axiom φ_1 has been removed, whilst φ_2 and φ_3 have been replaced by the new axioms ψ_2 and ψ_3, respectively. Thus, theory modification and theory replacement are completely specified by the above expression, in terms of axiom addition and axiom subtraction.

It is evident, therefore, that for the case of multiple theories, one simply replaces the single theory S, in the first instance, by the union of the component theories that constitute S, namely, the S_j;

$$S = S_1 \cup, \ldots\ldots\ldots, \cup S_m = \bigcup_{j=1}^{m} S_j = T_\cup S_m$$

for m theories S_j. Now, each of the theories S_j contains $\alpha_j = |S_j|$ axioms, which is the axiom number of S_j. Thus, each of these component theories can be expressed, as before, as the union of α_j single-axiom theories $\{\varphi_i\}$;

$$S_j = \{\varphi_1, \ldots\ldots\ldots, \varphi_{\alpha_j}\} = \{\varphi_1\} \cup, \ldots\ldots\ldots, \cup \{\varphi_{\alpha_j}\} = \bigcup_{i=1}^{\alpha_j} \{\varphi_i\}$$

Then, the theory S is written as a 'double' union;

$$S = \bigcup_{j=1}^{m} S_j = \bigcup_{j=1}^{m} \bigcup_{i=1}^{\alpha_j} \{\varphi_i\}$$

The total number of axioms involved is, clearly;

$$|S| = |S_1| + |S_2| +, \ldots\ldots\ldots, |S_m| = \sum_{i=1}^{m} S_i$$

Thus, for theory unification we would write;

$$S' = T_\sqcup T_= S = T_\square^k S = T_\square^k \bigcup_{j=1}^{m} \bigcup_{i=1}^{\alpha_j} \{\varphi_i\}$$

where, of course, the component theories S_i, must be mutually incompatible. Obviously, if $m = 1$, we have the single theory case, which is just theory reformulation, as we discussed earlier and because the domain is unchanged, no questions of compatibility arise and axiom combination takes place within the domain of the existing theory. We see that theory unification is nothing more than theory reformulation over the set-theoretic union of mutually incompatible theories. Considering the latter part of §4.3, we observe that through the process of axiom combination theory re-formulation and theory unification are equivalent processes. This is hardly surprising, since axiom combination is simply a manipulative process that applies, by virtue of that mechanical procedure, in both the case of a single theory and the case of several theories, the latter being considered as a single theory after the set-theoretic union of its components subject to the criterion of domain incompatibility. It is clear that the mechanisms of theory-change are not independent, but can all be expressed in terms of the fundamental operators T_- and T_+; by which means any change in a scientific theory may be described in a precise and mechanistic way. The scientific enterprise is thereby seen to be a definite and discernable process, which in principle could be performed by a suitably sophisticated machine. We shall examine this concept later.

4.5 Evolution equations

Since theory replacement is just a special case of theory modification and since theory unification is equivalent to theory reformulation, we need only consider the operators $T_\exists$ and $T_\equiv$ in expressing the notion of theory-change. Firstly, we consider the equation for theory modification. Given any theory $S = \{\varphi_1,..., \varphi_n\}$, it can always be written as the union of single-axiom theories $\{\varphi_i\}$. This, of course, is just a particular partition of the set S, of which there are many others. Thus, any theory is the disjoint union of 'sub-theories', which are just the members of the partition. If for an arbitrary partition of a theory S, there are m such 'sub-theories' S_i, each of axiom number $\alpha_i = |S_j|$, then, as with the case of theory unification, we can write;

$$S = \bigcup_{j=1}^{m} S_j = \bigcup_{j=1}^{m} \bigcup_{i=1}^{\alpha_j} \{\varphi_i\}$$

and we need make no distinction between single and multiple theories. We may now write the expression for theory modification as;

$$S' = T_\exists S = T_+^{k_i} T_-^{l_i} \bigcup_{j=1}^{m} S_j$$

$$= T_+^{k_i} T_-^{l_i} \bigcup_{j=1}^{m} \bigcup_{i=1}^{\alpha_i} \{\phi_i\}$$

We also have the equation for theory unification;

$$S' = T_\cup S = T_\cap^k S = T_\cap^k \bigcup_{j=1}^{m} \bigcup_{i=1}^{\alpha_j} \{\varphi_i\} = \bigcup_{j=1}^{m} S_j = T_\equiv S$$

These two expressions completely describe any and all of the mechanisms of theory-change that we have discussed. All the other operators are of the ordinary set-theoretic kind such as $T_\cup$ and $T_\cap$ and this also includes the fundamental operators T_- and T_+. Following our earlier discussion, it does not matter whether we are dealing with single or multiple theories in these expressions, provided that in the latter case the component theories, as we have seen, are mutually incompatible. Given that this latter condition is satisfied, then as far as the theory operators are concerned, we are dealing with a single theory and furthermore, every such theory can be expressed as the set-theoretic union of single-axiom theories. This is simply a consequence of the fact that any set can be partitioned into disjoint subsets and in this case, the subsets themselves are further partitioned into singleton subsets.

The two fundamental *theory evolution* equations;

$$T_\exists S = S'$$
$$T_\cup S = S'$$

can be combined into just one equation;

$$T_E S = T_\exists T_\cup S = S'$$

where $T_E = T_\exists T_\cup$ is the *theory evolution operator* and this expression is the *evolution* equation. The totality of all the mechanisms of theory-change is encoded in this simple equation, but that is inevitably so because of the simple formalism adopted. All our operations are essentially set-theoretic and they are so because of the choice of axiomatic structure for scientific theories. Essentially, the adoption of scientific laws as the primary axioms of a theory has allowed us to manipulate those theories in a way that would not be possible with an alternative approach.

Let us write the evolution equation in a more explicit form;

$$T_E S = T_{\exists} T_{\cup} S = T_+^{k_i} T_-^{l_i} T_{\Box}^r S = T_+^{k_i} T_-^{l_i} T_{\Box}^r \bigcup_{j=1}^{m} \bigcup_{i=1}^{\alpha_j} \{\varphi_i\}$$

$$= T_+^{k_i} T_-^{l_i} T_{\Box}^r \bigcup_{j=1}^{m} S_j = T_+^{k_i} T_-^{l_i} T_{\Box}^r T_{\cup} S_m = S'$$

With this expression for the evolution equation we can recover the theory-change operators:

Case 1; if $r = 0$, $l_i = 0$, $k_i = n$, then;

$$T_E S = T_+^n T_-^0 T_{\Box}^0 S = T_+^n S$$

Case 2; if $r = 0$, $l_i = n$, $k_i = 0$, then;

$$T_E S = T_+^0 T_-^n T_{\Box}^0 S = T_-^n S$$

Case 3; if $r = 0$, then;

$$T_E S = T_+^{k_i} T_-^{l_i} T_{\Box}^0 S = T_{\exists} S$$

Case 4; if $r = 0$, $k_i = l_i$, then;

$$T_E S = T_+^{k_i} T_-^{k_i} T_{\Box}^0 S = T_{\lor} S$$

Case 5; if $k_i = l_i = 0$, then;

$$T_E S = T_+^0 T_-^0 T_{\Box}^r S = T_{\equiv} S = T_{\cup} S$$

Case 6; if $r = k_i = l_i = 0$, then;

$$T_E S = T_+^0 T_-^0 T_{\Box}^0 S = T_+^0 T_-^0 T_{\Box}^0 T_{\cup} S_m = T_{\cup} S_m$$

We did not include $T_{\cap}$ here because it is not strictly an operator that contributes directly to theory alteration *per se*, but is more appropriate for the isolation of conceptual issues relating to a theory prior to theory-change. Of course, $T_{\cap}$ can easily be added if it is thought to be significant.

In this respect, we note that *any* additional operator can be included in the evolution equation, for the latter is simply of the form;

$$T_E S = T_1, T_2, T_3, \ldots\ldots\ldots, T_n S$$

and any T_i may be added if it is thought to be necessary. Thus, should the mechanisms of theory-change prove to be inadequate for some future case, or it appears that an important mechanism has been overlooked, then we can simply introduce an additional operator to take account of such a circumstance. Then, perhaps paradoxically, the evolution equation may itself evolve and in fact, will do so by the very mechanisms of theory-change that it purports to describe. This is not, however, of any real import and indeed, the evolution equation was obtained by axiom combination, as it had to be. There is no paradox here, the use of a technique to describe that technique does not preclude the technique from legitimacy, or introduce inconsistency of any kind; it is merely a property of language and meta-language, a distinction we have thus far, ignored.

Leaving this philosophical point aside, if such be possible in this context, we enquire as to whether the operations envisaged above completely describe the theoretical progress of scientific theories in the sense of the elucidation of natural laws or, as we continue to insist, scientific axioms. The evolution equation contains operators that act upon a theory considered only as a set of axioms and produces a new theory that may, or may not contain some or none of those axioms. It is to be considered as a purely mechanistic process. That is to say, that in principle, no directional agent is required in order to modify or change the content and structure of a theory. It is a random process that does not require the intervention of sentient beings. This is not to say such intervention does not occur, that would be absurd, for theories do not construct themselves. Rather, we wish to say that scientific theories, or for that matter, *any* theory which requires empirical verification, follows systematic laws that are divorced from the agent that may influence their progress or development. Thus, left to themselves, the operators will produce all possible theories and some of these will accord with reality. But of course, as indicated, a sensible agent is required to initiate the theory formulation and developmental procedures in the first instance. Indeed, one may continue, the operational entities in the evolution equation will not act without purposeful intervention by an intelligent agent and are therefore dependent on such. However, this is to miss the point. The mechanisms of theory-change describe only how theories must change and as such are entirely non-subjective. The essential point is that no ingenuity or intuition on the part of the investigating agent is, in principle, necessary, although in practice this may not be the case. Hence, theory evolution is governed by specific rules that are determined by the empirical process itself.

Thus, we need to specify precisely how this process occurs. In particular, we need a mechanism by which sensible beings can interact with the Natural World to produce theories thereof. This is not described by the evolution equation that we have posited. Nor is it clear that it is necessary for such an interaction between a sentient being and the empirical world to take place at all. There is a problem here, for an investigative agent, of whatever mechanical or biological form, appears to interact with the World through its sensible equipment and thereby discern the nature of physical laws, but reciprocally, the Natural world interacts with the investigative agent. Then, the intelligent agent is itself part of the process and no distinction between the observed and the observer is necessary. Hence, with such a participatory view, no specific mechanism of interaction is required and the investigative entity becomes an integral part of the investigative process. We are thus compelled to regard the experimental and observational methods of science as part and parcel of sensibility itself, which cannot be divorced from the environment in which sense perception takes place. In effect, the scientific method is nothing more than an inevitable consequence of the structure of the World and the empirical process exemplified by that method could not be otherwise.

All this may seem to be a little too vague and philosophical, so let us try to be more specific and perhaps, controversial. A theory, being nothing more than its axioms, is not actually constructed independently of Nature Herself. If axioms are postulated and accord with the observed properties of the World, then all is well, but of course, *any* axiom whatever can be so postulated. Yet, in practice this degree of arbitrariness rarely happens. When it does, however, it is usually due to some non-empirical, subjective influence, such as may be motivated by religious or similar irrationalities. Whilst such matters are extraneous to our context and thus fall under a quite different remit, they would still partially accord with our dictates, for the evolution equation allows axioms to be changed arbitrarily by any of the operators acting therein and if sufficient axiom modification is allowed, such speculations must eventually lead to a correct description. Since also, by the empirical method, a correct description must also be obtained and again, since the investigator is participatory in this process, then in all cases a valid and accurate rendition of natural phenomena must at some stage result. Hence, because the observer is an integral part of the process and he is subject to the same physical laws as apply to the external world, the only difference between these two approaches is one of mere expediency. The evolution equation does not care about this and operates independently of such methodology. Any number of axioms may be generated by the theory-change operators and sooner or later a correct axiom will be found, but the process is given a special impetus by the empirical method of science, as is clear historically, particularly since the more serious incorporation of such techniques since the time of Galileo. Then, theories can be obtained in an entirely random fashion and indeed, the evolution equation demands this, but the 'randomicity', to coin a neologism, is much reduced by the proper adoption of the scientific method. This is so because adherence to the gross facts of Nature effectively minimizes error. One would certainly expect this to be the case and it is easy to see how it works in practice. Assuming, for simplicity, that the number of possible hypotheses (axioms) are large, but finite, we may postulate n such axioms. The theory $S = \{\varphi_r\}$ containing any one of these axioms may be operated on repeatedly and the evolution equation takes the form;

$$\{\varphi_{r+1}\} = T_E\{\varphi_r\} = T_+T_-\{\varphi_r\} = T_\forall\{\varphi_r\} = T_E^r\{\varphi_1\}$$

where $1 \leq r \leq n$. Thus the initial hypothesis $\{\varphi_r\}$ is replaced by the next hypothesis $\{\varphi_{r+1}\}$ until a 'correct' axiom is found. Obviously, this is very time-consuming and hardly the most efficient method with which to extend scientific knowledge, but in principle it will work. If however, the correct form of the axiom can be determined directly from observation or experiment, then the above process is either not necessary, or only minimally so. In that case one may set $r = 0$, so that $\varphi_{r+1} = \varphi_1$. This is very satisfactory. The evolution equation accords with the scientific method as regards empirical laws. However, some degree of verification of an axiom will always be necessary, particularly as there is no time-dependency here. This brings us to the question of verifiability in general. As r increases; how does one know when the correct axiom has been obtained? Clearly, at every stage of replacement, empirical verification will be essential; otherwise the above 'axiom-finding' algorithm cannot halt. Then, testability is required for every value of r and this too, is part of the scientific method.

The question of what it means for a theory to be 'testable' or 'verified' is a complex one. It is intimately connected with the very notion of a scientific theory itself and the empirical process. The problems in defining these terms cannot be avoided and therefore, in order to reach a satisfactory understanding of how these different concepts interact with the observation process and the empirical methods of science, we shall have to deal with them. We shall attack these methodological and ontological issues in the next chapter.

Chapter 5

The Ontological Status of Scientific Theories

Philosophical questions relating to the nature of scientific theories and empirical knowledge are discussed. A brief revue of model theory is given. The ontology and epistemology of theories is investigated in that light and the idea of a 'natural interpretation' is postulated. We identify two types of scientific theory.

5.1 Validity of scientific theories

A scientific theory is never 'proved' in the mathematical sense of that term, for such would have no meaning in this context. Rather, scientific theories undergo a continual process of verification, falsifiability, or testability. These concepts are usually subsumed under the all-embracing, but not very enlightening term 'scientific method'. In practice, as is well-known, this merely consists in establishing whether a given hypothesis conforms with, or continues to conform to, the behaviour of the Natural World. If indeed, a theory is found to correctly describe the phenomena within its domain, by means of comparison with observed empirical data, then this is thought to constitute, 'evidence' for the 'correctness' of the theory and we say that the theory has been 'corroborated', or empirically 'tested'. Yet, still the theory is not proved. So this process continues. If no natural phenomena are found to be in conflict with the theory, perhaps over a considerable period of time and after much 'testing', then we may even say that the theory has been 'established', but still we do not say that it has been 'proved'. Then, it is clear that a scientific theory can never be proved in the mathematical sense of being completely beyond doubt, for the next series of empirical observations or experimental results may produce data that are inconsistent with, or contrary to the predictions of the theory and replacement or modification of the latter may then be required. In fact, as has been frequently pointed out, in principle it only takes one such observation to 'falsify' a theory, but in practice, it is likely that considerable reproduction of such a 'negative' result would be deemed necessary before radical theory modification or replacement is envisaged. And, if such drastic steps are indeed taken and a new or modified theory is adopted, then the whole procedure of empirical verification and testability also begins anew.

We may ask then, at what stage of this repetitive process would we consider a theory to be 'correct', or *valid*? Should a scientific theory be regarded as correct if it has passed all the empirical tests to which it has been so painstakingly subjected and yet still found to be in accordance with the empirical tests for perhaps a thousand times, or maybe ten-thousand or even a million times? Are we to assume that there is a 'degree' of 'correctness' in this sense; a theory being 'more correct' if it has been successfully tested one thousand times rather than one hundred times? Obviously, such a notion is absurd. Surely, a theory is either correct or it is incorrect, valid or invalid, with no variability in measure of these attributes? Or perhaps we could assign some sort of 'probability of correctness' that varies in direct proportion with the number of times the theory has satisfied the empirical criterion of testability. Clearly, that too would be equally absurd, for if one theory S is validated in such a way n times and another theory S' is validated $n+1$ times, but fails the $(n+2)nd$ empirical test, then S', on this assumption, is 'probably more correct' than S, even though S' has been 'falsified'. Then, we might attempt to refine our definitions of 'correctness', 'testability', etc, but we will still be faced with the same problem of exactly what is meant by 'scientific proof' and the 'correctness' of a scientific theory. Eventually, we may abandon all such attempts at clarity and simply agree to accept a scientific theory on 'faith' if it is well-tested and accurately describes the natural phenomena within its domain. But this most certainly will not do at all. Science is not a religion and cannot allow such subjectivity to enter its province or influence its methodology.

Nor need it do so. The philosophical and epistemological issues raised above primarily arise because no clear definition of a 'valid' scientific theory has been given. If we could specify some precise criterion by which we could call a theory 'valid' and 'invalid' otherwise, then many of these difficulties would disappear. If such were the case, then also the vague, ill-defined terminology invariably employed in this context could also be given a precise meaning. What is also clear is that the concept of 'validity' in relation to scientific theories cannot be based upon empirical notions of 'corroboration' or 'verification', for to do so simply invites the confusion that we have already alluded to. This is to say that any definition of the validity of a theory must be independent of the process by which the theory is verified, for if it is not, then since the verification process is never complete, no such definition could ever be precise.

In previous chapters, we have constantly emphasised the importance of the empirical method in the determination of the axioms of our scientific theories. We have even suggested that a theory is 'valid' if its axioms conform to the gross facts of Nature. Yet we have left this term undefined, assuming that some sort of meaning can be attributed to it in such a sufficiently precise way so as not to render our discussion vacuous, or even nonsensical. We see, in the light of the above, however, that if we are to understand the scientific process in empirical terms, then we must have some sort of working definition of validity. To this end, we need only reflect upon our conception of a scientific theory as consisting of just the axioms of which it is composed. Recall that the language in which the axioms are expressed is a first order predicate language and our well-formed formulas or axioms are already interpreted. Then, such an interpretation is just a 'model', in the conventional sense. This 'ready-made' interpretation of the theories and axioms is provided by the empirical process; by the very observation of natural phenomena that permits the variable and constant symbols in an axiom $f(x_i, c_j) = 0$ to be interpreted as measurable physical variables such as mass, velocity, or constants *a la* Boltzman, Planck, or the speed of light. It is *Nature* Herself that is providing the interpretation through the participatory act of observation. Of course, many other interpretations are possible, but such interpretations may not correspond to physical facts. Then we can imagine the set of all possible theories, with all possible interpretations. If we denote the latter by $\mathcal{T}$ then there is a subset $\mathcal{N} \subset \mathcal{T}$ that contains all the theories that describe natural phenomena; the theories or structures that are the embodiment of the Natural World. In effect the Laws of Nature under this 'natural' interpretation. Then, we can say that a theory S is *true* if it belongs to $\mathcal{N}$, i.e. $S \in \mathcal{N}$ and that it is *false* if $S \notin \mathcal{N}$. This does not mean that a theory is valid simply because it has passed all empirical tests, for that situation can easily change and in the history of science has often done so. On the contrary, a valid theory in this sense is valid for all time; it is a structure that exactly reflects the structure of the Natural World. Hence, a theory is valid if it is identical to one of the members of $\mathcal{N}$. The domain of the interpretation of a theory (model) is not the same as the domain of applicability, they are different entities and only in a platonic world will the latter be a subset of the former.

Using the ideas of model theory we can specify our theories more precisely and then define the notion of validity in terms of the more familiar idea of satisfaction within a model. We are not interested in the finer details of the formalism here; our primary purpose is of a more philosophical nature and we enter into just enough detail to enable the ontological status of a scientific theory to become sufficiently clear as to admit of an unambiguous presentation of that ontology within the current context. Hence, some 'primordial' excursion into the language and terminology of the relevant concepts of mathematical logic is necessary.

To begin with, we have a first-order language $\mathcal{L}$ that is rich enough to express any formula that may be required. Hence, $\mathcal{L}$ will consist of various types of symbols as follows:

A denumerable set $\mathcal{V}$ of variable symbols :
$$\mathcal{V} = (v_0, v_1, \ldots\ldots\ldots, v_n, \ldots..)$$
or simply variables.

Punctuation symbols such as parenthesis),(, and the comma ",".

The usual symbols of the logical connectives :
$$\neg, \wedge, \vee, \rightarrow, \leftrightarrow$$
negation, conjunction, disjunction, implication, and equivalence.

The universal and existential quantifiers $\forall$ and $\exists$.

The equality symbol " = "

These are common to all first-order languages with equality. In addition to the above, we require that the following sets be included:

A set $C = \{c_1, c_2, \ldots.., c_n, \ldots\}$ of constant symbols.

For every integer $n \geq 1$, a set $\mathcal{F}_n$ of n-placed, or n-ary function symbols.

For every integer $n \geq 1$, a set $\mathcal{R}_n$ of n-placed, or n-ary relation symbols.

These latter are often referred to as the *non-logical* symbols of the language. Instead of saying '1-*ary*', '2-*ary*' or '3-*ary*'; it is customary to say *unary*, *binary* and *ternary* respectively. We observe also that a constant symbol can be regarded as a 0-placed (0-*ary*) function symbol. Thus, the first-order language $\mathcal{L}$ is just the union of these sets;

$$\mathcal{L} = \mathcal{V} \cup \{),(,\neg,\wedge,\vee,\rightarrow,\leftrightarrow,\forall,\exists\} \cup C \cup \mathcal{F}_n \cup \mathcal{R}_n$$

This defines the 'vocabulary' of the language. The next thing to do is develop some sort of syntactic structure. Thus, one defines the *terms* of the language $\mathcal{L}$ recursively as follows:

> *Every variable v_i and every constant symbol c_j is a term.*

> *If $t_1,t_2,........,t_n$ are terms and if f is an n-ary function symbol, then $f(t_1,t_2,........,t_n)$ is a term.*

and a symbol-string is a term only if it is obtained by this prescription. Next, one defines the *atomic formulas* as follows:

> *If $t_1,t_2,......,t_n$ are terms, and $R \in \mathcal{R}_n$ is an n-ary relation symbol, then $t_1 = t_2$ and $R(t_1,t_2,......,t_n)$ are atomic formulas.*

Then, finally, one can define the *well-formed-formulas* or *wffs* thus;

> *Every atomic formula is a wff.*

> *If φ and ψ are wffs, then so are $\neg\varphi, \varphi \wedge \psi$,*
> *$\varphi \vee \psi, \varphi \rightarrow \psi, \varphi \leftrightarrow \psi$ and $\forall v_n \varphi, \exists v_n \varphi$ for*
> *any variable v_n and for any natural number n.*

and a symbol-string is a formula only if it is obtained by the application of the above rules. We then define a *theory T* as any set of formulas (*wffs*) of the language $\mathcal{L}$. Then, we have almost all the ingredients for our scientific theories. The formulas shall be the axioms of the latter. But, of course, the symbols included in the language and the axioms formed therewith have no intrinsic meaning; they are just strings of symbols and terms that are formed in accordance with the above rules. Since it would seem to be useful for the axioms of a theory to actually *assert* something, we need to assign specific meanings to each individual constant, variable, function and relation symbol. Then, an *assignment* or *valuation* $v(x_i)$ takes a variable x_i and gives it a value a_i. It is therefore a function with a given domain. Thus, if the domain is the set of integers, then the valuation $v(x_1)$ = 3, $v(x_2) = 4$ and $v(x_3) = 7$, makes the formula $x_1 + x_1 = x_3$ *true* in the sense that $3 + 4 = 7$ holds. In other words, this valuation *satisfies* the *wff* $x_1 + x_2 = x_3$. If a formula is true under *every* possible such valuation, then we say that it is logically *valid*. We can extend this idea as follows. Define an *L-structure* as a set D_I consisting of the *variables* of the language $\mathcal{L}$, together with a mapping that assigns a meaning or *interpretation* to the various symbols in $\mathcal{L}$. D_I is called the *domain* of the interpretation. Note that the interpretation of the logical connectives $\wedge, \vee, \neg, \rightarrow, \leftrightarrow,$ and the quantification symbols $\forall$ and $\exists$ retain, by convention, their usual meanings, namely 'and', 'or', 'not', 'implies', 'if and only if', 'for every' and 'at least one', respectively. Thus, an interpretation I of a first-order language $\mathcal{L}$ is given as follows:

A non-empty set D_I called the domain of the interpretation.

For each predicate or n-ary relation symbol R_n of $\mathcal{L}$, an n-ary subset R^I of D_I^n.

For each n-ary function symbol of f $\mathcal{L}$ is assigned an n-ary function f^I from D_I^n into D.

For each individual constant c_j of $\mathcal{L}$, is assigned an element c_j of $\mathcal{L}$.

These assignments are the interpretations of the symbols of the language, sometimes called a *model* of the language $\mathcal{L}$. More specifically, we consider an *L-structure* as the pair $\mathfrak{A} = (D_I, I)$. Then, if a formula φ of a language $\mathcal{L}$ is true under an interpretation I, we say that $\mathfrak{A}$ *models* φ, or that the $\mathfrak{A}$ *satisfies* φ and write $\mathfrak{A} \vDash \varphi$. As before, if a formula is satisfied under *every* interpretation of $\mathcal{L}$, then it is logically *valid*. In fact such a formula is a tautology. Thus, the notion of satisfaction merely states that under a given interpretation, a formula is true. Hence, a tautology is always true. For a set of formulas $T = \{\varphi_1, \ldots, \varphi_n\}$, if $\mathfrak{A} \vDash \varphi_i$ for each $\varphi_i \in T$, $(1 < i < n)$, then we write $\mathfrak{A} \vDash T$ and say that $\mathfrak{A}$ is a *model* of the theory T. This gives us an idea of when a theory is *true* for a given model, or interpretation. Since, we require our theories to be consistent, then, by the completeness theorem, they are satisfiable and have a model.

After this brief and not very rigorous excursion into model theory, we may return to the initial problem of this section, specifically, just what it means to say that a scientific theory is true. Firstly, we note that any theory T may be satisfied by more than one model, simply because, whilst the axioms of the theory remain unchanged, the interpretation of the language is different, yet still yield true statements for that interpretation. Thus, the concept of satisfaction is highly dependent on the interpretation that the model provides for the symbols of the language. Additionally, a theory may be true in one model, but not in another. Truth, in this sense of satisfaction in a model, is therefore relative. There are many examples of these ideas in the physical sciences and we can illustrate the point as follows. Consider the following expression;

$$\forall x_1 \forall x_2 \forall x_3 \forall x_3 R_1^2 (f_3^2 (f_1^2 (f_1^2 (c_2, f_1^2 (x_1, x_2)), f_1^2 (f_2^1 (x_3), f_2^1 (x_3))), x_4), c_1)$$

where the x_i are variable symbols, the c_i are constant symbols and the f_i^n and R_i^n are n-ary function and relations symbols respectively. If now, we interpret these symbols as follows:

R_1^2 is the equality symbol, so that $R_1^2 (x, y)$ means $x = y$

f_1^2 is standard multiplication, so that $f_1^2 (x, y) = xy$

f_3^2 is subtraction, so that $f_3^2 (x, y) = x - y$

f_2^1 is the inverse operator, so that $f_2^1 (x) = 1/x$

then with the interpretation of the variables x_i and constants c_i as:

$$x_1 = m_1, \quad x_2 = m_2, \quad c_1 = 0$$
$$x_3 = r, \quad x_4 = F, \quad c_2 = G$$

The above expression becomes;

$$\frac{Gm_1 m_2}{r^2} - F = 0$$

which is, of course, Newton's famous inverse square law of gravitation. However, if we now re-interpret the variable and constant symbols thus:

$$x_1 = e_1, \quad x_2 = e_2, \quad c_1 = 0$$
$$x_3 = r, \quad x_4 = F, \quad c_2 = k$$

then the expression becomes;

$$\frac{ke_1e_2}{r^2} - F = 0$$

which is Coulomb's law of electrostatics. The domain of interpretation, which is the range of the variables, is the same in both these cases, namely the positive real numbers. However, the domains of applicability are not the same and certainly bear no relation to the domain of interpretation. It is important to distinguish between the domain of interpretation within a given model and the domain of applicability which corresponds to that interpretation.

Now, suppose we have a theory T. We shall say that T is a *scientific theory S*, if there is a model of T such that, in addition to the constant and other symbols, the *variables* of T are interpreted as (in principle) measurable physical quantities. Thus, there is an interpretation I_S of T that assigns physical meaning to the variables within the domain of interpretation D. Then we can say that if;

$$\mathfrak{A} = (D, I_S) \vDash T$$

then $\mathfrak{A}$ is a scientific theory and we write $S = \mathfrak{A}$. Hence, what we have thus far called a scientific theory is in fact a model of the theory T under the interpretation I_S. It is convenient to continue this abuse and refer to S as a scientific theory. We shall say that S is *scientifically valid*, or just *valid* if the context is clear, if it is a member of the set N, i.e. if $S \in N$. If S is valid in this sense, then we regard it as a 'correct' and 'true' scientific theory. It correctly describes the Natural World within its domain of applicability. This does not imply that such a theory cannot be modified, for it is bounded by its domain of applicability and should the latter be altered, then so will the theory be accordingly modified. The interpretation I_S that makes this possible shall be called the *natural interpretation*. This will require some discussion.

5.2 The natural interpretation

If we select at random a theory from the set $\mathcal{T}$, then with the interpretation I_S it becomes a scientific theory, which may of course, be quite meaningless from a physical and scientific viewpoint. Thus suppose $T \in \mathcal{T}$ consists of the single axiom;

$$T = \{\forall x_1 R_1^2 (f_3^2(x_1, f_1^2(c_2, f_1^2(c_3, c_3))), c_1)\}$$

with the relation and function symbols interpreted as before. Then, setting $x_1 = r$, $c_2 = G$, $c_3 = e$, $c_1 = 0$, results in;

$$r - Ge^2 = 0$$

which is a scientific theory by virtue of the interpretation I_S, but does not appear to express anything physically meaningful. However, taking $x_1 = E$, $c_2 = m$, $c_3 = $ c and again $c_1 = 0$ gives;

$$E - mc^2 = 0$$

with which we are very familiar. Thus, we can say that in the first interpretation $T \notin N$, whereas in the second, it is. Both are scientific theories, but the latter is valid and the former is not. But the equation $r = Ge^2$ *does* assert something; it says that the distance between two points is constant and is equal to the product of the gravitational constant and the square of the value of the electronic charge. Clearly, there are situations when this is true. Yet, the formula says that it is true for *all* values of r and a geometry constructed on this principle would not correspond to the real world in the physical sense. However, with a suitable choice of units one may set $Ge^2 = 1$ and since $r = d(x, y) = 1$ if $x \neq y$ and is zero if $x = y$, then the axioms of a metric space are satisfied. Then, under this interpretation the formula is a model of those axioms. It is, of course, the discrete metric. This illustrates a number of simple facts.

Firstly, as we noted before, a well-formed formula is merely a string of symbols that is formed in accordance with the 'grammar' of the language and does not actually assert anything until those symbols are interpreted. However, because a *wff* is 'grammatically correct', once it has been interpreted, then it *must* make an assertion of some kind about the relationship between the specific variables within the *wff*. If I_S is this interpretation, then the assertion made is a scientific one. Similarly, a set of *wffs* T becomes the scientific theory S under the 'natural' interpretation I_S. This does not mean that S is scientifically valid; for that to happen it must belong to the set N. Nor does it mean that S cannot simultaneously be a theory about something else, whether valid in another domain of applicability or not, as we have seen above. This is so because the variable assignments in the natural interpretation are relevant to more than one domain of applicability. However, whilst the expression $r = Ge^2$ is physically meaningless in the sense that it is not realized in the physical universe, one cannot simply re-interpret the variable r in the context of another domain. The 'r' in metric topology is not the same as the 'r' as understood as a physical distance, even though it may well appear to be similar. The mathematical definitions of variables that interpret *wffs* of the language are of a different character than the physical variables, even of the same name, than are the variables in the domain of the interpretation I_S. The former are of a different ontological character, much like the abstract variables of the language $\mathcal{L}$, than is the physical interpretation we are considering here. We shall discuss the relation between mathematics and the physical sciences later, for the moment, however, we are concerned with the natural interpretation and the variables assigned thereby.

The question arises as to how the assignments of the natural interpretation come about and we are led to consider whether empirical observation of the Natural World is the only possible source of identification of the physical variables used in scientific models or theories. In this connection, we recall from §4.2 that scientific theories can be formulated from empirical and experimental observation, or by hypothesizing upon the nature of the axioms involved. In either case, however, it is necessary to prescribe the physical variables and constants of Nature that are to describe the natural phenomena under consideration. These are, evidently, the usual quantities that are familiar to us from the natural sciences, such as distance, area, velocity, acceleration, momentum, mass, temperature, pressure, entropy, the speed of light, the unit of charge, the Hubble constant, etc. Notice that these are of various kinds. Some are geometrical, some physical and some can only be defined in terms of more basic quantities and are not directly observable. Now, these quantities must be measurable if a theory is to be empirically tested and they must be in some sense *descriptive*, if we are to know what we are talking about. By this latter criterion, we mean that they must describe some property of the object under study that is a gross fact of Nature; and that without which the specification of the object would be incomplete. Such parameters, we have maintained, are provided by Nature, irrespective of whether we observe them directly or indirectly. Nor does it matter to us whether such variables are theoretical or mental constructs, for they describe Nature whatever their ontology. Obviously, the *names* that we ascribe to the variables and constants are of no consequence either; the important point merely being that they are attributable to physical properties of natural phenomena. One thing however is clear. If some measurable facet of the

Natural World changes with time, then it is a variable that we can use in our scientific theories, however specified. If it does not change with time, then it is a constant. Additionally, the natural interpretation demands that relation and function symbols shall be just those that are used in common mathematical practice. We therefore state that the natural interpretation of theories includes the following;

> A *natural variable p* is a measurable physical quantity that changes with time; i.e $dp/dt \neq 0$.

> A *natural constant c* is a measurable physical quantity that does not change with time; i.e. $dc/dt = 0$.

> All predicate or *n-ary* relation symbols are the standard mathematical symbols; i.e. $=, <, >, \cong$, etc.

> All the operator or *n-ary* function symbols are the standard mathematical operations; i.e. $+, -, \times, x^n$, etc.

and however philosophically inadequate this definition may seem, it is adequate for our purpose, which is essentially pragmatic and the cognitive status of the theories and variables contained therein need not at present concern us. Then, in a an arbitrary theory T, the natural interpretation turns T into a scientific theory S by assigning a natural variable p to each variable x of T and a natural constant to each constant symbol of T and interpreting function and relation symbols as standard mathematical operations. The logical connectives remain as usually employed and the variables shall range over the real or complex numbers, as required. We also say that any theory that does *not* satisfy these criteria, is not a scientific theory, for clearly, such a theory, by definition, cannot be relevant to the physical or Natural World in the sense that its variables and constant symbols are unable *ipso facto* to describe any property of a physical system. Evidently, if it did, then the measurability criterion would have to be satisfied. This excludes a large class of theories from achieving the status of being scientific.

It might be thought that there is some circularity here, for if a theory describes the Natural World, then it does so exclusively by the employment of the natural interpretation. Yet, it *must* be of this nature, again, by virtue of this fact. Hence, no theory can describe the Natural World unless it is interpreted accordingly; namely, through the natural interpretation. But, as has been seen, this is to miss the point. *Every* theory under the natural interpretation describes a physical or natural phenomenon, as was made clear from the foregoing, but that does not in any way imply that the theory is scientifically valid for the implication is merely that it is a scientific theory. Hence, any theory, or more properly any set of *wffs* that constitute a theory, when interpreted under the natural interpretation, will be a scientific theory; but there may be an infinity of such theories, only a finite subset of which will be scientifically valid in the sense defined. This is to say that the natural interpretation does not determine when a theory is scientifically valid; for that will be the case only if the theory is included in the set (class) N of valid scientific theories. Essentially, if a theory describes the Natural World then it employs the natural interpretation, but the converse does not follow, a theory that employs the natural interpretation does not necessarily describe the Natural World, for although it is a scientific theory, it may not be scientifically valid.

Now, if some property of Nature is observed to change with time, then it is quite appropriate to isolate that property by giving it a name. Such a designation will be a natural variable or 'named' symbol. Furthermore, if another property of the system under observation is also found to vary with time, then it too, shall receive a designation and if additionally, the manner in which these two natural variables change is observed to be related in some way - the variation of the one being dependent on that of the other - then one would certainly be inclined to state this relationship explicitly, perhaps in ordinary language. But, since one has presumably discovered the connection between the two natural variables by some process of measurement, even if only initially by casual and imprecise observation, then it is more expedient to express the relationship in mathematical form. Subsequent observations may refine the relationship, improve the precision and ultimately, almost without effort, an axiom of the form $f(x_1, x_2) = 0$ expressing the dependency of the natural variables is obtained. This then, is a scientific theory. It is a model of some *wff* of the language in which the variables x_1 and x_2 have been interpreted. But the observer has played no part in this interpretation; all he has done is to casually notice that certain properties of the physical world behave in a specific way. The interpretation has been 'compelled', or 'forced' in a most natural way, just because it has already been made, not by the observer, but Nature Herself. This is the natural interpretation. The fact that the resulting theory, containing here just the single axiom $S = \{f(x_1, x_2) = 0\}$ is a model for a *wff* in some language $\mathcal{L}$ is neither here nor there so that the existence of the theory and hence the natural interpretation, requires no explanation. Rather, it is the abstraction engendered by the model-theoretic approach that, by the very use of the seemingly vague term 'interpretation', seems to suggest some sort of subjectivity or arbitrariness, thus demanding an explanation of the particular interpretation. Yet, this is not so. Model theory makes no such stipulations; any interpretation, valuation or assignment is just as good as any other and there is no subjectivity and model theory *per se* is precisely defined. The problem arises when one enquires why one interpretation is apparently favoured over another. But we have observed that this is not the case. The problem is a philosophical issue which has nothing to do with model theory or logic. It is in fact, under another guise, the ancient question: Why is the Universe as it is? In the present context this reduces to the frequently posed question in the philosophy of science: Why are the variables and constants of Nature as they are? This can now be re-phrased as: Why does the natural interpretation pertain? In this formulation, we can see how nonsensical such speculation really is, for whatever interpretation is placed upon the abstract entities of a formal language, the same questions will arise. The situation becomes even more absurd when we realize that no matter which interpretation is currently extant, the resulting scientific theories will be the same, although perhaps expressed in unfamiliar terms. Thus, one may well read these pages in Mandarin but that will not alter the meaning contained therein, for clearly that must remain the same. We are saying that the natural interpretation will pertain irrespective and regardless of the names given to the symbols and constants, functions and relations of the formal language and that this is so because in consideration of all of the foregoing, it can hardly be otherwise. Hence, far from being arbitrary, the natural interpretation is the only possible interpretation in science and is quite independent of the practitioner. Indeed, the latter is often quite oblivious to the philosophical or epistemological processes that may be involved in his pragmatic investigations.

The above discussion rests upon an empirical-observational method of the determination of the relationships that may be ascertained with regard to the workings of the Natural World, specifically, the scientific method, although the latter has almost appeared to be accidental in its simplicity. If this is so, then it is because of the inevitability of the natural interpretation, as we have emphasised. But, in the absence of any empirical observation or experimentation, one may wonder if the natural interpretation would still impress itself upon the variables and constants that may not then be envisioned without the aid of direct access to the physical phenomena involved. In other words: Does the natural interpretation depend wholly upon the interaction of the observer with that which is observed? If it does, then it would be highly improbable for an investigator to obtain information concerning the Natural World in terms of axioms and theories by pure introspection or theoretical speculation. It is, of course, still possible to do so, for we recall from §4.2 and the end of §4.5 that the evolution equation allows, given sufficient time, the generation of all possible theories and clearly some of these will be scientifically valid. But, this does not address the present problem of whether the natural interpretation would be a participatory factor in such theory formulation. This is an important question which we must now consider.

There have been many instances in the history of physical science when theories have been formulated independently of and apparently without access to the empirical world and yet, such formulations have always made use of the natural variables and constants that the natural interpretation provides. We must therefore ask why this is so. For how is it possible for the natural interpretation to operate when there is no interaction with the physical world? And how are scientific theories formulated without such interaction? If they are so formulated and not just by the random process described earlier, then clearly a satisfactory explanation of such a fortuitous event is required. If this is not the case however, then the empirical mechanism by which the natural interpretation obtains without apparent interaction also urgently demands explication. Let us therefore briefly examine some historical examples.

Einstein formulated Special Relativity without recourse to the empirical world and so this was an entirely intellectual construct that had nothing to do with the scientific method, yet the variables and parameters employed in the development of Special Relativity were just those that would have been perceived through the natural interpretation.

Einstein later formulated the General Theory of Relativity, apparently employing geometrical methods that were hitherto completely alien to the (then) current physical (but not mathematical) conceptual framework. Yet, hid did so without the benefit of empirical interaction, but nevertheless, was able to express his theory in terms that the natural interpretation would have provided.

Maxwell initially formulated electrodynamics by analogy with Newtonian Gravitation and later, by analogy with Hydrodynamics. In the process natural variables were employed, but no empirical methods were used.

Lagrange and Hamilton reformulated Newtonian Mechanics in a novel way. New variables were introduced without recourse to the empirical method. Such variables seemed to be not only expeditious from an analytical perspective, but also 'natural', as though they had been received directly and empirically through the natural interpretation.

Aristotle decreed that the heavier a massive body, the faster it falls. Then, by the terms used, the variables employed are just those of the natural interpretation.

The list is endless and one could cite many such examples over the long history of science. But clearly, this is all nonsense. *Most* theories in physical science, which must include all of those which purport to be scientific, are formulated in this way. But we are not concerned with the formulation of scientific theories *per se*; we dealt with that problem earlier. Our main concern is the implication in the formulation of a theory with respect to the natural interpretation. Then let us rectify the above misconceptions accordingly.

Einstein was trained in physics. He therefore already had a very clear idea of the natural variables and constants that the natural interpretation imparts to a theory. He did not need to seek the necessary variables himself because of this familiarity. In the case of Special Relativity, the negative result of the Michelson-Morley experiment had empirically determined a constant of nature, namely the speed of light. Certainly, the natural interpretation is present here. In General Relativity, the same argument applies. For an empirical result of importance in this context we may refer to the apparently anomalous advance in the perihelion of Mercury. The geometrical framework of General Relativity consists simply in the ingenious utilisation of well known techniques of differential geometry in the physical realm. Whilst much conceptualising on the part of Einstein would have taken place, there is no conflict with the natural interpretation.

Maxwell was also a trained physicist and similarly well-acquainted with the natural variables used in that discipline. The formulation of Electrodynamics by analogy with Newtonian Mechanics or

Hydrodynamics illustrates the use of natural variables. The electric and magnetic natural variables had been meticulously and painstakingly identified earlier by Faraday in a *tour-de-force* of the empirical method. One can hardly say that the natural interpretation is absent here.

Again, both Lagrange and Hamilton were trained mathematical physicists. The 'new' variables employed such as the generalised position and momentum, action angle variables, etc. whilst not directly observable in the commonly understood sense, were clearly defined in terms of already familiar natural variables from Newtonian Physics. The introduction of the least-action principle and use of the Calculus of Variations was just theory re-formulation, which we discussed in §3.9 with this specific example in mind.

The case of Aristotelian physics is somewhat different, but it is clear that Aristotle had some idea of velocity, acceleration, mass and force that is, at least in the abstract, similar to the modern sense in spite of the peripatetic element. This can only have been obtained through familiarity with the Natural World, which thus provided natural variables, albeit with different names for some of them. The fact that his conjecture is false merely affirms the necessity of empirical verification.

Then, in all of the above, we can see that theories concerned with specific natural phenomena have always been expressed in terms of the natural variables imparted by the natural interpretation. This is why we have *defined* a scientific theory as any model of a theory in which the interpretation is the natural interpretation. There is no circularity here when we say that a theory is not scientific *unless* it is so interpreted, for the latter statement is just the inverse of the former and therefore equivalent to it. From the above discussion we therefore feel justified in asserting that *no scientific theory can possibly be formulated without the natural interpretation*. Every scientific theory is, then, formulated with the natural interpretation. This is again the definition of a scientific theory that we have been led to. Then, the implication is straightforward and the answer to the question we first posed in relation to the dependency of scientific theories on the natural interpretation is clear. All scientific theories and all investigators formulating such theories, do so with the assistance of the natural interpretation. Any theory formulated otherwise, cannot be a scientific theory. This, obviously, does not mean that, without empirical intervention, an arbitrary theory cannot be formulated that *has* the natural interpretation, for that is quite possible, it is just improbable. Note that there is no requirement for a scientific theory to be *testable*. If it is, then all well and good, the scientific method can then proceed and further refinements made to the theory. However, testability is not mandatory and indeed, there have been many examples of theories that *cannot* be empirically verified, either because they lie outside the scope of contemporary science, or because the predictions made are not currently verifiable with existing technology. A present example of the latter in modern physics is string theory, currently the best candidate for a 'Theory of Everything', where the predictions made are of such high energy and on such small scale that, at the time of writing, they are not testable with existing technology. This, however, may soon change as larger and more powerful particle accelerators come online.

We have not yet considered the second part of the natural interpretation; namely, the predicate and function symbols and their interpretation. We specified that the relation and function symbols will be the 'standard' mathematical notation, as is used in practice. In fact, we wish take a much broader, more linguistically encompassing view than this. Mathematics is a rich and ubiquitous language, which is capable of expressing any relation of any kind that we may envisage between any number of variables. It can, as we have seen, model structures in the real world in addition to an infinity of possible structures that have not (yet) been physically realised. It is free of the ambiguities of natural language and sufficiently abstract so as to be devoid of subjectivity, from which the latter may suffer. It is, therefore, the most appropriate language in which to express the properties of natural phenomena, especially in physics and the relationship between mathematics and the physical sciences has been, as the history of physics will testify, reciprocal; each contributing and extending the other, often in unexpected ways. Thus, mathematics evolves and as it does so, new structures, new ideas, concepts and symbols are introduced. It is the latter that concerns us here, for we do not wish to restrict the natural interpretation to the existing relation and function symbols employed in mathematics at any given time. Rather, we allow the natural interpretation to embrace all the symbology and expressive power of any mathematics that may in the future be found to be expedient or necessary. This is not to say that mathematics or the natural interpretation *changes* with time in the reformist sense, for each is fixed linguistically, but not expressively. Mathematics evolves through the introduction of new concepts, which are often found to be equivalent (isomorphic) to already familiar ideas, but the formal *language* of mathematics is still just that, a system in which relationships can be succinctly expressed. Hence, new symbols may be introduced as advancements are made and new theorems demonstrated, but that they can is merely due to the present state of ignorance, or current intellectual failings. We envisage that the natural interpretation is not so confined. Nor could it be so, for it would be patently absurd to discover that a physical process is not expressible in precise terms, in the linguistic terms of mathematics, because if it were not so, then mathematics would not be versatile enough to encompass Nature, which is contrary to the very conception of the natural interpretation itself. Hence, should a new kind of relation or function symbol be required, then it will be demanded by the natural interpretation and mathematics will accommodate itself accordingly. As mentioned, this has frequently been the case in physics; one may mention many examples, vector analysis, the Dirac δ-function, Virasoro algebras, non-commutative geometry, etc. Hence, we see that even though our mathematical knowledge may be limited at any given time, this does not place any such restriction upon the variables and symbols, relations and functions, of the natural interpretation, for which Nature will provide[*].

5.3 Theory type and empirical verification

Scientific theories, once formulated, are generally if it is possible, subjected to rigorous testing by the usual empirical method of comparison with observation and experimentation. If the axioms of the theory were first obtained directly from observation, then they will merely describe that which has been observed to already be the case in the Natural World and therefore must, at least in some sense, be considered valid, if only for the instant of time at which the observation took place. Since the natural interpretation pertains, then in this case the variables of the theory will necessarily be natural variables. Such theories can be called 'empirical' because their axioms are obtained directly through observation or experimentation.

However, when a theory is formulated from first principles, it does not possess this epistemic quality. Yet, as we have seen above, the axioms will still be framed in terms of natural variables and the natural interpretation still obtains. For want of a better term, we might call theories of this sort 'non-empirical', although they must necessarily contain an empirical element. In fact, we have already discussed these two types of theories in §4.2, where they corresponded to the two methods of theory formulation

---

[*] This does not of course imply that mathematics is empirical.

mentioned in that section. Then, the method of theory verification involves turning a non-empirical theory into an empirical one and this is done by comparison of the claims and predictions of the theory with observation. This distinction between empirical and non-empirical theories is often cited as being problematic in epistemic philosophy, for often non-empirical theories are formulated in variables *defined* in terms of the natural variables that are obtained by direct observation. But such variables are, by our former stipulations, still natural variables because either they can *in principle* be observed with appropriate instrumentation, or their *effect* can be observed through direct observation of measurable natural variables.

Thus, variables defined in terms of natural variables inherit the 'measurability' of the latter, provided that they *could* be measured with sufficiently sophisticated equipment. In the case when a *concept* is defined, such as entropy, we can still express this in terms of natural variables. We may illustrate this with a familiar example from physics. Boyle's law states that in an ideal gas at constant temperature T, the pressure P is inversely proportional to the volume V;

$$PV = RT$$

with constant of proportionality R, the universal gas constant. This is, of course, an empirically determined axiom and is valid only under restrictive conditions (low pressure). However, as is well-known, this empirical law can be derived from the kinetic theory of gases by making a number of assumptions concerning the molecular nature of matter. The principle assumptions are that the gas is particulate and each particle or molecule behaves as a point centre of mass, which then allows Newtonian mechanics to be employed. The molecules are sufficiently far apart so as not to exert forces of attraction and repulsion upon one another and yet they act in a continuous fashion, moving both linearly and randomly. With these and a couple of additional simplifying assumptions, one uses Newton's second law and considers collisions between the molecules of the gas and the forces involved in such impacts. The details are straightforward, readily available and need not be repeated here. Finally, one ends up with the equation;

$$P = nkT$$

where $n = N/V$ is the average number of molecules per unit volume and k is Boltzmann's constant. This is the same as the above with $R = nk$. The point here is that all the quantities involved are expressible in terms of measurable variables that can be obtained directly from observation. Furthermore, although at the time of the development of kinetic theory, individual molecules were not only hypothetical but certainly unobservable with the instrumentation of the time, that is no longer the case and hence variables relating to molecular motion can now be obtained directly from observation. Hence kinetic theory was testable in principle from the very beginning, but it is now testable in practice which has been done many times. Notice also, that Boyle's law (or rather the ideal gas equation of state) is now a *consequence* of another theory, namely kinetic theory. For this reason kinetic theory can be considered to be more fundamental and Boyle's law may be dropped, since our scientific theories are to contain axioms only and not derivable results. This is what happens when new assumptions are made, but that does not mean that the assumptions are valid, for other suppositions may well produce the same result. If however those assumptions can be directly observed then the theory will be empirical and hence all the variables are directly obtained through the natural interpretation.

There is an interesting point that is well illustrated by the above. When the axioms of one theory become the *deductions* of another theory, then the theory that can be deduced from the other is to be considered less fundamental, or less 'deep' than the new theory. Thus given two scientific theories S and S', if $S \vdash S'$, then S is to be regarded as the more fundamental of the two theories and the axioms of S' should not be included as fundamental axioms of S. Thus, in the case of kinetic theory K and the gas (Boyle's) law B, we have $K \vdash B$. Note that if S' consists of a number of axioms φ_n, i.e. $S' = \{\varphi_1, \ldots, \varphi_n\}$, then $S \vdash S'$ means that $S \vdash \varphi_i$ for each integer i with $1 \leq i \leq n$.

We noted above also that simply because a true theory may be deducible from another theory, this does not imply that the theory from which it is deduced is also true, for true theories can be deduced from false theories. Hence, if B is true, then $K \vdash B$ does not (necessarily) mean that K is true. However, the converse is not the case; if K is true, then B cannot be false, for B is now a *theorem* of K. Then, we see that no matter how 'true' the gas law may be, it does not imply the truth of kinetic theory. However, we may well consider that because the former is a consequence of the latter, this is *evidence* for the truth of kinetic theory. To take such a position is dangerous and certainly not justified. We would do better to follow the above and attempt to obtain the axioms of kinetic theory empirically.

Yet, in practice, as scientific theories develop and become more and more sophisticated, this somewhat dubious procedure is often followed and for good reason. If the axioms of one theory S' can be deduced from another theory S, then not only would the latter seem to be more fundamental, it would also seem to provide an 'explanation' for S'. This is generally taken to mean that the assumptions of the theory S must be correct, merely because there is good empirical evidence for S'. This supposition can lead to difficulties and, in fact, a great deal more, as we shall discuss at length in the next chapter. For now, there is still the question of how a scientific theory can be 'validated' by empirical means.

We agree, as we must, that an empirical theory must be scientifically valid at the time of observation. But this can only be the case if the measurements of the natural variables are completely error free. If they are, then the purely empirical theory so obtained, which is no more than a set of axioms, or a model with the natural interpretation, must necessarily belong to the set N of scientifically valid theories. If however, there are errors in the measurement of the natural variables, then all that can be said is that the theory holds at the instant of observation, but at best only approximately. Yet, as above, this is not justified either, for however small the observational inaccuracies may be, they may be sufficient to ensure that the theory does not hold at all, but merely *appears* to represent the workings of the Natural World when in fact errors of measurement are artificially excluding the correct theory. In principle, a small error in practice may lead to erroneous conclusions, thus negating the theory. The natural interpretation provides all that is required for the empirical determination of a scientifically valid theory, but in order to obtain such validity, perfect measurement is required. Clearly, this is never possible in practice and it would seem that the only way one may determine whether or not a scientific theory is scientifically valid, is to ascertain whether that theory is one of the members of the set N of valid scientific theories. But, such membership is just the definition of scientific validity in the first place, and the definition thus becomes circular.

It would seem that we are in something of a dilemma, for surely the scientific method is the 'Gold Standard', the cornerstone of investigative methodology and is thus appropriately sacrosanct. If it cannot be trusted to yield, unambiguously, an accurate description of the Natural World, then the scientific method can warrant no more credence than mythology or religion, which would be a very worrying situation indeed. But, of course this is not so. We are discussing epistemological and methodological issues here, which do not bear directly upon the facts of the situation. It is clear that science works and modern technology is ample testament to that fact. These words would not have been written without semiconductor technology, which relies upon many physical theories, especially quantum mechanics via the Schrödinger equation. Without Maxwell's equations we would not have the means to generate the electric power that supplies the computer upon which these words are written, nor would we have radio and television. Simple mechanical devices have been used for centuries in virtually all human activities, from manufacturing to warfare and are obviously dependent upon Newtonian theory. In medicine, tomography relies upon the Dirac equation, as do many other modern devices. Mobile telephones also depend upon both semiconductor technology and electromagnetic theory. Even General Relativity that not so long ago was thought to be without application outside of fundamental physics and cosmology, is essential for interplanetary navigation as well as the Global Positioning System which is used on a daily basis in major transportation systems and even by individuals. And so the list goes on.

There is no question of the utility of science, a fact that is often ignored by its opponents. Hence, if the difficulties mentioned above in relation to the empirical method, i.e. the scientific method, are correct, then it is legitimate to ask why knowledge obtained via that method has found success in practical applications, for clearly our machines *work* and they do so in accordance with scientific theories that have been painfully elucidated and rigorously tested over the last four centuries. This fact is also significant in relation to the 'psuedo-science' and 'anti-science' culture that still persists in spite of such evident success of the application of science. We return to this topic in Chapter 11 when we discuss it in that context.

Yet, a feeling of unease may still persist. It is true that our measuring instruments are not perfect and it is possible that such imperfections may result in the formulation of an incorrect axiom. However, it may be argued that such a situation could not obtain for long, since continued verification with ever increasing accuracy would soon discover such a discrepancy and indeed, that is the very reason why constant measurement repetition is so vital in science. Furthermore, we have no way of determining whether a theory is scientifically valid and neither can we unequivocally place a numerical value on any possible measure of probability of validity, for no degree of empirical verification can provide such information.

But, this is the whole point of the scientific enterprise, for if our instrumentation were perfectly precise and sufficiently extensive in its range then the natural interpretation would alleviate us from the need to continually test our theories. We would already have complete theories and possibly a Final Theory. It is because of these imperfections in the scientific method, together with a number of additional factors that are the reasons for the evolution of scientific theories in the first place.

Thus, although the scientific method is not perfect, one would be hard-pressed to find an equally effective alternative and with respect to the acquisition of gross knowledge of the physical realm, one is also forced to conclude that it is the best available epistemic method. Indeed, it would seem that any alternative epistemological procedure if we are to be at all confident in the data that such a substitute methodology might provide, would have to operate just like the scientific method and would probably have to *be* the scientific method.

However, before we can proceed further, it is necessary to discuss how the scientific method operates, not only as an ideal, but actually in practice on a day-to-day basis, i.e. how a working scientist *should* proceed and how he *does* so operate in practice. There are of course, many possible notions, views and opinions as to how the epistemological aspect of science is related to the empirical process and we have discussed some of them here, but we need to connect this with the method of science itself so that we may capture both the essential positive characteristics of the latter and isolate any difficulties that may be found in the technique. This is the subject of the next chapter where we shall elaborate upon and extend some of the points raised here.

Before proceeding, let us summarise the definitions of natural variables and constants for future reference:

Definition 5.2.1
A *natural variable p* is a measurable physical quantity that changes with time, i.e. $dp/dt \neq 0$.

Definition 5.2.2
A *natural constant c* is a measurable physical quantity that does not change with time, i.e. $dc/dt = 0$.

It will be important to bear these crucial definitions in mind throughout the sequel, for they are the foundations of the natural interpretation and the empirical determination of the axioms of a scientific theory.

Chapter 6

The Scientific Method

We discuss the practical formulation of theories and the relation thereto of assumptions made in the construction thereof. We further discuss the two types of scientific theories and examine there interrelation. Difficulties inherent within the scientific method itself are clarified. We conclude with a discussion of the "ideal' of the empirical method of science.

6.1 The method in practice

The progress of science essentially consists in the development and refinement of its theories, the whole point of which is to 'explain' the diverse nature of the physical phenomena that pervade our experience of the universe and in which we ourselves are willing participants. Thus, we observe, experiment and observe again in an attempt to arrive at some understanding of the workings of the Natural World. We have developed an empirical technique, which we call the 'scientific method' in order to facilitate and accomplish this grand goal that ultimately, is no less than a complete explanation of the entire universe of natural phenomena, including ourselves. The observations that we make disclose relationships amongst the quantities (natural variables) that we measure, thus revealing the detailed behaviour of various natural phenomena. Often tentative at first, we repeat our observations, perhaps under restricted or specialised conditions, until we are sure that we have ascertained the correct form of the relationship between the observed parameters. When we are satisfied, we propose a 'scientific law', which we write down in an appropriate way and which is usually of a mathematical kind. We continue to 'compare' our new-found law with more and more observational and experimental data, just to check if it is really correct, forgetting that it was obtained from observation in the first place and therefore, if our measurements were not initially flawed, then it *must* be correct and inevitably the new measurements will produce the same result. Then, when we find that our remarkable law still obtains, although it never occurred to us that it might

109

be time-dependent, we are nevertheless elated. We have discovered a 'Law of Nature' and we experience the profound satisfaction that accompanies such a revelation, rather as might an explorer finding a new, far and distant land, or a zoologist discovering a new species, except that our delight is all the greater, for what we have accomplished is an intellectual achievement, a triumph over Nature and furthermore, embraces the entire universe in its scope. But of course, it isn't and it doesn't. What we have done, we later realise, is merely to observe and experiment. True, we have been meticulous in our endeavour, but the fact remains that we have simply found something that was already there and plain for all to see had they taken the trouble to look. No invention or special ingenuity on our part has taken place. To make matters worse, we cannot legitimately say that our newly discovered Law of Nature is really universal, for its range of application, its domain of applicability, has been confined to the area and phenomena in which we have been able to empirically verify its validity in an operational sense. What we have obtained, as described in §5.3, is just an empirical theory, an empirical law, an axiom of the form $f(x_i, c_j) = 0$. What we really want to do is to understand *why* our new law should be the case at all. We wish to *explain* it and to do so we must build some sort of conceptual framework; a theoretical edifice upon which it can be founded. This will involve making assumptions of some kind, which we do carefully and minimally, for we wish our assumptions to appear 'reasonable' and at best, since we are on uncertain ground, to have as few of them as possible. If we are successful, then we will have indeed made a great intellectual step forward, for we will have provided the foundations for a Law of Nature and thereby obtained a deeper understanding of the universe and its physical processes. We may expect a Nobel Prize and a large cheque without further delay.

However, our expectations of scientific glory and immortality are short lived. When our results come before the Nobel Committee, whom we hitherto had little respect for, it is pointed out that although we may have discovered an 'experimental' or 'observational' relationship between various quantities, it only holds within the margin of error of our instrumentation and measuring devices and then just in the specific area of our observations. To compound our disappointment, it is made clear to us that our brilliant explanation of the phenomena described rests upon a number of assumptions and suppositions, the credibility of which the learned Committee is not convinced. To add fuel to the fire, we learn that there is a contender to the prize with an alternative explanation that is founded upon far more plausible assumptions. We are disheartened and do not expect to win the Nobel Prize.

In spite of this, we do not give up, for we are scientists and made of sterner stuff. We will not be thwarted by either the Nobel Committee, about whom we were obviously quite right all along, or the precocious young contender who thinks that he has a simpler and more intuitive explanation for our empirical law and whom we suspect is enjoying special privileges. We therefore refine our theory. Some assumptions are discarded and new, more plausible ones introduced. After many endless nights and much work we again present before the scientific community and our peers, the now new and improved, non-empirical model, from which the original law most elegantly follows. This time we are successful and we are told that the cheque is in the post. However, when the long-awaited cheque does arrive, or at least the promise of it, we are again disillusioned, for it is only for half the expected amount. Apparently, we are sharing our prize with our illustrious contender. Yet, we have nevertheless achieved something of scientific value and many accolades follow, but we do not attend the ceremony, after all, we are busy and Stockholm is such a long way away. There is still much science to be done and we have recently had some new, exciting ideas. The midnight oil continues to burn, for we now have some notion of the theory *underlying* our own theory which explained the initial empirical theory and between engagements this is what we shall work on, if only we are left alone to do so. Measurement techniques have improved and now modern technology has shown that the axioms of our non-empirical theory have been observationally verified, thus making it an empirical rather than a non-empirical theory. This is just as well, for our new theory contains additional assumptions that will explain exactly why the former theory is true. After nearly two years of intense work in which we traverse countless agonising and seemingly interminable blind allies, we are ready to publish. Strangely, however, the reception is somewhat less than we had hoped for. It seems that the young whippersnapper has also been busy and we hadn't noticed, so involved were we with our new ideas. It appears that he has revised his earlier theory and has produced a model that clearly fits all the available data. What's more, he has a more fundamental model that explains everything, albeit with many new assumptions. But then, we too, have made further progress and discovered new though perhaps untestable assumptions in formulating our own model; but surely that doesn't matter, for we have been down this path before. Technology will come to the rescue. Anyway, our contender's theory is also currently untestable with existing equipment and observational techniques. Therefore, we are not deterred and we soldier on with our research, delving ever deeper into the mysteries and forces governing our universe. We discover new ways to express the axioms of our model and exciting new mathematical methods for deducing the consequences of them, so creating a powerful, but now highly non-empirical theoretical edifice. Somewhere, however, in the back of our mind, we feel that we may be building our house upon foundations of sand, for not only is our theory now beyond the 'testability' of current technology, but it looks as though it might remain so for the foreseeable future. Additionally, there are now several similarly 'deep' theories that also explain our original model, which explained the model before that and the catalogue of assumptions that we have had to make grows ever longer and unverified. As we grow linearly older, so does our early contender and we learn that an even more fundamental theory has been formulated by a coalition of several young whippersnappers[*], and that this new model is hailed as the greatest step forward in the understanding of Nature ever devised. But, it too, is not testable and we begin to question the logic, at least as practiced, of the scientific method itself. In despair, we abandon practical science and decide instead, to write books on evolutionary epistemology.

The above narrative may seem to be somewhat scathing and derisive of the scientific method, but we do not mean to derogate the value of this technique, merely to illustrate many of the points raised earlier in §5.3, which this account is intended to clarify. In this respect we note several characteristics that we have identified earlier and which are specifically exemplified by the foregoing fiction. Let us number these lexicographically as follows:

a) Observations are made and/or experiments conducted. A relationship between various observable quantities is discerned. Thus a tentative empirical theory is formed.

b) The Law or axiom thus obtained is subjected to empirical tests. If it survives such tests, then it is accepted as a 'Law of Nature'. However, the observational tests may result in modification of the law, but still an empirical relationship exists.

---

[*] The writer is uncertain as to whether this is the correct collective noun for whippersnappers.

c) There is a clear distinction between empirical and non-empirical theories. Although both are merely sets of axioms, the non-empirical theory is deliberately formulated in order to 'explain' the empirical theory.

d) The non-empirical theory is constructed under additional assumptions or hypotheses that are designed to allow the empirical theory to be derived from it in a purely deductive manner. If such a deduction is successful, then that is considered to be sufficient 'explanation' for the empirical theory.

e) It is realised that the accuracy of the empirical theory and therefore its validity, is dependent upon the accuracy of any measuring devices used and also that the domain of applicability is limited to the area of investigation.

f) The hypotheses of the non-empirical theory are formulated in terms of natural variables and the natural interpretation is necessarily operational in both theories.

g) There is a competing non-empirical theory, also formulated in terms of natural variables but with different hypotheses. The non-empirical theory can also be deduced from this competitor.

h) The hypotheses of competing non-empirical theories are subjected to empirical testing. If only one survives this criterion, then that theory is now accepted as the 'correct' explanation of the empirical theory. This theory now becomes an empirical theory, since it has been empirically verified. The original empirical theory is now simply a consequence, or an inference that can be drawn from the new theory.

i) Another, deeper theory is formulated in order to explain the new empirical theory. This involves further assumptions, which are designed to ensure that the new empirical theory is a deductive consequence of this 'deeper' theory.

The process continues, until the point is reached at which the latest theory is untestable. When such a situation obtains it is always because existing apparatus is unable to obtain empirical data of sufficient accuracy to ascertain confirmation of the predictions of the theory. Then, since we require that a scientific model express empirically determined or verified axioms, any theory that does not satisfy that requirement cannot be called scientific in our sense of the meaning of that term. It might be thought however, that a theory that employs natural variables and is able to predict as yet unobservable phenomena *must* be more than merely speculative, but there is no basis for that conclusion by the criteria specified, no matter how sophisticated or mathematically elegant the theory may be. Indeed, theories of this speculative kind do occur in the practice of science and whilst they may be empirically confirmed eventually, they fail to meet the relevant qualifications unless their axioms are directly determined by observation and experiment, or verifiable by such methods. As prime examples of the latter in modern times, we may consider supersymmetric gauge theories and superstring theory as examples. Theoretical structures of this kind present philosophical and epistemological questions relating to the ontology of scientific theories about which one could debate at length and indeed we will return to the topic later. For the moment however, it is sufficient to note that by our definition of a scientific theory, no such problems arise, for we have essentially bypassed these issues through our attempts to actually define what we mean in the first place. Failure or lack of motivation to do so in the past may have contributed to the long and protracted debates concerning such matters, however inadequate our own definition may be.

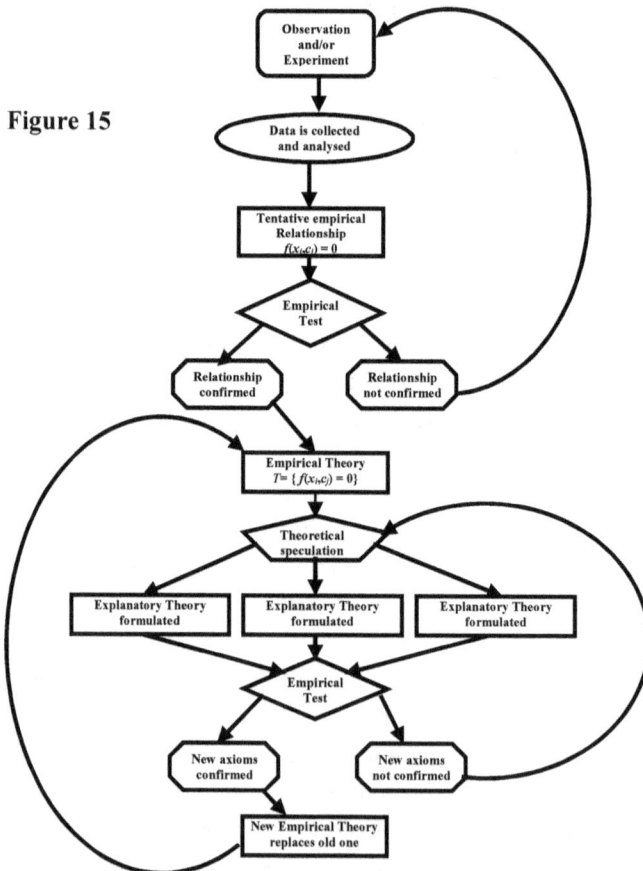

Figure 15

Note that the picture presented above of the method of science is only one possible scenario; there may not be any competing theories and the 'explanatory' theory may not be testable from the beginning. Clearly other variants of the procedure exist. The basics of the situation are illustrated in Figure 15, where we have shown three 'explanatory' theories, only one of which replaces the original empirical theory to become the new empirical theory. There are a number of points to note in relation to this figure that need clarification. Let us therefore follow it through from the beginning and see what we can discern.

We begin with an initial observation, or series of observations. This will result in a (presumably) fairly extensive data set in terms of natural variables, which are then analysed. It is found that a relationship amongst these variables can be extrapolated from the data and a preliminary hypothesis is formed. Clearly, this needs to be verified and to this end, further observations are made. There are now two possibilities. If the relationship is confirmed, then we feel that we are justified in expressing the newly discovered empirical relationship in a precise form such as $f(x_i, c_j) = 0$, which can thus continue to be subjected to empirical verification. This is then our empirical theory consisting in this case, of a single axiom. If the relationship is not confirmed however, then we continue to experiment and observe, collecting relevant data, which we again analyse in the hope of discovering some other relationship that will enable us to obtain an empirical theory. We note here that even if an empirical theory is confirmed, this does not mean that it is scientifically valid, for there is the question of the accuracy of the measurements of the natural variables. This inherent error may disguise a true empirical relationship that differs, perhaps only slightly from that which is observed. Additionally, there is the issue of domain of applicability, which is not *a priori* defined in practice and the operational validity of the observed relationship must therefore be considered only with respect to the specific area of investigation to which it pertains, although, if it is more universal, then such a greater encapsulation of the immediate domain will later become evident with further observational and theoretical developments.

The next stage in the figure concerns the formation of hypotheses, but not in an *ad hoc* fashion, for ideas and conjectures are now formulated only in relation to the empirical law hitherto determined. Thus, given the new empirical scientific theory, for it is now just that since theories consist only of axioms, then one may speculate upon the properties and underlying phenomena of Nature that are made evident through this empirical model. One may formulate several 'explanatory' theories based on various assumptions, from which the empirical theory may be deduced. Such speculative models can have no empirical basis *per se* and must be entirely hypothetical in that sense, although, doubtless there will be something of the nature of 'physical intuition' that will act as a guiding light in their formation. This is not a criterion for theory construction however, for we have seen from the evolution equation that a suitable theory can be obtained quite randomly given a sufficient period of time. But, we do not have the benefit of infinite time scales and therefore, a selection effect of this kind becomes operative, so we choose to formulate our explanatory theory with the natural variables available to us and within the domain of applicability of the empirical law obtained. We have imagined that we have three such theories in Figure 15, but there may be just one, or many more. The important point is that (a) they are all separate and with different assumptions and (b) from each of them the empirical theory may be deduced. The next stage is to test all of these non-empirical theories by comparison with observation and/or experimentation. The theories are expressed in terms of natural variables, indeed possibly new combinations thereof and will make some prediction concerning the behaviour of the physical system under consideration that should, if the theory is successful, be an observable consequence. This cannot be the original empirical law $f(x_i, c_j) = 0$, for that is a consequence of each of the explanatory theories, which is of course, the reason for their formulation in the first place. Therefore, some other prediction must be tested and for each of the theories it must be different, since otherwise the test will be inconclusive. We have envisaged in the figure that only one of the three theories is empirically confirmed in this way, but if this is not the case and in particular, if *none* of the theories are confirmed, then we return to the 'drawing board', so to speak and look for further predictions of the non-empirical theories that will allow the test to be conclusive. This will be the case when more than one or *all* of the theories pass the test. However, if only one of the theories is confirmed then the others are abandoned and this non-empirical theory becomes the new empirical theory. The assumptions of the non-empirical theory thus become the axioms of the new empirical theory. We then begin the cycle of theoretical speculation all-over and attempt to construct an even 'deeper' theory that explains the new empirical theory and produces the latter as a consequence. This is indicated in Figure 15 by the arrow leading from 'new empirical theory' at the bottom, to 'empirical' theory in the centre of the diagram. It is clear that we could follow this procedure of theory formulation and empirical observation and testing indefinitely, each time introducing new assumptions that then become new axioms for ever deeper and deeper theories. At some point however, we will inevitably reach the stage where empirical confirmation is no longer possible, either because the theory is inherently 'untestable', or it is not testable with the existing measuring apparatus. This need not deter us, but if we so continue with theoretical speculation without empirical confirmation, we may ask whether we are doing science at all rather than pure philosophy. In practice, this sometimes happens and in modern times String Theory and more especially, 'M-Theory' are examples of this.

6.2 Difficulties of the method

Whenever a new theory emerges, whether empirically or non-empirically, assumptions are made. The assumptions involved in purely empirical theories are of a different character than those of non-empirical theories. They tend to be more of a philosophical and ontological kind, relating to the existence of those very fundamental attributes of the physical world that are usually taken for granted, such as space, time and motion and the continuity of the variables employed therewith. Many of these assumptions are taken into account via the natural interpretation, which allows us to regard them as merely an interaction between the observer and the observed. In this way they become part of the fabric of the universe, in which the observer himself participates and with which he is inexorably intertwined. Thus, the natural interpretation does indeed allow us to take such assumptions for granted, for they are subsumed thereby, which enables us to discard them. We therefore *do* discard them, for now their epistemic importance is further removed from practical considerations of the scientific method.

On the other hand, the assumptions made in formulating non-empirical theories are primarily of a physical nature and tend to be concerned with the properties of physical systems and their fundamental constituents. It may appear that there is some overlap here, for non-empirical theories often relate to the ontological nature of space and time, or similar 'basic' properties of the universe, but there is a fundamental difference. Non-empirical theories treat such notions as already given, without questions of existence or ontology, but only with respect to the *nature* of such conceptual frameworks as realised in the Natural World, which is essentially the natural interpretation. The epistemic properties of the concepts and variables that denote and describe physical reality or the Natural World are of no concern in the theoretical consideration of the latter in relation to scientific speculation and theory formulation as we have described it. This is why, once a theory has been tested and without ontological reservations, we may take the assumptions made in a non-empirical theory to be the axioms of the new empirical theory, for there is no epistemic or philosophical 'baggage' that is thereby transmitted.

The scenario of the investigative technique of science depicted in Figure 15 and discussed above is clearly not inclusive of all possibilities, but nevertheless, portrays most explicitly a number of salient facts in relation to the scientific method as it is actually practiced. In this regard, we are able to adduce the following potential sources of difficulty in the method that may lead to error in theory construction:

a) In the first instance, an empirically determined theory is entirely dependent upon the accuracy of the observational or experimental data obtained. Thus, this margin of error may lead to inaccurate formulations of empirical axioms.

b) Secondly, the construction of a non-empirical theory in order to explain an empirical theory does not preclude the formulation of alternative theories that are equally explanative.

c) It is often assumed that if an empirical theory can be deduced from a non-empirical theory, then the non-empirical theory must be correct, for otherwise no such deduction would be possible.

d) When non-empirical theories are untestable, it is nevertheless the case that further formulations of non-empirical theories of an 'explanatory' nature are in practice pursued, thus adding greatly to the assumption set of the existing theoretical structure. Let us consider each of these in turn.

With regard to (a), it is clear that any empirical law is obtained within the bounds of existing observational accuracy and techniques, which evidently, can easily lead to an equivalently confined empirical law. This is to say that, assuming that gross errors have not been consistently made by the observer, then the empirical law may be only an approximation to the 'true' law that *would* have been obtained had the measuring process or instrumentation employed been more sophisticated. Indeed, in the history of science, there are numerous examples of this. By way of illustration, consider the Newtonian inverse-square law of gravitation; $F = Gm_1m_2/r^2$. Within the accuracy of the time and for a considerable period thereafter, this law seemed to be perfectly accurate and in fact 'universal'. However, as the accuracy of scientific instrumentation increased, so discrepancies to this well-established law were subsequently found, most notably and famously in the advance of the perihelion of the planet Mercury. Then, suppose that in Newtonian times more accurate and sophisticated measuring apparatus had been available and that, as a result, instead of the inverse-square law, the following expression had been found to fit the available data;

$$F = \frac{Gm_1 m_2}{r^2} + \frac{K}{r^3}$$

Which, with hindsight, we recognize as the appropriate modification of Newton's law required by General Relativity in order to account for the anomalies in the perihelion of mercury (with a suitable choice of K). But such a retrospective view is irrelevant. The point is that *had* the above law been empirically verified, then it would have been due only to the observational accuracy and competence of the observer, now endowed with scientific equipment of far greater sophistication, thus enabling the investigator to render the effects of General Relativity as empirical axioms. There exist no non-Newtonian conceptions of space and time in this process, for it is purely empirical and at least for a while, no necessity to depart from those foundations is immediately apparent. Thus, the importance of accuracy and precision of observation is illustrated as being an essential part of the scientific method, but is also a limiting criterion that constrains the success of that method. The essence of the matter is that given an empirical physical law of the form $f(x_i, c_j) = 0$, then the latter has been determined within the constraints imposed by observational accuracy. This law will continue to be empirically verified until the instrumentation improves and non-empirical theories will be formulated in order to 'explain' this axiom and from which it can be derived. Yet, the correct empirical law may be of a different form, say $g(x_i, c_j) = 0$, which may be only slightly different. The first law is only an approximation to the correct law. It is by virtue of this circumstance that the current accuracy of measuring apparatus may cause the investigator to explore several 'blind alleys' until such time as more precise observation reveals the true situation. But, as we have seen above, this inherent inaccuracy is a double-edged sword, for we may imagine that the modified expression for Newtonian gravitation given above would probably not have led to the 'true' axioms (Einstein's equations) that were later made evident with the advent of General Relativity, but then again, perhaps they *could* have done. Hence, imprecision in observation may obscure the real situation, but greater precision may do likewise. As mentioned before, only *full* precision is free of these difficulties.

With regard to (b), the construction of an explanatory theory of a non-empirical kind consists in making unjustified assumptions in relation to the fundamental constituents of Nature, or at least as far as the physical system under consideration is concerned. The fact that this can be done in several ways with varying underlying assumptions and hypotheses would not initially be a cause of conceptual difficulties if it were always the case that the scientific method could quickly determine, through observation or experimentation, precisely which assumptions are justified and which are not. Unfortunately, the procedure is not as discerning as this. There may be an infinity of assumptions that can be made and within each particular model, or non-empirical theory, the axioms of the empirical theory may consequentially follow and observation alone may not then be able to discriminate. Whilst eventually, it may be possible to eliminate a given theory through empirical means and perhaps after much arduous theoretical and observational 'fine-tuning' it is never the case at the moment of hypothesising, that this is actually possible, for otherwise there would be little point of such formulations in the first place. This may seem to be obvious and to be just a fundamental attribute of the scientific method, but again, that would be to miss the point. The fact that it is even possible to construct more than one non-empirical theory for the sole purpose of explaining an empirical theory is itself a weakness in the procedure, since such theoretical formulations necessarily demand new and to some extent arbitrary assumptions, which may again impede the progress of scientific knowledge. Thus, essential as this process may be to the scientific method, it relies upon 'physical intuition' and *ipso facto*, the imperfections of the latter are restrictive. Then, this results in another inherent difficulty in the method itself.

With regard to (c) however, the situation is entirely different and was mentioned in §5.3. A non-empirical theory may be considered 'successful' if the axioms of the empirical theory can be derived therefrom. But that does not mean that the hypotheses of the non-empirical theory are in any sense justified or valid. We saw an example of this with kinetic theory and the perfect gas law earlier. Simply because the gas law may be derived from an 'atomistic' theory of matter does not imply that atoms exist, it merely indicates that certain assumptions lead to a conclusion that has been observed to be true and does not preclude the possibility that alternative assumptions may also lead to the same conclusion*. Indeed, if it did then one would never be able to formulate more than one non-empirical theory in order to explain an empirical theory, which obviously has not been the case throughout the history of science. Even if two theories (as models) are isomorphic (or elementarily equivalent) in the mathematical sense, this does not mean that they are the same theory in the scientific sense, for interpretation of the symbols is all important for the latter and the natural interpretation admits of several possibilities. We saw an example of this in §5.1 in connection with the inverse-square laws of Newton and Coulomb. This is because the natural interpretation is actually many interpretations in terms of the natural variables, whatever the strings of symbols or *wffs* of the language may be. In actual scientific practice, this is obviously well-observed. Nor can it be said that if an empirical theory follows from a non-empirical theory, that the former is 'corroborated' thereby, or that the latter is 'evidence' for the former. No such circumstance obtains as long as alternative non-empirical theories are possible which in principle, is always the case. The best that can be said is that the fact of consequence is sufficient to warrant further investigation of the non-empirical theory in question. Then, one may ask whether such judgmental errors are actually made in practice and the answer is most certainly in the affirmative. However, this is a difficulty not with the scientific method *per se*, but with the investigator and his adopted logic, or lack thereof. Nevertheless, it is still problematic from a purely pragmatic perspective, i.e. in relation to the scientific method as it is sometimes practiced. The issue is further compounded in the last of our lexicographical considerations here.

Then, with regard to (d) we may say that when a non-empirical theory is formulated then the hypotheses of the theory are, by definition, untested, for otherwise it is an empirical theory. Hence, forming a 'deeper' theory that 'explains' the first one serves only to multiply the number of untested hypotheses. Thus, suppose we have an empirical theory S_0 such that $S_1 \vdash S_0$. If the hypotheses and assumptions of S_1 are never empirically verified, then clearly S_1 can never become the new S_0, or the new empirical theory. The loop at the bottom of Figure 15 is broken and we are left with theoretical speculation only and although S_1 may be aesthetically pleasing,

---

* The writer possesses a text on Statistical Thermodynamics that specifically states that the explanatory success of kinetic theory in deriving the ideal gas law is indicative of its reality.

it is still purely hypothetical. This 'one-level' situation, by which we mean just one step removed from an empirical theory (hence S_1) rarely happens in practice, simply because (usually) the non-empirical theory S_1 is not yet 'deep' enough to prevent empirical verification or falsification. This is because, generally, such a one-level theory will be constructed in terms of natural variables that may easily be tested and the predictions made therewith should, probably, be readily accessible with current instrumentation or measuring devices. We have adopted the adverbial qualifications here in order to cover those infrequent cases where this is not so. However, one may well construct a now deeper, more fundamental theory S_2 that explains S_1, so that $S_2 \vdash S_1$. This implies the addition of further untested assumptions, even if S_2 is still more aesthetically satisfying than S_1. The loop in Figure 15 is now well and truly broken and we are now two steps removed from the empirically verified theory S_0. Then, we may go further to the third level S_3 and then to S_4 and we may think to continue *ad infinitum*, at each stage adding more and more untested hypotheses. But this can never happen in practice, so let us confine ourselves to S_n, with $n < \infty$. We might think that we are now dealing with a theory that has a great number of assumptions that have accumulated additively at each level as follows: The first level S_1 has, say, h_1 assumptions (or hypotheses), the second level S_2 has h_2 assumptions and the nth-level has h_n assumptions. Then the number of assumptions for the nth-level S_n is h_n which is then the algebraic sum of these. However, this would be quite incorrect. Each theory S_i is derivable from S_{i+1}, i.e. $S_{i+1} \vdash S_i$ and therefore the hypotheses of S_i, which are the axioms thereof, are not contained in S_{i+1} because we exclude axioms from a scientific theory that are logical consequences or deductions from that theory. Therefore, given a sequence of theories $S_1, S_2,..., S_n$, where for each i we have that $S_{i+1} \vdash S_i$, then there is no change in the axiom number of any member of the sequence on account of being derivable from another member of that sequence. This is to say that if;

$$S_n \vdash S_{n-1} \vdash ,............, \vdash S_1 \vdash S_0$$

then it is not the case that the axiom numbers $h_i = |S_i|$ are additive, i.e.

$$|S_n| = h_n \neq h_{n-1} + h_{n-2} +,........, + h_1 + h_0$$
$$= |S_{n-1}| + |S_{n-2}| +,..........., + |S_1| + |S_0|$$

for this is only the case with set-theoretic union, which is quite different from our present considerations. In other words, by formulating a hierarchy of non-empirical theories in this way, we are not forming the set-theoretic union of their axioms, but merely replacing, at each level, one set of axioms for another such that the old axioms are derivable from the new. In fact, there is no telling what the axiom number of a non-empirical theory may be, for it is dependent upon the ingenuity of the individual theorist but, ideally, we would like it to be less than that of the theory which it is supposed to explain. However, it is clear that the domain of applicability of any S_i in the non-empirical theory sequence cannot be less than that of the original non-empirical theory, i.e. $\forall i \; \mathcal{D}(S_0) \leq \mathcal{D}(S_i)$. This, of course applies generally to all scientific theories that are consequences of others, whether empirically verified or not.

What is of greater concern however, and our primary point here, concerns the credibility and consequential validity or otherwise of an untested scientific theory that is formulated either as an initial or subsequent member of a sequence of non-empirical theories that are themselves untested. The essential point and the matter in hand, is not the nature or multiplicity of the hypotheses involved, for the latter is of only minor importance since, as observed above, such assumptions do not increase with the level or degree of removal from the original empirical theory. Rather, the issue of concern is the extent of detachment from the empirical realm that such a procedure necessitates. It may be argued that the continual formulation of non-empirical theories in this sequential fashion, however intellectually stimulating, will ultimately reach a stage that cannot be properly called scientific in its methodology, but must be considered to be little more than mere philosophical speculation and therein lies the epistemic danger. Yet, we can clearly address this point, for *any* theory that has not been subjected to empirical tests is automatically speculative and until such tests are made; only hypothesis prevails.

Therefore, if the scientific method is to be practiced in a manner that is not merely speculative; that does not consist of endless hypothesising, then it must retain a direct and intimate relation with the empirical world. Evidently, this should be accomplished at each stage in the process of axiom acquisition, the latter being *either* direct observation as is the case in the original empirical theory, or what amounts to the same thing, the verification of the axioms of the non-empirical theory that thereby elevates the latter to the status of an empirical theory. In this way, at each stage of the investigation and at each level within the succession of theories that are correspondingly formulated in the sequential manner hitherto described, the non-empirical theory gains credibility *before* further non-empirical theories are envisaged. We have emphasised earlier that in practice this may not always happen and we shall later encounter important examples of this phenomenon in connection with the problem of a 'Final Theory', although in principle it should be the 'ideal' of the scientific method to proceed in a strictly empirical and methodological manner that is free from unnecessary or extraneous hypotheses that may subvert that process in the first instance. This brings us to the question of the correct way to proceed in practical epistemology, so that we may avoid the difficulties here envisaged. Hence, let us consider what this putative 'ideal' may be construed to be.

6.3 Ideal science

It is quite possible to attain complete knowledge of the external world by means of speculation and theory formulation. The evolution equations of §4.5 make this clear, for random theorising, if continued indefinitely, must result in a correct theory at some point and there is no doubt that this will work. All knowledge of all things can thus be obtained by this simple process, providing sufficient time is available. But obviously and perhaps unfortunately, time is not available in so great a measure for mortal sentient organisms with a limited life-span in spite of their scientific motivation. Then, it might occur to such entities that an appropriate way to proceed under such a restriction would be to actually *observe* the external world in order to see whether the countless theoretical speculations that may have been arbitrarily postulated are actually the case in the real world. Thereby, it is hoped, many of the more tentative hypotheses concerning the alleged structure of the external world may be eliminated through an appeal to Nature. Thus a selection process is introduced.

The only possible selection procedure that can be objectively employed is therefore the confirmation of the hypotheses obtained from empirical data by observation or experimentation. In this way, needless hypothesising is avoided and Nature Herself becomes the ultimate arbiter of what may or may not be admissible within a given understanding of the physical world or physical phenomena described by a particular set of hypotheses. Under this restriction the complexity of the situation is much reduced and Figure 15 of the last section takes the following simple, but highly idealised form;

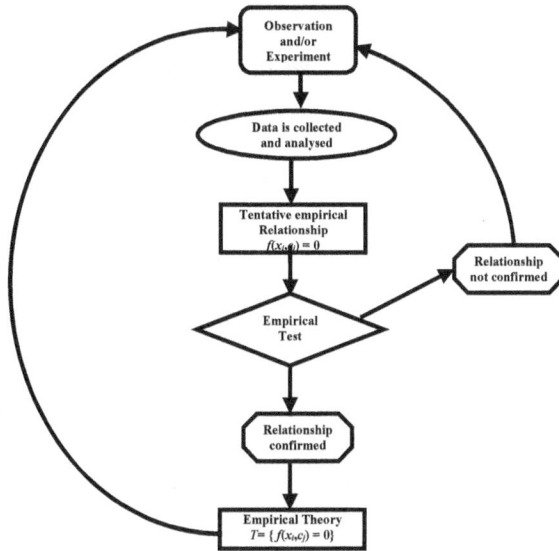

Figure 16

and we see here that there is never any 'theoretical speculation' as before. Thus, in this scenario there exist no non-empirical or 'explanatory' theories. Everything is then thoroughly empirical and no extra assumptions are introduced. The situation depicted in Figure 16 suffers from only one of the difficulties described in §6.2, namely the question of the accuracy of the measuring apparatus which again entails the possibility that the subsequent empirical theory may be unrepresentative of the 'actual' theory that would have been obtained if greater precision of observation had been available. But this is now considered unproblematic because it is part and parcel of the scientific method; it is *incorporated* into the method itself. Any observational discrepancies will, it is supposed, be resolved either by further observation and empirical tests with the extant scientific apparatus, or will be refined at some future time when more sophisticated equipment has been developed. Hence, there is a steady, continuous and gradual accumulation of scientific knowledge, the validity of which rests entirely upon the quality of the empirical process, which leaves little room for error. This 'ideal' scientific method, it must be admitted, does *prima facie* seem to present us with a virtually infallible epistemic technique of discovery and it would appear that given sufficient patience, with which the practising scientist is traditionally attributed in copious quantities, then it should be possible to systematically acquire greater and greater knowledge of the external world, thus the progress of science is thereby assured.

Certainly, the above epistemic set-up, or minor variants thereof, has often been portrayed as the *de facto* empirical method that is invariably employed in the investigation of Nature by practitioners engaged in that endeavour and has additionally been praised for both its efficiency and objectivity as a practical epistemological procedure for the acquisition of human knowledge, so much so that we take pains to impart the technique to our children, that thereby they may also be rational observers of the Natural World. In fact, much of this is justified, for it *is* the case that if one observes a given physical process to behave in a certain way, then it is the case that it behaves in that way and the way it behaves is expressed as an empirical axiom that states this fact, namely that it is the case that it so behaves! But the circularity here actually *guarantees* that the method is successful, for it could hardly be otherwise and it is this fact that accords merit to the 'ideal' as we have here envisaged it. Then, objectivity is almost a *sine qua non* of the scientific method. Efficiency however, is another matter, but only when an alternative and equally objective method can be displayed which is more epistemologically economical in the acquisition rate of raw data and the determination of physical laws.

Given such an idealistic epistemological methodology we may enquire as to how such a procedure can actually contribute to the growth of knowledge in the absence of theory formulation or the proposition of testable hypotheses, for these apparently have no place and nowhere to participate in the above scenario as illustrated in Figure 16. This in turn begs the question and demands an answer as to whether scientific knowledge can be obtained without theoretical speculation and if empirical means are entirely adequate for the continued acquisition and perpetual expansion of such knowledge of the World, for this is the claim that is made. This is a question of sufficiency. We have already agreed that empirical techniques such as observation and experimentation are necessary in the scientific enterprise since, as we have alluded to several times, little more than mere speculation results otherwise. Now, the practice of science will always involve natural variables, for the natural interpretation of all models of physical systems automatically pertains. Mere empirical interaction on the part of the observer will provide the appropriate natural variables and these will be generally of a macroscopic nature such as velocity, pressure, temperature, viscosity etc., at least initially. Often however, new natural variables are constructed from others, or at least defined in terms of them. A familiar example of this is kinetic energy $E = \frac{1}{2}mv^2$ (we ignore the definition of mass which is already provided by Newton's laws). In such cases, the variables are measurable and empirically observable just because their constituent components are, and relationships may be obtained therefrom by the empirical methods described above. But what of unobservable quantities, or variables that are defined under given hypotheses? Individual photons may register on a suitable photon counter or a charged couple device, but one cannot see them directly. The angular momentum of elementary particles is not directly observable for that matter; the wave functions of such particles are no more accessible to observation than is the wave function of macroscopic entities, or indeed, that of the universe itself. One cannot argue here that the *effects* of such quantities are empirically observable and testable because the variables themselves are defined within the domain of a *non-empirical* theory, whereas our context here is the ideal scientific method that admits of no such speculative entity. Nor can we say that with the advent of infinitely sophisticated measuring apparatus, such microscopic quantities may eventually be empirically obtainable as natural variables or measurable quantities, for it is not possible to directly measure anything below the Planck scale in principle and string theories of quantum gravity would fall into this category. One may argue that we are discussing quantities that are defined by non-empirical theories as though the latter are already established, while the ideal method we are describing cannot countenance such definitions or theoretical constructs. Hence, no problem arises in the method because no such situation can ever pertain; the method as described simply does not deal with hypothetical formulations at all and therefore, any consideration of non-measurable variables cannot arise. But this, of course, is the whole point. The 'ideal' in the form given places so stringent a restriction upon the formulation of speculative hypotheses that it is difficult to see how such a simplified procedure can achieve very much more than the presentation of a mere catalogue of empirically determined axioms, without depth or insight. Yet, there is a sense in which we *should* be able to elucidate the structure of the World entirely by empirical means. This entrenched feeling has no logical foundation; it is merely an expression of faith in the method itself. That is not to say that the method is flawed in any way, just that it is not entirely adequate to the enormous epistemic task that it has hitherto been imagined to embrace. This is why in practice, the ideal can never be attained and that in practice also, the scenario of

Figure 15 is more closely adhered to rather than that of the more simplistic depiction exhibited in Figure 16 and *that* is why scientific theories must often undergo fundamental change. But sometimes the 'ideal' is pictured in an even simpler way as in Figure 17 as follows;

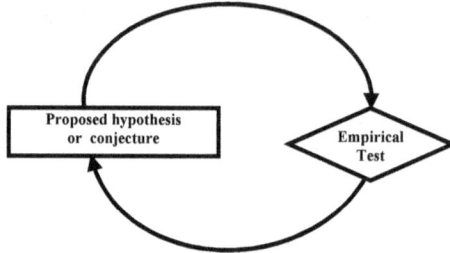

Figure 17

where hypotheses are proposed, empirically tested, then further hypotheses are proposed and the procedure continues *ad infinitum*. Here, the empirical test incorporates observation and experimentation and new hypotheses are formed whether or not the test is successful. This scenario suffers from the difficulties of observational accuracy in addition to the problem of hypothesis construction, except that a hypothesis is tested at each stage of formation. It does not matter whether we consider Figure 16 or Figure 17 to be our ideal scientific method since they are essential equivalent and both are incorporated within Figure 15 which we adopt as the *pragmatic* method that is actually employed in the scientific investigation of the physical world. Obviously, we need to examine the connection between the mechanisms of theory-change, the evolutionary aspect of scientific theories and the scientific method in this entirely pragmatic sense that we have found must now be the case. This is the concern of the next chapter.

Chapter 7

Scientific Revolutions

We consider the cumulative growth of scientific knowledge and the difficulties compounded in that process. We examine the status of scientific theories at any given time and suggest that a point is reached when reform is inevitable. We isolate the causes of theory change in science and suggest that the latter lie in the nature of the scientific method itself. We define and classify nine types of scientific revolution and discuss examples of each from the various sciences.

7.1 Cumulative progress

There is no question that scientific theories undergo modifications and refinements from time to time. They are required to meet the constraints of empirical verification and testability which, in the first instance, justifies their very existence. Usually, any modification in the axioms of an empirically verified theory will be a slow process, involving much painstaking observation on the part of the practitioner and will most likely be a consequence of improvements in observational techniques, or new innovations thereof that now reveal the necessity of correction to the original data. Such work must, therefore, necessarily be incremental in relation to the modification process and progress will be gradual and steady, proceeding at a nearly constant rate. A glance at any scientific journal will make this abundantly clear, for therein one may experience the diligent efforts of dedicated individuals whose sole purpose is the pursuit of knowledge through observation and experimentation, with ever greater precision and ever more meticulously than before, in order to thereby establish or alter the axioms of the scientific theory under enquiry. From an operational standpoint this is clearly commendable. But, such changes in physical laws are relatively minor and unless there has been a serious error in the observational or experimental methods on the part of the practitioner, the axiom under consideration will simply be modified to accord with any new data that may arise. It will then comply with this new empirically obtained information and the problem is resolved. Hence, scientific investigation may now proceed without hindrance until further such modification becomes necessary. *The mechanism employed here is theory modification and it becomes necessary because of imperfections in observational accuracy.* For evidently, if the observations were perfectly accurate in the first instance, then no subsequent measurements could reveal a discrepancy. Then, it is these very discrepancies that bring about theory modification in this sense, which is thus caused by the limitations of the scientific tools available at any given time. This is one fundamental way in which scientific theories are gradually refined. That such change is incremental results from the circumstance that improvements in the measuring apparatus are themselves incremental. Indeed, perhaps one measurement may, at best, be only two or three per cent more accurate than the last, possibly resulting in no axiom change at all. But, as empirical techniques advance in this manner and at each stage greater and greater precision of measurement is attained, then the affect is cumulative and theory modification inevitably follows. In the language of Chapter 4, let us consider the action of the theory modification operator T_3 on a simple theory S_0 consisting of just one axiom; $S_0 = \{\varphi_0\}$. This theory has been empirically obtained and, hopefully, verified with the existing measuring devices. As measurements improve and fresh empirical data is obtained, the axiom φ_0 is modified to say φ_1, but not in such a way as to change the theory S_0 in any radical or foundational way. Yet, although modification of a theory via T_3 is technically axiom replacement, this need not mean that there exists a great difference between the given axiom and that which replaces it. The difference may be minor and as in this case, just a slight alteration as a result of improved measurement. We need to distinguish between the two cases where axioms in which the values of any constant symbols contained therein are altered and the case when the mathematical form of the axiom itself is changed. We consider the former first. Thus the axiom φ_0 is replaced by a new axiom φ_1 that is of the same form as φ_0, except perhaps for a slight change in the value of any additive or multiplicative constants, or even a small correction term, none of which require a fundamental re-evaluation of the underlying principles involved. Thus, no assumptions of the theory are brought into question and the domain of applicability remains the same, i.e. $\mathcal{D}(S_1) = \mathcal{D}(S_0)$. The process continues and, with even greater accuracy of measurement φ_1 is replaced by φ_2 and in each case the domain of applicability is unchanged. In this type of theory-modification the form of the axiom is not altered and we can illustrate this with a very simple example taken from cosmology.

Suppose φ_0 is a model under the natural interpretation of the following *wff* in some formal language $\mathcal{L}$;

$$\forall x_1 \forall x_2 R_1^2 (f_3^2 (x_1, f_1^2 (c_2, x_2)), c_1)$$

where we employ the notation of §5.1 and use the interpretation given there for the function and relation symbols. If we now make the identifications $x_1 = v$, $x_2 = r$, $c_2 = H$ and $c_1 = 0$, then the above becomes;

$$v - Hr = 0$$

where v is velocity, r is distance and H is a constant (Hubble's constant). We immediately recognise this as a particularly simple form of Hubble's law for the velocity of recession of distant galaxies in the expanding universe. The accurate determination of the value of the 'constant' H had been a pre-occupation of observational cosmology for the best part of the twentieth century and continues to this day. Every new measurement of H therefore results in a new axiom, but the mathematical form of the axiom is not changed and neither is its domain of applicability. This type of activity roughly corresponds to what we may describe as 'everyday science', in which slow and gradual progress is made, knowledge of the Natural World is steadily increased thereby, but remains within the existing conceptual framework, for no fundamental assumptions or important principles are challenged. Thus, for each new empirically obtained value of H, made possible by either an improvement of observational techniques or the development of measuring equipment of greater sophistication, the constant c_2 in the *wff* above is re-interpreted with this new value, which then results in a new axiom φ_{k+1} that now replaces the axiom φ_k with the former interpreted value of c_2. However, since the syntactic form of the *wff* in the language $\mathcal{L}$ is left unaltered, no foundational issues in the physical system concerned, in this case the entire observable universe, are confronted. The process is therefore simply one of continual refinement and represents gradual, incremental growth of scientific knowledge.

There is then the case when the form of the axiom *is* changed. Considering the above example again, it is well-known that Hubble's constant is not actually constant and in fact varies with time. Then, the constant symbol c_2 in our *wff* above becomes a variable symbol, say x_3 and the *wff* now takes the form;

$$\forall x_1 \forall x_2 \forall x_3 R_1^2 (f_3^2 (x_1, f_1^2 (x_3, x_2)), c_1)$$

where we now need to quantify over the new variable $x_3 = H(t)$ say, (where t is cosmological proper time) which incorporates the newly discovered time-dependency. In this case an assumption has been modified, namely that of the constancy of H, but still the domain of applicability is unaffected. The fact that a constant has now become a variable may well change our *perceptions* of the theory and may compel us to consider the ramifications of that new insight, but our axiom is nevertheless empirically determined and what is more, under the new interpretation, specifically, the (observed) recognition of the variability of H, the *algebraic* expression of Hubble's law is of a similar form from a mathematical viewpoint;

$$v = H(t)r$$

although this is not strictly the case formally. The modification of the original axiom is more far-reaching than before when mere numerical values for the constant symbol were the subject of investigation, but all that has happened is that the rate of variability of H is now the object of enquiry. The empirical investigation from the standpoint of the practitioner has taken a new 'turn' so to speak and perhaps in a more intellectually stimulating way, but apart from this apparent 'transition', subsequent progress will remain of an incremental nature. In this particular case, in relation to the time-variability of H, it will presumably follow the same empirical, observational and scientific epistemic progression as did the prior case of when H was assumed to be constant. In fact, we may regard the two expressions as being of an identical linguistic form if we take the constant symbols to be special cases of variable symbols. Since subsequent observation *changes* the value of the constant symbols, the latter are evidently variable within the scientific-empirical context of theory modification. Simply because a re-interpretation of a constant as a variable alters the ontological nature of the theory does not necessarily imply that the syntactic form of the theory must thereby change as a result of such re-interpretation. A constant that does not change is thus a variable with zero rate of change and in general, all constant symbols that are not specifically stated to have precise time-independent values in the reals may be considered to be variable symbols. With this proviso, which is admitted by the natural interpretation, the latter modified example of Hubble's law is equivalent to the former and no formal distinction need be made in this context. Let us be clear, in a formal language with a constant symbol c, one cannot have both c and $\neg c$, for c is either constant or it is not, but we are not dealing in the present context with the formal language *per se*, but with its interpretation. The 'constancy' of the constant c is dependent upon the interpretation, which is entirely empirical and subject to modification in that epistemic sense. No part of the formal language is affected thereby and no contradiction results therefrom. Thus, here again, at each stage of the process empirical observation brings the operator $T_{\exists}$ into play and the initial axiom φ_0 is replaced by φ_1, φ_1 is replaced by φ_2 and so on as follows;

$$S_1 = \{\varphi_1\} = T_{\exists}S_0 = T_{\exists}\{\varphi_0\} = \{T_{\exists}\varphi_0\} = \{\varphi_1\}$$
$$S_2 = \{\varphi_2\} = T_{\exists}S_1 = T_{\exists}\{\varphi_1\} = \{T_{\exists}\varphi_1\} = \{\varphi_2\}$$
$$\vdots$$
$$S_n = \{\varphi_n\} = T_{\exists}S_{n-1} = T_{\exists}\{\varphi_{n-1}\} = \{T_{\exists}\varphi_{n-1}\} = \{\varphi_n\}$$

or in general simply $S_n = T_{\exists}S_{n-1}$ for the case of a theory with a single axiom. Since $n \in N$, where N is the set of natural numbers, the recursion above is sequential and the theories S_i change incrementally, as we have observed. Hence, in both of these cases we may consider that a slow cumulative expansion of scientific knowledge obtains and we may visualize this diagrammatically;

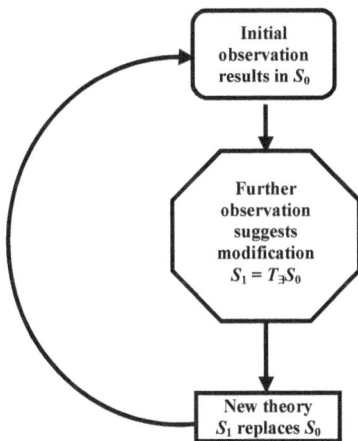

Figure 18

Then, as we cycle through this process, we obtain a succession of theories $S_0, S_1, \ldots, S_n$, beginning with the initial theory S_0 and ending with the final theory (to date), S_n. This final theory in the sequence represents the most accurate data currently available, but although it is different from the initial theory, it is not so very different as to be unrecognizable, by which we mean that the similarities of S_n to S_0 are such that it is quite evident from their mathematical or linguistic form that both theories are dealing with the same physical phenomena and with identical domains of applicability. Thus, this type of theory evolution $S_n = T_E S_{n-1} = T_3 S_{n-1}$ is a special case of the more general situation and one in which slow sequential progress occurs. Note two important characteristics here, firstly, the theories S_i are all empirical theories and secondly, at each stage the form of the *wff* of which the theory is a model under the natural interpretation, is unchanged. This latter in particular, is the hallmark of cumulative growth in the sequential sense that we have described.

It is clear that the sequence of theories that are obtained as above by purely empirical means are of an epistemic nature that grants them particular attributes in the growth of scientific knowledge. Each theory is obtained from the preceding theory by the process of observation and experimentation, in fact by standard scientific empirical methods. It is useful therefore to have a symbol to denote this. We shall denote such an 'empirical deduction' by '$\triangleright$'. Then, $S_{k-1} \triangleright S_k$ shall mean that the theory S_{k-1} has been modified by empirical means into the theory S_k, which corresponds to a special action of the theory evolution operator T_E in which only observation and/or experimentation has been involved and additionally, the domain of applicability has remained constant. Thus, the situation above is expressed as the sequential empirical process;

$$S_0 \triangleright S_1 \triangleright, \ldots \ldots \ldots, \triangleright S_n$$

The symbol '$\triangleright$' is nothing more than arbitrary notation for observation and experimentation, it signifies the empirical process and is the 'act of observation' or measurement of the physical world, or more specifically, the variables of the physical system under investigation. Note that there is no subjective component implied here. Whilst the empirical process can be carried out by a sentient biological organism, the same results can be achieved just as easily with a suitably equipped and appropriately programmed mechanism capable of receiving and analysing data from the empirical or external world.

Theory modification of this sort is relatively uneventful and consists essentially in the mere refinement of existing physical laws, which therefore both lends and substantiates a degree of stability to those axioms. However, as noted in §6.1, the empirical process is very much confined by the restraints placed upon the accuracy of the measuring apparatus available to the observer. Yet, we have implicitly assumed above that subsequent measurements with ever more precise equipment will permit the continued development of the sequence $S_0, S_1,\ldots\ldots, S_n,$ Usually, at least for a limited but reasonably long period this is the case, but there are two instances where the cumulative process may not hold. Firstly, the time-period between successive measurements may be inordinately long, thereby rendering the empirical process temporarily inoperable. This may occur when there is a delay in the development of scientific instrumentation of increased precision. The existing measurements may then be so imprecise that, at the current epoch, they are no longer trustworthy. In such case, *very* different results may be obtained from the data at hand and the scientific process reaches an impasse in that the testability of a theory is not possible at that time. Secondly and more seriously, it may no longer be feasible to make further measurements because that possibility is denied by the very nature of the physical system itself, in that the required accuracy of such measurements lies far beyond the foreseeable capabilities of any future apparatus. In practical terms, these two situations seem to be equivalent, for both pertain to observational accuracy. In the first case because of a time delay in obtaining results and in the second case because of an impossibility in obtaining those results, which is effectively an infinite time delay in the acquisition of empirical data. Thus, the cumulative process is halted and we must consider the effect of this upon the scientific endeavour in such a situation.

7.2 Minor revolutions

An empirically determined axiom or theory will have been tested and found to hold within the limitations set by the accuracy of the measuring apparatus available. But if that degree of precision is insufficient to distinguish between one empirical law and another, then since the empirical process is halted in the absence of greater precision, a choice may be made as to which axiom is adopted as the primary scientific model within the context of existing knowledge. Whilst this may often be the scientific *status quo*, it may not always be so. In the last section we considered something like this with the illustration of Hubble's law, which we may now imagine to have been determined initially with H constant, followed by a considerable delay in determining the variability of the latter due to insufficiently accurate scientific apparatus. When, eventually it is found that H is in fact variable, a significant length of time may have passed during which the constancy of H has become the foundation of further theoretical developments. Thus, the realisation of the non-constancy of H, after such confidence in the contrary view, represents and is the cause of a non-major but physically non-trivial revision of the expanding universe model. A similar situation has arisen recently when it was gleaned from supernovae and other observations that the expansion rate of the universe was actually accelerating. This was something of a surprise, brought about entirely by improvements in measuring techniques and instrumentation. Yet, in neither of these examples is the domain of applicability changed. However, scientific knowledge has nevertheless taken a small but significant step forward that transcends the steady and gradual cumulative 'everyday' progress that we discussed earlier. Such discoveries may lead to further, more important developments, particularly in the theoretical realm and the formulation or modification of non-empirical theories and we will refer to these kinds of theory modifications as 'minor revolutions'.

Then, we see that minor revolutions can occur because of an improvement in observational methods or scientific instrumentation, or both, but also that a *failure* of the measurement technique to provide new data may prevent scientific progress in an empirical sense that thus results in the continued acceptance of an inaccurately determined axiom, so that when superior equipment becomes available, the illusion of a minor revolution obtains. As mentioned above, the latter circumstance may lead to the pursuit of theoretical consequences of an incorrect axiom or theory which results in non-empirical and non-testable hypotheses that present a view of the physical world that is entirely false in that it can never be scientifically valid in the sense of §5.1. However, the similarity of these two situations in relation to accuracy of measurement suggests that we regard them as essentially the same thing from the pragmatic standpoint, since the success in obtaining new data with new equipment may be regarded as a failure of having been able to obtain that data prior to the development of the new instrumentation. Hence, minor revolutions occur because of the inadequacy of the measuring process. In fact, minor revolutions *must* occur, simply because of these imperfections in the methods of data acquisition. They are then, part of the scientific process itself and the existence of such phenomena requires no special explanation. Note that in all of the cases discussed so far, the domain of applicability remains unchanged, but the linguistic or algebraic form of the axioms of the theory may or may not be altered. We use the former as a criterion for the definition of a minor revolution and the latter to distinguish between various types of such revolutions;

Definition 7.2.1

A (minor) revolution of the first kind consists in the assignment of a new value to one or more of the constant symbols of the language.

Definition 7.2.2

A (minor) revolution of the second kind consists in the re-interpretation of one or more constant symbols of the language as variable symbols in which the algebraic or linguistic form of the axioms is unaltered, but the domain of applicability remains unchanged.

Definition 7.2.3

A (minor) revolution of the third kind consists in the re-interpretation of one or more constant symbols of the language as variable symbols in which the algebraic or linguistic form of the at least one of the axioms is altered, but the domain of applicability remains unchanged.

In all of the above, the theories are empirical and the natural interpretation obtains. The reason for placing the adjective 'minor' within parentheses will become clear in the sequel and for the remainder of this section we shall dispense with this qualification. Let us say a little more in connection with these apparently arbitrary definitions.

Revolutions of the first kind are not, of course, revolutions at all and in fact they are nothing more than the 'everyday science' that we mentioned earlier; just the routine accumulation of scientific data in the usual empirical way. They are a necessary part of the scientific method only because the observational nature of the latter demands that the determination of the values of the constants of a theory must be dictated by all the criteria of testability and verifiability that the method provides through the interaction of the observer with the physical world through the natural interpretation. It should be clear in this respect that any system of units employed by the investigator is quite irrelevant and it is merely the 'constancy' of the constants of the language that is of concern, rather than their numerical scaling. As far as we are concerned the constants are 'dimensionless', but their magnitude is of considerable importance in the determination of the structure and stability of the physical world. In this sense therefore, if one wishes to impose it, revolutions of the first kind can indeed be considered to significantly alter the status of a scientific theory, although they should be regarded as 'null' revolutions in the above classification.

Revolutions of the second kind are more significant, but do not cause a major upheaval in scientific thinking. As discussed above, they are the result of new measurements, as in the variability of the Hubble constant and the later recognition of accelerated expansion. The form of the axioms is not changed, but of course the re-interpretation of constants as variables means that some assumptions are questioned. In modern times this situation would be realised if certain physical constants such as the gravitational constant, the fine structure constant, or the speed of light were found to be variable. This would be quite far reaching, but easily incorporated within the framework of present theoretical models and the domain of applicability would be the same. The importance of such revelations lies in the consequences that follow therefrom, rather than the structure of the theories or axioms themselves. For example, the discovery of the accelerated expansion of the universe does not change the form of Hubble's law but is of major significance in relation to the future evolution of the universe, for it suggests that the universal expansion will continue indefinitely and that time has a beginning, but no end. These however, are *consequences* of the axioms and whilst the former are of enormous scientific interest, we are concerned in this context only with the latter and in this regard we recall that our theories are axiomatic and that which may be deduced from the axioms is not of immediate concern in this context. Revolutions of the first kind have not been uncommon in the history of science, ranging from the non-constancy of the Earth's rotation, the precession of the equinoxes, to the more extreme proposals mentioned above in relation to the fundamental constants of Nature. Thus, in revolutions of the second kind an axiom of the form $f(x_i, c_j) = 0$ is acted upon by the theory modification operator to produce a new axiom $f(x_k, c_l) = 0$ such that;

$$f(x_k, c_l) = f(x_{i+m}, c_{j-m}) = T_3 f(x_i, c_j)$$

where there are now $k = i + m$ variable symbols and $l = j - m$ constant symbols, i.e. exactly m constant symbols have been turned into variable symbols. The functional relationship $f(x_k, c_l) = 0$ is unchanged, as is the domain of applicability. Observe that this process is reversible; we could have m constant symbols replacing m variable symbols.

Revolutions of the third kind are much more fundamental. When the mathematical or linguistic form of an axiom is changed, the scientific model is thereby altered in a way that modifies the theory significantly, but that change does not extend to the entire domain of the scientific field concerned. For the latter to take place, the domain of applicability would need to change also. Hence, in this case the axioms of the theory are changed and a new theory results, but the effect of that change is still confined to the specific area in which the theory was originally applicable. Evidently, this condition places severe restraints upon the nature of the admissible modification, for the latter cannot be allowed to influence the axiomatic structure of any other extant theory that is a part of the general scientific discipline under enquiry. We are therefore dealing with sub-domains that are specifically designed to function entirely in relation to one specific area and modifications of which are strictly confined to that area and have no effect upon the greater area of which that sub-domain is a part. This is obviously necessarily the case when the domain of applicability remains constant under theory modification and is the reason that we refer to such axiom modification as of the 'minor' type. The essence of the matter here is that within a restricted domain of a particular scientific discipline, it has been found by empirical means that one or more of the axioms of the theory pertaining to that sub-discipline are incorrect and that new, empirically established axioms must now replace them. Of course, this again would not have been necessary if the scientific measuring apparatus or the techniques employed therewith had been sufficiently precise in the first instance. Thus, in revolutions of the third kind an axiom of the form $f(x_i, c_j) = 0$ is acted upon by the theory replacement operator to produce a new axiom $g(x_k, c_l) = 0$ such that;

$$g(x_k, c_l) = g(x_{i+m}, c_{j-m}) = T_\forall f(x_i, c_j)$$

Again, there are now $k = i + m$ variable symbols and $l = j - m$ constant symbols, i.e. exactly m constant symbols have been turned into variable symbols in the new functional relationship $g(x_k, c_l) \neq f(x_k, c_l)$ and, as expected, the domain of applicability is unchanged. Essentially, revolutions of the third kind amount to axiom replacement and the modification operator $T_\exists$ is replaced by the operator $T_\forall$ as in the expression above. We note that this process is also reversible.

In the foregoing, the discussion has been confined to empirical theories only and we may wonder how non-empirical theories fare in this modification process. However, we have already noted that non-empirical theories become empirical theories under circumstances in which the empirical process is able to operate, i.e. where measurements are still possible. This was illustrated in Figures 16 and 18 and discussed therewith. Then, subject to the continued validity of empirical methods, no difficulty in this regard arises and we may consider *all* our theories to be empirically testable models, providing this condition obtains. This should be evident, for clearly, if a non-empirical theory is not testable or empirically verifiable, then it is merely a set of hypotheses and can have no effect upon any existing empirical theory. Thus, if a (minor) revolution is to occur at all, then it does so via empirical processes. Considering that such revolutions are the result of the failure to make accurate measurements, this is hardly surprising. Then, we are led to categorically state the following;

> **For a scientific revolution to occur at all, it is**
> **necessary that the empirical process be operative.**

which can be stated in many ways, for example: *A scientific revolution cannot take place outside of the empirical method*, or we might state, *non-empirical methods can never lead to a scientific revolution*. Clearly, we can develop this further, for if we define 'empirical' to mean interaction with the external world, i.e. measurement of physical variables, then, since the latter is accomplished via the natural interpretation, we see that scientific revolutions imply the natural interpretation. This leads us to identify the latter with the empirical process itself so that the two, at least from a purely pragmatic viewpoint, are indistinguishable. We may ask whether the converse of the above holds: Does the empirical process imply scientific revolutions? Clearly, at least on the basis of the foregoing, it cannot, for it is imperfect measurement that results eventually in (minor) scientific revolutions and there is no reason why such measurements cannot be continuously improved over sufficiently small time intervals so as to prevent a revolution, or indeed, why measurements may not be very nearly perfect in the first case. Then, although the empirical method is a *necessary* condition for scientific revolutions to occur, it is not *sufficient*. Before we can go further with this analysis we need to consider other types of scientific revolution.

7.3 Intermediate revolutions

The essential defining characteristic of what we have termed 'minor' scientific revolutions is that the domain of applicability is unchanged in the process of axiom modification. We now consider the case when this condition is relaxed. The first thing to note is that the domain of applicability may increase or decrease. Now one may reduce the domain of applicability in several ways and we discussed these in Chapters 2 and 3. But, in the present context we are dealing with *scientific* theories in the sense of §5.2, whether they be scientifically valid or not. Thus, the natural interpretation demands that such theories are empirical theories and have therefore been subjected to the usual battery of empirical tests. Hence, if the domain is decreased, then the axioms of the theory were never applicable to the larger domain in which they were thought to originally apply and were therefore empirically invalid, which contradicts their initial verification. Hence, we need not concern ourselves with domain reduction, for the domain of applicability of the modified axioms would need to be a subset of the domain of applicability of their unmodified form and this is only possible if that original form is not empirically verified. But, the option exists that new measurements admit this possibility; that the domain of applicability was inappropriate in the first place. This cannot be so, because the axioms had been empirically verified within that domain and the domain of an axiom is just the area in which that axiom holds whilst the domain of a theory is the intersection of the domains of its axioms. Thus, we concern ourselves only with the case of domain enlargement. This, of course, does not preclude the possibility that a particular formulation of a theory may be found to be applicable within a sub-domain of that theory, but that such special circumstances cannot negate the original scientific theory that has, by definition, been empirically verified within the larger domain of applicability.

Then, by an intermediate scientific revolution, we shall in the first instance, mean theory modification, or theory-change in which the domain of applicability is extended either by modification or replacement of the axioms of the theory. But this cannot be the only criterion, for there is more than one way to increase the domain of a theory. Firstly, the axioms of a theory may be modified so that they extend beyond their original domain. This must necessarily involve changing the form of the axioms so that they may accommodate the new, now extended domain. Secondly, the axioms of one theory may be combined with those of another theory by the process of theory unification, thus extending the domain of applicability to that of both theories. At the same time, we do not wish to alter the fundamental principles and assumptions of the theory or theories involved at this stage, for that would entail a new discovery. Hence, as before, we encapsulate this in the following definitions;

Definition 7.3.1
An (intermediate) revolution of the first kind consists in the modification or replacement of one or more axioms of a theory where the algebraic form of the latter is altered such that the domain of applicability of the theory is increased, the variables remain the same, but no assumptions of the theory are altered.

Definition 7.3.2
An (intermediate) revolution of the second kind consists in the modification or replacement of one or more axioms of a theory where the algebraic form of the latter is altered such that the domain of applicability of the theory is increased. New variables replace old variables but no assumptions of the theory are altered.

Definition 7.3.3
An (intermediate) revolution of the third kind consists in the unification of two or more theories such that the domain of applicability is the set-theoretic union of the domains of the component theories. The variables of all components are unaltered, but no assumptions of any of the component theories are altered.

The essential point here is that the underlying assumptions, the fundamental basis on which the theory is founded, shall not be modified by this process, although the axioms will be. Thus, the theory will be applicable within a larger domain than was hitherto the case, but there has been no *conceptual* leap in the sense of a new understanding of the fundamental principles upon which the theory or theories were initially founded. This latter criterion is important, for we envisage intermediate revolutions in science to be more groundbreaking than minor revolutions, but not so revolutionary as to change irrevocably the nature of the relevant science *per se*, yet such change may nevertheless be of a magnitude that significantly transforms a particular scientific field or area of enquiry in such a way as to contribute considerably to the understanding and development of the theory of that special sub-discipline, or by the unification of several of the latter, which was hitherto unavailable. Thus, revolutions of this sort will alter the face of a sub-discipline, or set of sub-disciplines, but will not be of so great philosophical depth in that foundational issues are brought into question, or the ontological status of the formalism and basic tenets of the theories concerned will require major revision. Hence, intermediate revolutions operate within the existing framework of current science, but admit into a sub-discipline of the latter non-trivial modifications that represent significant and substantial advances within the area concerned, without the major upheaval that would result from a re-writing of the fundamental physical and philosophical basis of any of the theories involved. Then, it will be clear that intermediate revolutions constitute a far greater change than their minor counterparts. Let us consider initially such a modification as presented in the first of the above definitions.

It might be thought that it would be possible to extend the domain of applicability of a theory simply by extending the domain of its axioms. However, the domain of an empirical theory is defined in terms of the domain of its axioms and these have already been empirically determined, a point which we have alluded to several times above. Therefore, for intermediate revolutions of the first kind it is clear that the axioms must change. In fact, new axioms are required that extend the domain of applicability of the old axioms. Hence, as with minor revolutions of the third kind, the theory replacement operator is effective. Since the domain of applicability of a scientific theory is the intersection of the domains of its axioms, removal of the axiom with the smallest domain or replacement of the latter with an axiom of larger domain will result in a theory with greater or equal domain of applicability and this is the minimum requirement for the attainment of the latter. We discussed axiom subtraction in §5.2 and we observe now that it can never lead to a scientific revolution simply because no new axioms can arise through this process and the domain of applicability is already contained within those of the remaining axioms. Hence, axiom replacement is the only possibility. In that case there are a number of considerations. Obviously, if the axiom with the least domain, say φ, is replaced by the new axiom ψ, we require $\mathcal{D}(\psi) > \mathcal{D}(\varphi)$ and in general for a theory S_2 to replace a theory S_1 we must have $\mathcal{D}(S_2) > \mathcal{D}(S_1)$. Also, the natural variables may or may not change and whilst the form of the axioms may be altered, the fundamental assumptions are not. Then, the process is just axiom replacement with the appropriate condition on the domain. In terms of the physical theories involved we can envisage this as the action of the theory replacement operator $T_\vee$ on the theory $S = \{\varphi_1,....,\varphi_n\}$ as in §4.5;

$$S' = \{\psi_1,...,\psi_n\} = T_E S = T_+^{k_i} T_-^{k_i} S = T_\vee S$$

but with $\mathcal{D}(S') > \mathcal{D}(S)$. However, this is just theory-change and does not indicate how each individual axiom itself is modified. However, the latter may be made more explicit if we write, for each axiom $\varphi_n(x_i,c_j)$ an expression for the new axiom;

$$\psi_n = g(y_k,c_l) = T_E f(x_i,c_j) = T_E \varphi_n$$

with $g(y_k,c_l)$ of a different functional form to $f(y_k,c_l)$ and where the new variables y_k have replaced the variables x_i and the constants c_l have replaced the constants c_j. This is essentially a mapping;

$$Z_{ij}^{kl} : f(x_1,.....,x_i,c_1,.....,c_j) \rightarrow g(y_1,.....,y_k,c_1,.....,c_l)$$

for each axiom $\varphi_n(x_i,c_j) \equiv [f(y_i,c_j) = 0]$. So, the mapping Z_{ik}^{kl} turns the variables x_i into the variables y_k and the constants c_j into the constants c_l. Thus, within a given axiom φ_n of the theory S, some or all of the variables are replaced by new variables and similarly for the constant symbols. Additionally, the functional form of the axiom is altered and the domain of applicability increased. This essentially encapsulates Definition 7.3.1 and 7.3.2, but some clarifying remarks are necessary.

Intermediate revolutions of the first kind are similar to minor revolutions of the second kind in that the algebraic or linguistic form of an axiom(s) is altered but without the re-interpretation of constants as new variables and with an increase of the domain of applicability. Thus, the expression of a law $f(x_k,c_l) = 0$ takes a new form $g(x_k,c_l) = 0$ where $\mathcal{D}(g) > \mathcal{D}(f)$ for this particular axiom. Exactly the same variables and constant symbols pertain and, of course, the fundamental ontological or epistemic principles underlying the model are not thereby changed. Thus, this corresponds to the simple case when a physical law is modified to encompass situations in which the original axiom was inapplicable whilst retaining the latter as a special case and clearly if such a modification is to be of any merit, then the domain of applicability must increase. Essentially, a 'correspondence principle' is operating here, which must be so since the original axioms were empirically verified in the first instance and the modified axioms must incorporate that fact. In terms of the above mapping Z_{ij}^{kl}, this corresponds to the case where $k = i$ and $l = j$ and for each k, $y_k = x_i$. Needless to say, there are many examples of this sort of modification in the physical sciences. An obvious and familiar illustration is again the perfect gas law $PV = RT$ which we discussed §5.3. This axiom applies only to 'ideal' gases and is only an approximation to the real case, which is better described by the well known van der Waals equation;

$$\left(P + \frac{a}{V^2} \right)(V - b) = RT$$

where the constants a and b are specific to the particular gas under consideration. Clearly, in the limit $a = 0$ and $b = 0$ we recover the perfect gas law. Notice that we have the same variables, P,V,T as before and the same constant R. However, there are now the new constants a and b, however, definition 7.3.1 says nothing about constants, but only requires that the variables be unchanged. In this case we have gained two constants, but this need not be so; we could easily have found examples without this property. Evidently, the van der Waals equation (with suitable values for the constants) describes gaseous phenomena to which the perfect gas law is inapplicable; it therefore has a greater domain of applicability than the latter but still retains that of the former axiom. It might be thought however, that new fundamental assumptions have been introduced because the van der Waals equation is derived through considerations of the molecular nature of gaseous matter, whereas the perfect gas law is entirely empirical and requires no such assumptions. We saw in§5.3 that this is not so, for we described there that $PV = RT$ can also be obtained through kinetic theory on similar assumptions. But even if that were not the case it wouldn't matter anyway, since both equations (axioms) become empirical once observationally or experimentally verified. Furthermore, the criterion that fundamental assumptions are not altered does not refer to such notions as the constitution of matter or similar verifiable gross facts, but rather to conceptual ideas and pre-suppositions of a more fundamental nature that demand a re-evaluation of the very nature of space, time and the structure thereof, i.e. issues of a philosophical, existential and ontological kind. Many more examples of the sort illustrated here could be given, especially in relation to the numerous equations of state in the physical sciences. We see that intermediate revolutions of the first kind are quite common and an essential part of the scientific process brought about by the need for empirical precision.

Intermediate revolutions of the second kind take those of the first kind just one stage further. The algebraic form of at least one axiom is altered and the domain of applicability is increased. It is not, of course, this new functional relationship that is the cause of domain increase, but rather it is the introduction of the new variables that must carry that responsibility. More precisely, the replacement of the old variables by new variables admits of a larger domain because the new functional relationship is demanded by those new variables. If the functional form of the axiom(s) were unchanged then this would not be possible, for the natural interpretation had already determined the old variables, which are thus natural variables and the latter were empirically established within the mathematical, functional or linguistic relationship that was empirically found to obtain between them and by definition to hold within the domain of applicability of that relationship. Hence, the new variables cannot exist in that functional form if the domain of applicability is to be extended. In this connection, recall that when an axiom is determined it is by empirical means, for mere hypothesis does not, for us, constitute an axiom or physical law. The natural interpretation *defines* the variables that will partake of a given functional relationship and this will establish the relevant domain of applicability. Hence, the functional or algebraic form of the relationship between the natural variables is a result of the empirical process itself and again, the reliability and accuracy of the methods employed in that enterprise. Thus, if new variables are to replace old variables, then they cannot stand in the same relation to one another as was formerly the case.

Then, an intermediate revolution of the second kind is just axiom replacement in which new variables are introduced into one or more of the axioms of the theory. It is not required however that *all* the variables are renewed in this way, but only a sufficient number of them to ensure an increase in the domain of applicability and as mentioned above, the functional form must also change. This corresponds to axiom replacement where the mapping Z_{ik}^{kl} replaces some of the variables x_i in the original axiom $f(x_k, c_l) = 0$ with new variables y_k in a different functional relationship $g(y_k, c_l) = 0$ as discussed above. It is possible that constants may also be replaced, but again, this is not demanded. Thus, some of the y_k will be equal to the x_i in most cases and therefore be the same variables. If none of the y_k are equal to any of the x_i then the new axiom $g(y_k, c_l) = 0$ has nothing in common with the old axiom except possibly with respect to a few constant symbols. This will result in a completely new theory, but the underlying assumptions and conceptual framework is still the same as before. When some of the variables of the old axioms are retained and the domain of applicability increased, it is again reasonable to expect that the new theory should be reducible in some limit or other to the old theory, so that the later theory may embrace the earlier one. This however, may not be the case when all variables in all axioms are replaced, for in that case the old theory has been essentially abandoned and the new theory replaces it in its entirety. This is rarely the case in practice. Let us therefore consider a famous example from physics in which a relativistic wave equation is formulated. It is assumed that Special Relativity and non-relativistic quantum mechanics is already given, so that no fundamental conceptual issues are questioned. Then, a possible relativistic wave equation is;

$$\left(\partial_\mu \partial^\mu + m^2\right)\psi = 0$$

which is the Klein-Gordon equation for the free field $\psi(x,t)$. The notation is standard. As was realised, this equation suffers from a number of deficiencies such as negative energy solutions and probabilities. An equation of the first order was needed and subsequently obtained by Dirac;

$$\left(i\partial/\partial t + i\boldsymbol{\alpha} \cdot \nabla - \beta m\right)\psi = 0$$

where $\alpha = (\alpha_1, \alpha_2, \alpha_3)$ and α_1, α_2, α_3 and β are matrices. The details need not concern us here, but the solutions of the Dirac equation also satisfy the Klein-Gordon equation, which is what we require. In fact if we take;

$$\beta^2 = 1, \quad \alpha_1^2 = \alpha_2^2 = \alpha_3^2 = 1$$
$$\alpha_i \alpha_j + \alpha_j \alpha_i = 0, \quad i \neq j$$
$$\alpha_i \beta + \beta \alpha_i = 0, \quad i = 1, 2, 3$$

the two equations become identical and therefore the Dirac equation encompasses at least the domain of applicability of the Klein-Gordon equation. That in fact the domain is larger is obvious from the observation that the Klein-Gordon equation describes particles of spin zero, whereas the Dirac equation additionally describes fermions of half-integer spin. New 'variables' have been introduced in the form of the matrices α_i and β and clearly the form of the two expressions differ considerably; one is second order and the other first order. As mentioned above, we emphasise that since we took Special Relativity and quantum mechanics as our starting point in the form of the Klein-Gordon equation which is *already* relativistic, no additional assumptions of a fundamental nature have been made, so that even though these two disciplines contain non-classical ideas, that fact is of no concern here since relativistic and quantum mechanical notions were already incorporated into the initial axiom. We shall see however, that the Dirac equation also provides an example of revolutions of the third kind when this is not the case.

A more simple and familiar example is afforded by β-decay in nuclear physics. If we consider the decay of a neutron (n) into a proton and an electron (β-particle, e^-) or a proton (p) into a neutron and a positron (e^+), one would write;

$$n \rightarrow p + e^-$$
$$p \rightarrow n + e^+$$

which, as is well known, violates conservation of angular momentum. Instead, the existence of the neutrino and anti-neutrino is postulated to carry away the 'missing' energy, respectively v_e and $\overline{v}_e$, one then has;

$$n \rightarrow p + e^- + \overline{v}_e$$
$$p \rightarrow n + e^+ + v_e$$

which was eventually verified experimentally when sufficiently accurate equipment made that possible. Here the 'new variables' are the neutrinos which appear in the full mathematical formulation as variable replacement.

Intermediate revolutions of the second kind therefore represent a significant advance within the field of scientific enquiry concerned and generally result in a new physical law, or axiom. The domain of the latter, however, is confined to the area of investigation only, which may be greatly extended thereby, but has no affect *per se* upon unrelated scientific disciplines. We therefore see the particular field advance in a very non-trivial way, with accompanying new discoveries and revelations, but also a distinct absence of any modification to fundamental principles. Thus, such revolutions are truly revolutionary within their respective domains, but do not extend to the domain of applicability of other areas or sub-disciplines of the more general scientific enterprise. Now, it might be thought that a new physical law or axiom should have a significant affect upon the whole of science and indeed sometimes this has been true, but that is not our meaning here. A new law may revolutionise a given field, but nevertheless have influence on other domains merely by virtue of application only. By this we mean simply that the new axiom or theory may be *applied* within domains of applicability other than its own in a purely practical sense, but this does not in any way *contribute* to, or *extend* those extraneous domains to which it is applied in the manner of theory-change. Nor are any assumptions of a fundamental nature in any domain altered by such applications. Thus, the Dirac equation has direct applications in medicine through Nuclear Magnetic Resonance Imaging (NRMI), because one is measuring the spin states of protons, i.e. a fermion governed by that equation – but this has no bearing whatsoever in relation to the biology of living organisms. However, when a physical law or theory is combined with an otherwise distinct and separate theory, the effect on both may be substantial. Hence, we consider intermediate revolutions of the third kind.

In fact, we have already discussed intermediate revolutions of the third kind in §3.10, where we gave an example of what we mean by this process in terms of the Hydromagnetic and the MHD approximation. Thus, we envisage revolutions of this sort to be nothing more than theory unification as described in chapters 3 and 4, but with the important proviso that fundamental philosophical or ontological assumptions must not be altered. Furthermore, no *new* variables are introduced, but axiom combination – which is a fundamental part of this process – may alter the form of the physical laws or axioms involved. Obviously, the domain of applicability is extended to that of each of the component theories in the set-theoretic sense.

Recall from §2.2 that the unification of two (or more) theories requires that they be incompatible in the sense that the domain of each theory is distinct as far as applicability is concerned. Essentially, the one theory cannot encroach upon the domain of the other, for it is simply not applicable there. We are, it seems, dealing with two different animals. But we have a mechanism by which this situation may be rectified and by which new revolutions may ensue. It is a twofold process as we have earlier described and consists of ordinary set theoretic union together with axiom combination. Then, theory unification constitutes a revolution because new insights are revealed merely by the process of combining, or conjoining two or more disparate fields. They are however disparate only in the sense that it was hitherto not directly realised that appropriate algebraic or linguistic manipulations would allow both theories to embrace the same, but larger domain of applicability. Yet, in so doing, new axioms or new physical laws *appear* to be obtained, but evidently they were already there in the first instance. This is only so because no assumptions of a foundational kind have been changed, which is our requirement for this sort of revolution, for if the fundamental assumptions of either theory were altered in any way, then the physical laws that were so combined in the unification process would need to take a different form and possibly contain new variables. The essence of the matter here, is that at least in the *original* formulation of the axioms of the component theories, no alteration is necessary, but in the unified theory, clearly the axioms adopt a new functional form but without affecting the original domains of the component theories, which still independently stand on their own in relation to their respective domains of applicability.

Thus, in the MHD example of unification of §3.10, Maxwell's equations were combined with the fluid equations and this admits of new insights and applications, such as in plasma physics, stellar atmospheres, geophysical fluid motion, etc. However, each of the individual theories are not themselves altered, but new revelations emerge from the combination of the two and the domain of applicability of the new axioms is just that of the union of the domains of the component theories. So it is also with relativistic wave equations such as the Dirac equation above. The latter combines non-relativistic quantum mechanics with Special Relativity, thus uniting the domains of both. Obviously, the same applies to the Klein-Gordon equation and indeed the relativistic Schrödinger equation, together with a myriad of other similar equations. In each case, the new domain becomes the set-theoretic union of the component theories and in each case also, no *fundamental* assumptions are changed in any manner that might affect the philosophical or ontological foundations of either component theory.

Perhaps the most famous example of an intermediate revolution of the third kind and arguably the most profound is Maxwellian electromagnetic theory itself, which we have mentioned several times. The unification involved here is that of the seemingly disparate phenomena of electricity and magnetism, although it was already clear from experimental evidence, principally best documented by Faraday, that the two were intimately connected. In the standard physics textbooks Maxwell's equations are easily and quickly derived, but whilst this understandable sacrifice to expediency has merit it fails to capture the essential features of the scientific development of physical theories, which in the case of sentient beings such as ourselves is often more intuitive and analogical than it might otherwise be. Frequently, the painstaking work and insights of previous workers in a given field will have reached the point when a major development *should* be inevitable if only a suitable intellect were available to complete the process. This is seen to hold in almost all revolutionary scientific advances that lead to fundamental changes in the overall picture of the world and electromagnetic theory is no exception in this respect. We shall therefore adopt a *semi*-historical approach in our description of electromagnetic theory as a revolution in this category in the hope that some perspicuity may thereby ensue with regard to the points alluded to in the foregoing remarks.

Much of what are now called Maxwell's equations were of course already known and experimentally verified prior to Maxwell's own contribution. However, there was an inconsistency that Maxwell identified and was able to rectify. Faraday's meticulous experimental work was well known to Maxwell and he drew an analogy between 'lines of force' in relation to magnetism and the motion of fluids. We begin with the continuity equation in fluid mechanics for incompressible flow that we quoted in §3.10. Let us recall the simple heuristic arguments used in the derivation of this equation. Firstly, if ρ is the density at a given point and $\mathbf{v}$ the velocity, then the mass flow per unit time through a surface element $d\mathbf{S}$ is $\rho\mathbf{v}\cdot d\mathbf{S}$. Hence the total mass flux through the closed (bounded) surface S is $\int_S \rho\mathbf{v}\cdot d\mathbf{S}$. This must be equal to the rate at which mass is lost from the volume V bounded by S, therefore;

$$\int_S \rho\mathbf{v}\cdot d\mathbf{S} = -\frac{d}{dt}\int_V \rho\,dV$$

By application of the divergence theorem to the left hand side this can be written as;

$$\int_V \nabla\cdot\rho\mathbf{v}\,dV = -\int_V \frac{\partial\rho}{\partial t}\,dV$$

which leads to the continuity equation;

$$\nabla \cdot \rho \mathbf{v} + \frac{\partial \rho}{\partial t} = 0$$

and for an incompressible fluid the density does not depend upon the co-ordinates, thus;

$$\nabla \cdot \mathbf{v} = 0$$

Now Maxwell drew an analogy between Faraday's magnetic lines of force and the above streamlines in fluid flow. He imagined 'tubes of magnetic force' by this hydrodynamic analogy. Thus, replacing the velocity $\mathbf{v}$ by the magnetic flux $\mathbf{B}$ gives the Gauss theorem in electrostatics;

$$\nabla \cdot \mathbf{B} = 0$$

which expresses the absence of single magnetic poles. Then, in exactly the same way Maxwell also writes for the electric field $\mathbf{E}$;

$$\nabla \cdot \mathbf{E} = 0$$

which is valid in free space in the absence of charges. However, since $\nabla \cdot \mathbf{E} = 4\pi\rho$, where ρ is now the electric charge density distribution and $\nabla \times \mathbf{E} = 0$ the electrostatic field is derivable from a potential Φ so that $\mathbf{E} = -\nabla\Phi$. Combining these expressions we have Poisson's equation;

$$\nabla^2 \Phi = -4\pi\rho$$

which means that for the charge density ρ we should have;

$$\nabla \cdot \mathbf{E} = 4\pi\rho$$

and Maxwell was well aware of this. Next we express Faraday's law of electromagnetic induction in suitable mathematical form. We consider the total magnetic flux through a surface S bounded by a circuit C and write;

$$\int_C \mathbf{E} \cdot d\mathbf{S} = -\frac{d}{dt} \int_S \mathbf{B} \cdot d\mathbf{S}$$

much as we did earlier. If we now use Stoke's theorem on the left- hand side of the above we obtain;

$$\int_S \nabla \times \mathbf{E} \cdot d\mathbf{S} = -\frac{d}{dt}\int_S \mathbf{B} \cdot d\mathbf{S}$$

from which we obtain;

$$\nabla \times \mathbf{E} = -\frac{\partial \mathbf{B}}{\partial t}$$

The field **B** is the magnetic *induction*, but we have also the magnetic *intensity* **H** which is interpreted as the magnetic force and in terms of this Maxwell wrote Ampère's circuital law as;

$$\oint_C \mathbf{H} \cdot d\mathbf{S} = \int_S \mathbf{J} \cdot d\mathbf{S}$$

for a *steady* current density **J** flowing through the surface S bounded by the circuit C. If we now use Stoke's theorem again, we get;

$$\int_S \nabla \times \mathbf{H} \cdot d\mathbf{S} = \int_S \mathbf{J} \cdot d\mathbf{S}$$

and therefore;

$$\nabla \times \mathbf{H} = \mathbf{J}$$

Thus, we have the following four equations for the electromagnetic field;

$$\nabla \times \mathbf{E} = -\frac{\partial \mathbf{B}}{\partial t}$$

$$\nabla \times \mathbf{H} = \mathbf{J}$$

$$\nabla \cdot \mathbf{E} = 4\pi\rho$$

$$\nabla \cdot \mathbf{B} = 0$$

Then, can we say that electric and magnetic phenomena have been unified? Not quite. The above set of equations is inconsistent and it took

Maxwell to realise this. The problem is with Ampère's law which pertains in the above for a steady current only such that $\nabla \cdot \mathbf{J} = 0$, but for the general case the movements of each charge when it is displaced from its equilibrium position creates a small current, the *displacement current* which is proportional to the electric field strength $\mathbf{E}$. We denote this by a new field vector $\mathbf{D}$, as Maxwell knew. Now, the current $\mathbf{J}$ and the charge ρ should therefore satisfy the continuity equation;

$$\nabla \cdot \mathbf{J} + \frac{\partial \rho}{\partial t} = 0$$

and with $\nabla \cdot \mathbf{D} = 4\pi\rho$ we can write this as;

$$\nabla \cdot \mathbf{J} + \frac{\partial \rho}{\partial t} = \nabla \cdot \left(\mathbf{J} + \frac{1}{4\pi} \frac{\partial \mathbf{D}}{\partial t} \right) = 0$$

and so Maxwell replaced the current $\mathbf{J}$ with the quantity in parentheses in Ampère's law for time-dependent fields. Then the latter becomes;

$$\nabla \times \mathbf{H} = \mathbf{J} + \frac{\partial \mathbf{D}}{\partial t}$$

In suitable units and in terms of the electric displacement $\mathbf{D}$, the full set of Maxwell's equations now reads;

$$\nabla \times \mathbf{E} + \frac{1}{c} \frac{\partial \mathbf{B}}{\partial t}$$

$$\nabla \times \mathbf{H} = \mathbf{J} + \frac{1}{c} \frac{\partial \mathbf{D}}{\partial t}$$

$$\nabla \cdot \mathbf{D} = 4\pi\rho$$

$$\nabla \cdot \mathbf{B} = 0$$

where c is a constant (which subsequently turns out to be the speed of light). Of course we have not been very careful in this discussion and indeed Maxwell later derived these field equations in a much more rigorous fashion, but our point was to show that intuition and insight can play a significant role in theory development and this is well-illustrated here with regard to unification and the ingenuity that the latter may require.

Then is definition 7.3.3 satisfied? Clearly, electric and magnetic phenomena have been unified and the Maxwell equations cover the domain of applicability of each, which is now the set-theoretic union of both component domains. Furthermore, the same variables in each of the component theories are used and thus no new assumptions have been introduced or indeed, old assumptions altered. Hence, Maxwellian electrodynamics is an intermediate revolution of the third kind.

The unification of electricity and magnetism is very similar to the hydromagnetic example given earlier, yet few would doubt that the former is of much greater significance from the point of view of scientific advancement. The Maxwell example therefore illustrates two interesting points; firstly, that intuition and analogy with existing theories and models can serve as a guiding hand in initiating progress in science and secondly, that revolutions of the same kind can differ greatly in importance. We shall return to both these points later.

There are numerous examples of intermediate revolutions of the third kind throughout the scientific realm and not just in physics. In evolutionary biology, Neo-Dawinism (but not Dawinism *per se*), or the 'modern synthesis' is an illustration (primarily linguistic) of this sort of scientific revolution. Here, the Darwinian thesis of species development, adaptation and natural selection is combined with Mendelian genetics (or its modern counterpart), thus increasing the domain of applicability to that of both theories, even on the molecular level and additionally providing new insights into the causes of species evolution. No fundamental assumptions are altered in either theory, which independently stand alone, but the unification provides a more complete representation. One may ask how axiom combination applies to this particular example, for few clearly defined mathematical formulae (apart from those of population genetics) exist in this area. There seems to be no substantial theoretical structure such as occurs in physics or similarly well developed fields that may be manipulated, at least from a mathematical point of view. But this does not matter. We recall that the mathematical expression of the axiom content of a given model does not preclude linguistic formulations of the relevant axioms, since the formal language in which any *wff* is given of *whatever* nature, requires only a meaningful interpretation and the natural interpretation does not care about the particular language used.

Mathematical expressions may be more succinct, but ordinary language is no less applicable. Then, if a purely linguistic axiom P from one theory is to incorporate a sentential axiom Q from another theory, axiom combination cannot simply be the conjunction or concatenation P ∧ Q of these two sentences, for that would achieve nothing. Rather, a new linguistic expression, say P □ Q results, which is the linguistic form of what would have been axiom combination in the mathematical and algebraic sense had the original formulation not been expressed in natural language. This is then, a *new* sentence of ordinary language that is of a different grammatical and syntactic form hitherto unknown and which expresses the axiomatic combination of P and Q that is not at all what would obtain from simple axiom concatenation, for now bears only a passing, if any, semantic resemblance to the original forms of P and Q. We have mentioned earlier that any meaningful expression, whether mathematical or otherwise is perfectly acceptable as far as the formal language is concerned, for in this sense mathematics is just another language. Hence, the formulations of the axioms in ordinary language are of no concern from an interpretive standpoint. All meaningful statements, if the symbols of the language are so interpreted, will be equivalent, but of course, it is understood that the natural interpretation obtains, as always must be the case. All this means is that it doesn't matter whether we use the symbol 'm', or the word 'mass' to denote a physical quantity, since they signify the same thing.

To briefly return to the physics of the latter half of the twentieth century, we would be amiss if we failed to mention the Weinberg-Salam electroweak model of the unification of electromagnetic and weak nuclear interactions. This is the theory with gauge group $SU(2) \otimes U(1)$, where $U(1)$ corresponds to electromagnetism and $SU(2)$ to the two-dimensional symmetry group of the weak interaction, describing leptons and quarks. If we also include the 'colour' degrees of freedom for quarks as in quantum chromodynamics with gauge group $SU(3)$ we may construct the so-called 'standard model' of particle physics with gauge group $SU(3) \otimes SU(2) \otimes U(1)$, thus unifying three of the four fundamental forces. Specific details would be out of place here and the point we wish to make is that these unified theories represent (somewhat complicated) examples of intermediate revolutions of the third kind, for each of the component theories are unaltered and the resulting new theory, after unification, does not entail the modification of any fundamental assumptions that were made, or formed the basis of any of the component theories. Any assumptions in the new theory are carried over directly from the component theories without alteration and no new fundamental notions of a philosophical nature are made. The same applies to the variables of the unified theory, but this does not mean that new results or variable combinations (thus defining new parameters) cannot occur in the new theory. Clearly, if that were the case there would be little point in unification in the first instance.

We therefore see that intermediate revolutions in this category, as is similarly the case with the aforementioned classification scenario that has been described, can be relatively simple or highly complex. Evidently, the complexity of a unified theory is directly dependent upon the complexity of the component theories and this is inevitably so since axiom combination will be a more involved process with mathematically more sophisticated theories. We observe also, that when unification of theories takes place, it must necessarily be a scientific revolution simply because a multiplicity of domains are combined which must *ipso facto*, clearly constitute substantial scientific advancement in all of the domains in question merely in virtue of the unification process *per se*. Thus, an intermediate revolution of the third kind is the most dramatic change that can occur within the scientific enterprise that does not bring into question foundational, ontological or philosophical issues. This is the criterion that places such revolutions firmly within the conceptual framework of existing science, but as we have observed, still allows scientific advancement to admit of new notions and ideas, such as the Higgs field in the standard model example, which introduces major change and offers new perspectives, but are not of themselves concerned with the deep philosophical assumptions of any of the particular theories involved. This again is an expression of a kind of 'correspondence principle' which could not apply if fundamental assumptions were modified for any of the component theories partaking in the unification process, for the new unified theory must reduce to each of the component theories under suitably restrictive conditions relating to their respective domains of applicability. Should any of those components be altered in the philosophical manner envisaged, then they would themselves partake of and be the subject of a scientific revolution. This brings us to the next and final category of revolutionary change in scientific theories; namely those in which the fundamental assumptions themselves undergo radical revision. Such changes are of an altogether different magnitude and lead to major upheavals in scientific thinking. This is the concern of the next section.

7.4 Major revolutions

Throughout the history of science there has, from time to time, emerged theories which, by their very nature, introduce such radical conceptual changes that previously held ideas and assumptions are no longer tenable. The *status quo* is overthrown and a new and different perspective on the nature of the physical world is obtained. However, this does not always mean that earlier hypotheses were grossly incorrect, although that may well be the case, but rather that new observational data and experimental results have revealed that only a partial truth in the original assumptions obtained. Then, we are dealing with two possible scenarios here: Firstly, the case in which the assumptions of a theory are modified or re-interpreted to such an extent that they can no longer be considered to be of the same nature, or in any relation of equivalence to their earlier original form and secondly, the case when the underlying assumptions of a scientific theory are completely replaced. This is not the same as theory modification and theory replacement, as it may on first contemplation appear to be, for when fundamental principles are changed or modified something far more radical and deep-rooted is taking place. Thus, it is because the nature of the changes here are of a foundational and conceptual kind that the theory is replaced or modified, rather than the converse. Replacing a theory does not necessarily entail the replacement of the underlying philosophical and ontological assumptions of that theory, as we have seen, but if those assumptions themselves are replaced, then so must the theory in question. Similarly for assumption modification which, from a purely pragmatic viewpoint, may be regarded as theory replacement. The difference between these two cases lies in the fact that with assumption replacement the resulting new theory may bear little or no relation to the old theory, whereas with assumption modification a sufficient relaxation of aspects of the new assumptions may allow recovery of the earlier theory in special circumstances in a kind of 'correspondence principle'. Evidently, in either of these processes new variables may be introduced and new relations between them defined. The new theory will therefore be mathematically or linguistically very different in syntactic form, even when some of the old variables are retained. Thus a new view of Nature is provided. However, the case of modification is less radical than total replacement, for it alone allows existing theories to serve as approximate descriptions of the physical world, which can never occur when an entire set of assumptions are abandoned in favour of a new and alternative philosophical foundation. There is also the question of domain applicability. When the assumptions of a theory are replaced, it is expected that the domain of the new theory be at least that of the old. However, when assumptions are modified, the new theory must have domain of applicability greater than that of the old theory. This latter circumstance arises because the new theory must cover the domain of both the phenomena described by the former theory, in addition to the phenomena with which the old theory was unable to adequately describe and which necessitated assumption modification in the first instance. This now ensures the correspondence principle and the new theory is reducible to the old theory just when the modifications of the new assumptions are removed. Then, as before, let us make the following definitions for such *major* revolutions;

Definition 7.4.1
A (major) revolution of the first kind consists in the modification or replacement of at least one of the fundamental assumptions of a theory such that that the assumptions that are replaced are logically inconsistent with those that replace them and the domain of applicability is increased.

Definition 7.4.2
A (major) revolution of the second kind consists in the replacement of all of the fundamental assumptions of a theory such that the assumptions that are replaced are logically inconsistent with those that replace them and the domain of applicability is not decreased.

We note that in the first case it is not necessary that *all* the assumptions of a given theory be modified or replaced. The alteration of *any* fundamental assumption in a physical theory is sufficient to generate major theory-change and thus instigate a major revolution. We observe also the importance of the domain of applicability yet again in this type and indeed all other kinds of revolution discussed thus far, as well as in the basic mechanisms of theory-change. Obviously, if *all* the assumptions of a theory are modified or replaced as in the second case, then a more drastic or more 'major' revolution occurs than if only some of the fundamental assumptions partake in this process. Thus, there are degrees of major revolutions which are directly related to the number of assumptions altered or replaced. This need not concern us however and anyway, similar circumstances obtain also for both minor and intermediate revolutions. The essential point is that it is the change in the fundamental assumptions that primarily characterises this sort of revolution and from that all subsequent theory-change follows. We therefore see no reason to further refine this category. There remains however, something of a bone of contention. We have continually referred to 'fundamental assumptions' throughout this work and sometimes ambiguously. Before we can properly describe major revolutions we need to clarify the meaning of this rather vague term.

Scientific theories often contain a number of assumptions of various kinds and we have encountered many of these in the present discourse. Some assumptions are of a relatively simple variety and relate to suppositions concerning the mathematical or linguistic form of the relations between the natural variables of a given theory. These are hypotheses that conjecture some relationship of the form $f(x_i, c_j) = 0$ and are empirically testable. They therefore need not concern us in this context and indeed, have been discussed at length earlier. Then there are assumptions that suggest a modification of the theoretical structure of a physical theory that entails a new concept upon which that theory may be built, or from which the observed empirical laws may be derived. Again, we saw this earlier; the molecular hypothesis of kinetic theory was our example then. This also is empirically verifiable and therefore equally of no import in this discussion. Both of the latter kind of hypothetical conjectures are part of the process of science and intimately connected with the scientific method as may be practiced by the sentient investigator.

Since our present preoccupation is with major revolutions, our concern is, on the contrary, with far deeper issues, specifically those that are not part of the theoretical or axiomatic structure of a scientific theory, but without which no such theory could even be formulated in the manner in which it is given. Hence, although we operate on a deeper level, the empirical process is still the determining criterion. This must always necessarily be so since no scientific progress, if it is not to be merely mythological, could ever otherwise be made. We have used terms such as 'philosophical', 'ontological' and 'fundamental' when referring to 'assumptions' of this kind and until now it has not been required of us to be particularly specific as to the precise meaning of these adjectival modifiers. Yet, if major revolutions are to be defined in terms of the modification of such notions as we have envisaged, then we need to consider their nature. This is not an easy thing to do, for each scientific theory will, in general, have a very different set of assumptions which seem to bear no relation to those of another discipline. Thus, assumptions in physics will have little in common with those in the biological sciences, except in a reductionist sense. Nevertheless, for a given scientific theory it is usually possible to identify the basic, often unspecified assumptions that precede the development of a theoretical structure. When we have in the prequel referred to an assumption of a theory as being 'philosophical' or 'ontological', we do not mean this in a particularly deep sense that would involve us with questions relating to the nature of reality or the existence of the physical world, for our domain is the scientific one and we must accept that our theories pertain to existent objects or the natural variables described by the axioms, even if this does not preclude a non-realist position. The assumptions that we wish to identify are specific to the theories upon which the latter are built and is therefore a relative matter. Hence, assumptions will have 'depth' in direct proportion to the sophistication of the scientific theory under consideration and this will further depend upon how well-developed the scientific discipline in which the theory is found actually is, at any given time. Evidently, an assumption of a given theory is some hypothesis that has not been empirically verified, for otherwise it would be an axiom of the theory. Hence, it is a supposition that lies at the foundations of a scientific theory and whilst it may not be generally of a philosophically deep nature or of ontological significance in the wider sense outside of the domain of the theory, it is so *relatively* to the particular model concerned and in that sense is fundamental. Then, in consideration of the foregoing, we may attempt a definition as follows;

Definition 7.4.3
A *fundamental assumption* of a scientific theory is a hypothesis without which the theory could not be formulated and the negation of which would negate the theory.

Whilst this is hardly an entirely adequate characterisation, it will serve our purpose for the practical aspects of major revolutions. We emphasise that although a fundamental assumption is a *hypothesis* and has not been empirically verified, or falsified, that does not mean that it cannot be so discredited. Indeed, major revolutions depend upon this vulnerability of fundamental assumptions, as is clear from definitions 7.4.1 and 7.4.2.

Now, a given theory S may have several fundamental assumptions upon which it depends for consistency, say $a_1,......,a_n$. What we require therefore is that for each a_i we have $\neg a_i \rightarrow \neg S$ so that the theory becomes untenable if any one of the fundamental assumptions is negated or removed. Conversely, $S \rightarrow a_i$ which merely asserts the assumptions upon which S is founded, as we expect. Then, we see the difference between the *fundamentality* of the assumptions in major revolutions and those that we considered earlier in relation to intermediate revolutions. In the latter case, the same scientific theory could be constructed from more than one set of hypotheses, so that the negation of such assumptions would not necessarily negate the theory, implying that the assumptions under consideration were not fundamental in this sense. In the present context however, this is not the case and the assumptions are truly fundamental, which is why revolutions in this category are the source of such major upheavals within the scientific enterprise. The relativity of the fundamentality of the assumptions is already incorporated within the definition, which thus refers to a specific scientific theory rather than a class of such theories. Clearly, the way to identify fundamental assumptions is to deny them and then see if the theory still holds. If it does not, then evidently the assumption is a fundamental one. Thus, if one denies absolute space and time, or Galilean relativity, then Newtonian mechanics fails, at least if one retains the usual definitions of the natural variables employed within that model. Having identified a fundamental assumption for a given theory, one may ask how it may be justified, for it is a hypothesis that cannot be empirically verified. We are not interested in suggesting that an assumption seems to be 'reasonable', that it may be 'justifiable' or 'naturally assumed' in some vague sense, for the history of science is littered with examples of assumptions once considered to be virtually *a priori* that have subsequently turned out to be demonstrably false. But it is this latter criterion that holds the key in definition 7.4.3, for although empirical verification of a fundamental assumption is by that definition impossible, it may nevertheless be empirically falsified. This is different from the situation with the axioms of a theory, which enjoy the pleasure of being either empirically verified or not falsified, or are just falsified. The distinction is an important one because its non-observance has led to much confusion in the philosophy of science, especially during the twentieth century.

Then, how might a fundamental assumption be empirically falsified? Evidently, we cannot write $\neg S \rightarrow \neg a_i$ and then simply declare that the fundamental assumption is invalid because the theory is invalid, for this obviously may not be the case. In fact, if it was so and the equivalence $\neg a_i \leftrightarrow \neg S$ obtained, then science would progress rather rapidly and major revolutions would never occur. Suppose then, that we have a theory S that is subsequently found to be in conflict with new, more precise observational or experimental data. The theory thus evolves under the action of the theory modification operator T_3 to produce the new theory $S' = T_3 S$. But S was founded upon some set of fundamental assumptions and hence $S \rightarrow A = \{a_1, ..., a_l, ..., a_n\}$. Now of course S' also has a set of fundamental assumptions, $S' \rightarrow A' = \{a_1, ..., a_m, ..., a_n\}$, where (for the sake of simplicity we consider only a single assumption) the assumption a_l has been replaced by the new assumption a_m. This does not necessarily mean that S cannot be formulated upon the assumptions A', for that may well be possible. However, if a_l and a_m are inconsistent, i.e. $\neg(a_l \wedge a_m)$, then since S' has now replaced S and been empirically verified, then a_l has been empirically falsified, but a_m is still not empirically verified for it remains an assumption. In this way, fundamental assumptions are empirically falsified, but only when the more accurate observations allow this to take place, for clearly the original assumption was perfectly adequate and did not warrant serious consideration prior to the arrival of new empirical data. Thus, major revolutions occur when inconsistencies are found within the hypothetical structure of the theory in question. Yet these are of two kinds and again this circumstance is intimately connected with the domain of applicability.

In major revolutions of the first kind, one or more of the fundamental assumptions are modified (effectively replaced) because empirical observations have found that they are incompatible with the new more reliable data subsequently obtained. This data has forced the axioms of the theory to be themselves modified, with the special effect, not present in lesser revolutions, that the assumptions that are required of them are now incompatible with the old assumptions; hence the replacement of the latter. As with all scientific progress, this is a consequence of more sophisticated observational and experimental techniques that admit of investigation into physical realms that were hitherto inaccessible. Then, the domain of applicability of the new theory must include that of the old theory since the latter was already well-established, but yet extend further into the area that the new, more precise empirical data, has revealed. Hence, the domain of applicability is increased because it includes the domain of the old theory. This is therefore the operation of a correspondence principle.

However, the situation is quite different with major revolutions of the second kind. Here, new axioms are empirically obtained and are of such a different nature that their fundamental assumptions are incompatible with those of the former theory. The result is a completely new model of the physical phenomena involved in which none of the original assumptions are retained. Consequently, in this case a correspondence principle cannot obtain, for the new theory completely replaces the old theory and bears little relation to it. However, the same natural variables may still pertain and in general will do so, but although the domain of applicability may not necessarily increase, it certainly cannot decrease, since the new theory must replace the old and therefore at least retain the same domain of applicability of the latter if it is to accomplish the same objective. Obviously, this does not preclude the possibility that the domain may actually increase, but such would be a rare event indeed and in most cases the domain of applicability will be unchanged. Once again, it is the empirical methods of science that initiate revolutions of this type, as must ultimately always be so.

We shall also need a further variant of the above definitions of major scientific revolutions for a number of reasons. Recall that a scientific theory is the concrete realisation of a set of axioms in the sense that it is a model under the natural interpretation. Hence, major revolutions of the first and second kind assume that some pre-existing model already obtains and the assumptions of which are already available to be modified or replaced. This in turn requires that any such prior model also is expressed in terms of axioms containing natural variables in some well-defined way. However, this may not always be the case. Natural variables are measurable quantities and a given set of hypotheses and concepts may be so vague as to render statements about the world, or axioms of a theory quite meaningless simply because no natural variables are adequately defined in this sense. Thus, in ancient Greek philosophy the four 'elements' of earth, air, fire and water with their accompanying attributes could not be considered to be measurable physical quantities or observables and consequently are devoid of meaning in terms of natural variables. The present writer is perhaps not as astute as Aristotle or his predecessors for he does not know how to measure the values of these quantities and although he can make seemingly axiomatic statements involving them such as 'earth is drawn to earth because it is of its nature to be so', he is unable to comprehend the meaning of such assertions or understand the model-theoretic sense by which such a theory obtains in relation to our earlier conceptual notions of a scientific theory. This being the case, we have little choice but to dismiss such ideas as entirely unsuitable for our purposes insofar as they do not satisfy our understanding of what it is to be a scientific theory as described in chapter five of this essay.

Hence, we are sometimes faced with the situation in which no pre-existing clearly defined theory is available for modification, either with regard to fundamental assumptions, axioms or indeed empirical content. In this circumstance a theory is developed *ab initio* with no former empirical foundation. In short, it cannot be built from an earlier theory because the latter does not exist in a form in which any systematic progress would be possible. Then, in such cases a theory must be formulated 'from scratch' and the only way this can be achieved if the natural interpretation is to obtain is via the empirical process. We have discussed similar situations in chapter six, but now observation and experimentation must take special precedence over theoretical constructs and natural variables need to be identified thereby so that meaningful axioms may be formulated. We therefore require a third category of major revolutions that will properly accommodate such situations as these. Hence we make the following definition:

Definition 7.4.4

A major revolution of the third kind consists in the determination of new natural variables and the formulation of fundamental assumptions that may or may not incorporate existing ideas into an integrated scientific theory such that at least one new axiomatic expression of a relation between the natural variables can be constructed.

This definition requires the notion of 'fundamental assumption' because it is not always the case that an axiom is obtained directly from observation and experimentation. Since we are dealing here with a first application of the scientific method and no suitable conceptual precursors have been identified, there is no question of the domain of applicability, for the latter is defined only when the theory is obtained. However, we have allowed in this definition that prior notions and ideas may be included in the new theory, but not scientific theories, for by assumption, there are none. We do not wish to exclude any useful pre-existing concept from our definition since to do so would seem to be unnecessarily restrictive. By their very nature, revolutions of this kind are maximally empirical and demonstrate more clearly than any other the effectiveness of the scientific method.

Major revolutions are relatively infrequent in the development of empirical knowledge, or knowledge of the external world in the sense of the natural interpretation, but occur periodically, which may be centurial or multi-centurial and are of such significance and of such a transformational nature that they have in philosophical discussion, tended to subordinate minor and intermediate revolutions. Yet both the latter play a precursory role that eventually precipitates major revolutions, entirely through the empirical process. We shall discuss these points later. However, in order to make such a detailed analysis it is helpful to clarify the foregoing definitions and speculations by the illustration of specific examples from various areas of science. Let us begin, somewhat arbitrarily, with those two great bastions of twentieth century physics; relativity and quantum mechanics. First, we look at Special Relativity.

Prior to relativity Newtonian mechanics prevailed. Space was three-dimensional and absolute in the sense that it was the geometrical background or arena against which physical phenomena took place. The physical processes of such phenomena were described as evolving through time, which was also absolute and was unrelated to the three spatial dimensions except in this dynamical way where a particular configuration of a mechanical system was described precisely at each point in time in a linear fashion. A crucial concept in both classical and relativistic mechanics is the existence of *inertial* frames, which are preferred reference systems in which Newton's first law holds. The *principle of Galilean relativity* asserts that the *form* of the laws of physics is the same in every inertial frame. In other words, the form of the equations or axioms $f(x_i, c_j) = 0$ cannot distinguish between any two inertial frames, as long as there is no acceleration. We are thus dealing with uniform rectilinear motion, i.e. motion of constant velocity. Then, given two inertial reference systems $R(x,y,z,t)$ and $R'(x',y',z',t')$ in rectangular Cartesian co-ordinates x,y,z and x',y',z' in which the system R' is moving at constant velocity v along the x axis with respect to R, we have the familiar Galilean transformation between the two systems;

$$x' = x - vt, \quad y' = y, \quad z' = z, \quad t' = t$$

which illustrates Galilean relativity. Note that the fact that $t' = t$ asserts that the time is the same in both systems, i.e. in all inertial systems and this absoluteness of t is fundamental to Newtonian mechanics. Observe also that the equation for x' is just the usual 'addition' property for relative velocities. If, for example, we consider a particle in the unprimed system moving along the positive x-axis with speed u, then we may compute its speed u' in the primed system;

$$u' = \frac{dx'}{dt'} = \frac{dx'}{dt} = \frac{dx}{dt} - v = u - v$$

and we find that $du'/dt' = du/dt$, showing that the acceleration is the same in both frames, as we would expect.

When we consider the four co-ordinates x,y,z and t it is easier to speak of an *event* rather than a point at time t. Now let us take two such events (1) and (2) with spatial co-ordinates x_1,y_1,z_1 and x_2,y_2,z_2 respectively and at times t_1 and t_2. The time interval, $\Delta t = t_2 - t_1$ and the spatial interval;

$$\Delta r^2 = (x_2 - x_1)^2 + (y_2 - y_1)^2 + (z_2 - z_1)^2$$
$$= \Delta x^2 + \Delta y^2 + \Delta z^2$$

are each invariant under the above Galilean transformation, clearly indicating the fact that in Newtonian physics, space and time are separate entities. The form of the above spatial distance is obviously the usual Euclidean generalisation of the Pythagorean Theorem, as would be expected, for this is the Euclidean 3-space that we are all familiar with. Hence, among the fundamental assumptions of Newtonian mechanics are the absolute property of space and time and the distinction or separability between them, together with the metric;

$$ds^2 = dx^2 + dy^2 + dz^2$$

which endows this space with its Euclidean topology (E^3). Evidently, if any of these fundamental concepts are replaced or modified, then a major revolution of the first kind is inevitable.

Of course that revolution came in 1905 with Einstein's publication of the Special Theory of Relativity. We shall briefly sketch the main ideas. The essential point was to challenge the assumption of absolute time and replace it with the constancy of the velocity of light c. More precisely, the second principle of Special Relativity asserts that the speed of light c is the same in all inertial frames. The only way this could be achieved and Maxwellian electrodynamics and Newtonian mechanics reconciled, was to include the time variable t into the distance interval. This was motivated by the Lorentz transformation;

$$ct' = \gamma(ct - \beta x)$$
$$x' = \gamma(x - \beta ct)$$
$$y' = y$$
$$z' = z$$

where $\beta = v/c$ and $\gamma = \left(1 - \beta^2\right)^{-1/2}$ as usual. But to do this would require a new variable which includes time, yet has the dimensions of length. Hence, one is led to define a spatial variable $\tau = ct$ so that the distance interval includes the time and is thus now a *space-time* interval Δs ;

$$\Delta s^2 = c^2 \Delta s^2 - \Delta x^2 - \Delta y^2 - \Delta z^2$$

which is accordingly the new invariant under the Lorentz transformation in analogy with the Galilean case and represents the distance or *interval* between two events in *space-time*. The new metric or *line-element* describing the infinitesimal interval between two space-time points is then;

$$ds^2 = c^2 dt^2 - dx^2 - dy^2 - dz^2$$

which is no longer a Euclidean but a 'pseudo-Euclidean' geometry. This is the four-dimensional Minkowski space-time of Special Relativity. Note however that this space still has a Euclidean topology, namely the four-dimensional topology of E^4.

Indeed, this is a radical departure from the Newtonian picture. Time and space are no longer absolute and we cannot speak of them separately. We must consider time and space as a single entity, space-time, in a four dimensional pseudo-Euclidean (flat) manifold. A particle in this space describes a trajectory or 'worldline' in space-time. In particular, the metric is not positive definite as in the Euclidean case, for we now have $ds^2 > 0$ (*timelike* interval), $ds^2 = 0$ (*lightlike* interval) $ds^2 < 0$ (*spacelike* interval). This clearly has no analogue in Newtonian physics.

Why is Special Relativity a major revolution of the first kind? According to definition 7.4.1, we have to satisfy only two criteria: Firstly, a fundamental assumption must be modified or replaced and, secondly, the domain of applicability must increase. Furthermore, the modified assumption should be logically inconsistent with that which it replaces. Well, obviously this condition is fully met by Special Relativity since one clearly cannot have both absolute and relative time or separable space and time with co-joined space and time. Nor can the speed of light be both finite and infinite. But definition 7.4.3 requires that Newtonian mechanics could not be formulated without the assumption of absolute time and that Special Relativity can similarly not be constructed without its fundamental assumption of the constancy of the speed of light in all inertial frames. Clearly, this is also satisfied and, in effect, is really included in the earlier definition 7.4.1 for reasons of consistency. Note, that the principle of relativity, namely, that the laws of physics are the same in all inertial frames applies to *both* theories and hence Special Relativity is not new in this respect. This may lead one to wonder why modern relativistic theory had not been formulated earlier, but in the light of the above, the problem was evidently an inability to depart from the Euclidean regime.

The next criterion relates to the domain of applicability. From the Lorentz transformation it is immediately clear that in the limit $c \to \infty$, the Galilean transformation results. Then we note that when $v \ll c$ Newtonian mechanics is incorporated into the relativistic model as a limiting case and hence, also the domain of applicability of classical mechanics. Thus, Special Relativity has a larger domain than Newtonian theory, but the latter is still, as might have been anticipated, included within it. There is, therefore, a correspondence principle in operation, as indeed would be expected for this revolution category, which certainly admits of an increase of domain of applicability, but it should be clear from the definitions that it is not an essential requirement of a new theory to be reducible to that of the old, for it is quite possible for the domain to increase without any correspondence between a given theory and that which supersedes it.

The geometrical change with Special Relativity also requires some alteration of the natural variables. Thus, in Newtonian mechanics the instantaneous velocity of a particle is defined as the tangent to the trajectory and is a vector quantity in 3-space. In Special Relativity, the analogous velocity is a 'four-vector' in space-time and is tangent to the worldline of the particle, it belongs essentially to the tangent space of a differentiable manifold, and the same applies for other quantities in that regime. These are, of course, still natural variables, but some of them, such as, for example, the energy $E = mc^2$ have no direct counterpart in classical mechanics. This is hardly surprising considering the conceptual change involved.

The need for Special Relativity arose through the empirical process, but it was not because of the negative result of the Michelson-Morely experiment to measure the motion of the Earth through the ether, even though Lorentz had found the correct mathematical expression for the length contraction of material bodies in the direction of motion that would explain the failure of this famous experiment and which is now part of special relativistic mechanics.. The anomalous advance of the perihelion of Mercury was also not a primary motivation as far as Einstein was concerned, although that discrepancy seemed to indicate that something was wrong with Newtonian mechanics. Rather, Einstein was concerned with the incompatibility of Galilean relativity and Maxwell's electrodynamics. Both of these theories had a wealth of experimental and observational data to support them, but they clearly could not be simultaneously correct. Einstein chose to modify the Galilean principles rather than Maxwellian theory. With great insight and intuition, he was able to see that Special Relativity was required by Maxwell's electromagnetic theory. But the revolution provided by Special Relativity was still precipitated by empirical observation, as must be so with all revolutions.

It is often said that Einstein 'unified' Newtonian physics with Maxwell's electrodynamics and to some extent this is true, but it is not theory unification in the sense of Definition 7.3.3 and hence an intermediate revolution of the third kind, because fundamental assumptions are not altered in the latter type of revolution. Nevertheless, there is a sense in which unification has taken place which must not be confused with our use of that term. In fact, in major revolutions of the first kind, there will always be unification of two theories in this way, simply because the domain of applicability must increase and a correspondence principle of some sort may apply.

Non-relativistic quantum mechanics offers another perspective on major revolutions of the first kind and perhaps in a unique way in the history of the physical sciences. The theory is transformational with regard to the fundamental assumptions of preceding scientific models, yet contains and is motivated by empirical criteria that bear no relation to pre-existent ideas. The resulting conceptual change is of such a devastating nature that it did at first, appear to deny any rational understanding in terms of classical physics and even to this day is counter-intuitive in that sense. Be that as it may, this can have no bearing in relation to scientific revolutions, which must remain impartial to such observer-dependent idiosyncrasies or investigator specific limitations. The fact that quantum mechanics is so *very* different from its classical counterpart is not relevant *per se* for our concerns, for such matters lie within the philosophical foundations of the quantum theory itself and whatever interpretation of the axioms of that theory one may choose. However, that the theory is almost entirely of empirical origin is of special significance in our context, for that suggests that it is divorced from the objectivity which would result from mere speculation and, therefore, is thus objective in some entirely empirical sense yet to be specified. We shall have much more to say about this ontological connection in the sequel.

Quantum mechanics deals with the micro-world of atoms and molecules and the particle constituents thereof, where it is found that classical theory fails. But that need not be so. Whilst domain of applicability of Newtonian classical dynamics cannot include that of quantum theory, the converse is not the case, for again we have a correspondence principle and, in fact. it was in this particular field that the very notion of the latter first arose. The final formulation of quantum mechanics – as it currently stands - allows the theory to be expressed 'axiomatically' in the form of simple postulates, not all of which are independent and some of which may be taken as axioms or physical laws in our sense, but others in a more fundamental way that directly challenges the Newtonian picture. Since we are concerned here primarily with the fundamental assumptions of the classical theory prior to the advent of quantum mechanics that are at once inconsistent with those of the quantum revolution, it is necessary to take a historical perspective, at least insofar as that provides some insight into the conflict and incompatibility between these two theories. Recall that from thermodynamic considerations, Planck obtained the following empirical formula for the energy spectrum of the radiation from a black-body;

$$\rho(v,T) = \frac{Av^3}{e^{Bv/T} - 1}$$

where A and B must be empirically determined, ν is the frequency and T the absolute temperature. Although this expression agreed very well with experimental data at the time, Planck was determined to find a theoretical justification for it. The details don't matter, but after careful consideration and much physical insight he obtained the following well-known expression;

$$\rho(\nu,T) = \frac{8\pi\nu^2}{c^3} \frac{h\nu}{e^{h\nu/k_B T} - 1}$$

where of course c is the speed of light and k_B is Boltzman's constant. But Planck derived this expression only on the assumption that the energy is a multiple of $h\nu$, with some constant (Planck's constant) h, i.e. $E_n = nh\nu$. This was the famous quantum hypothesis; that the energy came in discrete 'packets' called *quanta* instead of a continuous distribution as in the classical theory. Clearly, this is a major departure from the then prevailing view and a fundamental assumption has been replaced. A few years later, during his famous *anus mirabilis*, Einstein considered the photoelectric effect and postulated that light consists of 'particles' or 'photons' and the energy of an individual photon (i.e. with $n = 1$ above) should be $E = h\nu = \hbar\omega$, where $\hbar = h/2\pi$ and $\omega = 2\pi\nu$ is the angular frequency. Now, the relativistic energy of a particle is given by the well-known formula;

$$E^2 = (m_0 c^2)^2 + p^2 c^2 = h^2 \nu^2 = \hbar^2 \omega^2$$

where m_0 is the rest-mass and p the momentum. But for a photon travelling at the speed of light the rest mass must be zero. Hence, we have in terms of the wave-number $k = \omega/c$;

$$p = \hbar k = \hbar\omega/c$$

or in vector notation;

$$\mathbf{p} = \hbar\mathbf{k} = \frac{h}{\lambda}\frac{\mathbf{k}}{|\mathbf{k}|}$$

with the appropriate direction of momentum and photon. But the photon is travelling at the speed of light c and therefore has zero rest mass. Together with $E = \hbar\omega$ these constitute the famous De Broglie relations. Then we have;

$$p = \frac{E}{c} = \frac{h\nu}{c} = \frac{h}{\lambda}$$

and a wavelength λ is associated with the photon, considered as a particle. These are the Planck-Einstein relations and show that photons should behave as both particles and waves. Again, we have a major departure with classical theory which had insisted that electromagnetic radiation was either one or the other. But if waves can behave like particles, then the converse is also true and this is what De Broglie suggested; that to each particle (of non-zero rest-mass) the same relationship holds and massive particles are assigned a 'guiding wave' of wavelength λ as above. Thus matter behaves like waves and waves behave like particles. Thus was born the 'wave-particle duality' that again had no correlation in classical mechanics and seemed to be so paradoxical when first proposed. Bohr then applied the quantum principle to the hydrogen atom, in which 'energy-levels' of bound electrons were introduced so that only discrete values of the energy were allowed and in accordance with $E = h\nu$ so that the permitted 'orbits' of the electrons surrounding the nucleus were compatible with the quantum principle. This resolved several problems in atomic spectroscopy in addition to the inevitable decay of an electronic orbit if the electron emitted continuous radiation due to its acceleration as it must do in classical theory, thus making the atom unstable. The latter is similar to the problem of the infinite energy density in a cavity (black body) in relation to our earlier remarks concerning the Planck formula. All of this was required by empirical considerations and found to be in excellent agreement with the experimental data, which is important from our viewpoint, for the revolution exhibited by quantum mechanics, illustrates well the empirical process, as we require.

However, quantum mechanics introduces a further 'non-classical' notion, which is in fact necessitated by the foregoing. In Newtonian theory a material particle is described precisely at each point in space and time by its position and velocity, or its position (x,y,z) in some co-ordinate system (usually Cartesian) and its momentum p at each instant of time t. This works wonderfully for point particles of mass m, but cannot be applied to wave motion, for then there is no particular point to isolate and this is the principle dichotomy between the Newtonian and quantum mechanical picture alluded to earlier. If a material particle exhibits wave-like properties in the classical sense and also exhibits particle-like properties, again in the classical sense, then a new approach is required if some degree of reconciliation between these two conflicting notions is to be realised. One is therefore compelled to consider a statistical and probabilistic view of the situation, for no other option is available. Thus, classical physics describes a particle by giving its position vector $\mathbf{r} = (x,y,z)$ and its momentum $\mathbf{p} = (p_x, p_y, p_z)$. However, in quantum physics the state is described by a complex valued *wave function* $\psi(\mathbf{r})$ or *state-vector* of either the position or the momentum $\psi(\mathbf{p})$, at any instant t, but not both. These are referred to as the *co-ordinate* and *momentum* representations respectively, yet although this is convenient for the present discussion, it is realised that many other, in fact an infinity of other representations are clearly possible. The wave function is envisaged as containing all information concerning the state of the particle and the two representations are equivalent in this respect. If φ and ψ are two wave functions, then an inner product is defined thus;

$$\langle \varphi | \psi \rangle = \int_{-\infty}^{\infty} \varphi^*(x)\psi(x)dx$$

where of course φ^* is the complex conjugate of φ and in fact the wave functions are (unit) vectors in a Hilbert space. An important principle in quantum mechanics is the *superposition* principle. If φ and ψ are two states then $\lambda\varphi + \mu\psi$ is also a possible state, where λ and μ are arbitrary complex numbers. This is just a property of the linearity of the Hilbert space and the other vector space axioms also apply as usual, but that need not concern us here. Attempting to capture the wave and particle nature of matter into a single concept, one introduces in elementary quantum mechanics the *wave packet* as a superposition of plane waves of the form $e^{i(\mathbf{kr}-\omega t)}$. In one dimension the most general form for a wave packet with amplitude $A(k)$ is;

$$\psi(x,t) = \frac{1}{\sqrt{2\pi}} \int_{-\infty}^{\infty} A(k)e^{i(kx-\omega t)} dk$$

Then, the wave function $\psi(x)$ in the co-ordinate or x-representation is related to the wave function $\psi(p)$ in the momentum representation insofar as one is the fourier transform of the other. Thus;

$$\tilde{\psi}(p) = \frac{1}{\sqrt{2\pi\hbar}} \int_{-\infty}^{\infty} e^{-ipx/\hbar} \psi(x)dx$$

and conversely;

$$\psi(x) = \frac{1}{\sqrt{2\pi\hbar}} \int_{-\infty}^{\infty} e^{+ipx/\hbar} \tilde{\psi}(p)dp$$

The idea and primary motivation behind the construction of a wave packet is to describe the particle in such as way that there is a region of 'enhanced' amplitude which dominates over the distribution; the centre of the wave packet and this is why it is necessary to construct it as an infinite superposition of plane waves. A wave packet is supposed to look and behave like a wave and also a particle. Thus, there is dispersion (standard deviation) about this point, Δp or Δx, an inevitable 'spread' in the wave function which by definition leads to the familiar inequality;

$$\Delta p . \Delta x \geq \tfrac{1}{2}\hbar$$

which is one form of the well-known Heisenberg uncertainty principle. It might be thought that the latter is a fundamental principle of quantum mechanics that has no classical analogue, but that is not the case. The uncertainty principle is simply a consequence of the formalism. We are dealing with square integrable functions in a Hilbert space which admits of the above fourier representation. This necessarily leads to an uncertainty principle and in fact such notions also occur in communication theory, particularly signal theory. But that does not mean that it is not real, for indeed it is and if quantum mechanics is a correct description of Nature, then the uncertainty principle follows. In fact, the uncertainty principle ultimately stems from the particle and wave nature of matter, *wave-particle duality* and the way we have chosen to describe that phenomenon mathematically, in particular as wave functions in a linear space, as we have seen. It is the superposition principle that is of primary importance and that too is simply a reflection of the vector space description of a quantum system, yet is a consequence of empirical considerations in the sense that, at least historically, such a formalism seemed to be required by the observational and experimental data.

Probability enters into the picture by interpreting the square of the absolute value of the wave function $|\psi(x,t)|^2$ as the probability $P(x,t)$ of finding the particle at position x at time t. Recall that the state vector is a single-valued smoothly varying and positive-definite function and obeys the normalisation condition;

$$\int_{-\infty}^{\infty} |\psi(x,t)|^2 = \int_{-\infty}^{\infty} \psi^*(x,t)\psi(x,t) = 1$$

which meets the requirement that;

$$\int_{-\infty}^{\infty} P(x,t) = 1, \text{ and } P(x,t) \geq 0$$

so that $P(x,t)$ in fact behaves as a probability should. Probabilistic and statistical methods are hardly new in physics and ensembles of classical particles are frequently treated with such techniques, but in this case it arises as a property of the system itself and therefore of a characteristic of the Natural world, specifically and yet again, because of the formalism of the wave-packet description of particles which makes it impossible to isolate both position and momentum in accordance with the uncertainty principle. There is no classical analogue to this and indeed it is often taken as a postulate of non-relativistic quantum mechanics. We would expect the wave function to satisfy an equation of motion and in particular a differential equation as in Newtonian mechanics. Thus classically, for a free particle of mass m in an external field of potential $V(\mathbf{r})$ and kinetic energy T, the Hamiltonian is given by;

$$H = T + V = \frac{p^2}{2m} + V(\mathbf{r})$$

which is the total energy E of the system. In quantum mechanics the wave function obeys the equation;

$$-\frac{\hbar^2}{2m}\nabla^2\psi(\mathbf{r},t) + V(\mathbf{r},t)\psi(\mathbf{r},t) = i\hbar\frac{\partial}{\partial t}\psi(\mathbf{r},t)$$

which is the famous time-dependent Schrödinger equation. We note that this equation is linear and homogeneous, as it must be if its solutions are to satisfy the superposition principle. Now if one makes the following identifications;

$$\mathbf{p} \rightarrow -i\hbar\nabla$$

$$E \rightarrow i\hbar\frac{\partial}{\partial t}$$

$$H \rightarrow -\frac{\hbar^2}{2m}\nabla^2 + V(\mathbf{r})$$

the Schrödinger equation can be written;

$$H\psi(\mathbf{r},t) = \left(-\frac{\hbar^2}{2m}\nabla^2 + V(\mathbf{r})\right)\psi(\mathbf{r},t) = i\hbar\frac{\partial}{\partial t}\psi(\mathbf{r},t) = E\psi(\mathbf{r},t)$$

so that $H = E$ and we observe that the identifications of the momentum and energy with the differential operators made above allow the quantum variables to pass into their classical counterparts, as required by the correspondence principle. Hence, the Schrödinger equation is essentially the quantum version of the classical equation $E = p^2/2m$. Thus, natural variables are represented by (self-adjoint) *operators* as above, not all of which have a classical analogue. An example of the latter kind is the (necessarily quantised) spin of fermionic particles, which is hardly surprising. Thus, to any self-adjoint or Hermitian operator there corresponds an *observable* or natural variable if it is measurable, but the value obtained by empirical measurement will be an eigenvalue of that operator and that this too may be taken as a postulate of quantum mechanics. From all of this one may construct the full formal mathematical apparatus of non-relativistic quantum theory and although we have adopted the Hilbert space approach here as our example, we need not have done so; quantum mechanics can also be constructed algebraically via von-Neumann algebras such as C^*- algebras or by the Boolean lattice approach. This need not concern us, for we have isolated the central issues we need in order to see that quantum mechanics does indeed represent a particularly interesting example of a major revolution of the first kind. Then, let us summarise the revolutionary aspects of quantum mechanics in relation to Definition 7.4.1 and particularly the classical *status quo*.

In the first instance, the quantum hypothesis that the energy and in particular electromagnetic radiation is of a corpuscular nature is clearly at odds with the classical Maxwellian picture of continuous wave-like propagation and even though the definition of a wave-packet seems to unite these two disparate concepts it does so only at the expense of both ideas which are anyway dependent upon the superposition principle. The simple brute fact remains that the discreteness of quantisation is not compatible with the continuity of the classical formalism, where waves do not behave like particles, whatever mathematical device or conceptual cloaking may be employed. Definition 7.4.1 is therefore fully satisfied in this regard.

Secondly, particles do not behave like waves in classical physics in any individual sense. It is true that fluid particles may *en mass* exhibit wave like properties but this is nothing more than the collective motion of classical particles conforming to Newtonian dynamics, which is a quite different issue and unrelated to the de Broglie conception of 'matter waves' that is of such fundamental concern in the quantum assumptions. Thus, once again Definition 7.4.1. is appropriately satisfied.

We therefore regard the Einstein postulate and the de Broglie hypothesis as the fundamental assumptions of quantum mechanics and historically this was certainly the case. From these basic ideas, albeit with a little ingenuity, eventually follows the whole mathematical edifice of non-relativistic quantum mechanics and the negation of either one of them would also negate the theory in accordance with Definition 7.4.3. Yet, there is a strong empirical foundation in the formulation of the early quantum theory that reflects in great measure the importance of the scientific method and its interaction with the Natural World and which exemplifies most significantly the empirical basis of scientific revolutions. For indeed, quantum theory is perhaps above all scientific endeavours, the most empirically instigated and experimentally motivated of theoretical structures and could not have been realised in the absence of the observational and experimental techniques which had reached a stage of development that made such innovation possible. Then, the question of natural variables arises, for they would seem to be of a different nature in the quantum mechanical formalism. It is certainly true that the 'observables' or 'physical variables' of measurable quantities appear to have a different epistemic standing in quantum theories than that which is envisaged in the classical scenario, but on closer examination the ontological status of the natural variables, as we have defined that term, is co-incidental with both descriptions in relation to either model. We recall from §5.2 that all that is required of a natural variable is that it be measurable and time-dependent. Classically, these are just the familiar real variables such as velocity, momentum, temperature, etc. However, quantum mechanically they are the result of an action of a hermitian operator acting within a complex Hilbert space and thereby specific values of such variables are realised as the eigenvalues of that action during the measurement process. There is no dichotomy here as far as our definition of natural variables is concerned, for no such contrary criteria appear in the definition. In standard quantum mechanics each self-adjoint operator corresponds injectively to a physical or empirical observable and measurable quantity, even statistically and our definition of a natural variable makes no distinction in relation to this. Essentially, the fact that a physical quantity, or rather the value thereof, may be a particular eigenvalue of the operator that is identified with the natural variable concerned therefore has no relevance for us *per se* and the natural interpretation obtains merely because it must so do by its very nature and in this sense – especially pertinent in the present context – because the empirical process, *ipso facto*, necessitates that this shall be the case. Then the latter once again is the final arbiter in these matters if the scientific enterprise is to be of value.

We must therefore now consider the domain of applicability of non-relativistic quantum mechanics, for the definition requires that it be greater than that of the classical theory. In the former discussion of Special Relativity it was a simple matter to ascertain the domain of the new theory by application of the correspondence principle, which made it abundantly clear that the classical theory was a particular case of special relativistic mechanics and therefore that the latter encompassed the domain of both models. This was done simply by taking the limit by supposing $v/c \ll 1$ and thus reducing the relativistic equations to the familiar Newtonian ones. Hence, we may wish in the present context to appeal to the correspondence principle and argue that the replacement of the natural variables of classical mechanics by Hermitian operators on the Hilbert space demonstrates that quantum mechanics would necessarily embrace the domain of classical theory in an analogous manner and specifically in the limit $\hbar \to 0$, the equations of classical dynamics result. Then, the domain of quantum mechanics would include the entire domain of the macro-world of classical mechanics in addition to the micro-world for which it was originally intended to be applicable. However, whilst this is possible, it is unsatisfactory for a number of reasons. In the first instance, quantum and classical theory are quite different beasts, with distinct evolutionary phylogeny. Quantum mechanics is manifestly a linear theory in which the superposition principle is of fundamental importance to the entire formalism. But superposition is not a general property of classical Newtonian physics, where non-linearity is often evident and the theory contains no superposition principle *per se*. Let us look a little more closely at the classical limit of quantum mechanics as it is often portrayed in the literature.

We consider a particle in a potential V(**r**) as above. To this end, we may write the wave function in the form;

$$\psi(\mathbf{r},t) = A(\mathbf{r},t)\exp\left(\frac{i}{\hbar}S(\mathbf{r},t)\right)$$

where $A > 0$ and S is a real function. We are therefore separating the modulus and the phase and this can always be done for any complex function. If we now insert this expression into the Schrödinger equation then on separating real and imaginary parts we obtain the two equations;

$$\frac{\partial S}{\partial t} + \frac{1}{2m}(\nabla S)^2 - \frac{\hbar^2}{2m}\frac{\nabla^2 A}{A} + V = 0$$

$$\frac{\partial A^2}{\partial t} + \nabla\cdot\left(\frac{A^2\nabla S}{m}\right) = 0$$

The usual procedure is now to observe that $\hbar$ appears only in the third term of the first of these equations and when we take the limit $\hbar \to 0$ we obtain;

$$\frac{\partial S}{\partial t} + \frac{1}{2m}(\nabla S)^2 + V = 0$$

which 'looks like' the classical (non-linear) Hamilton-Jacobi equation with S as Hamilton's principal function. This is all very dubious of course, but is often presented as an application of the correspondence principle. However we can go further; if to the above equation we apply the operators ∇ and $\partial/\partial t$ and interpret $\dot{\mathbf{r}} = \nabla S/m$ as some sort of velocity we obtain;

$$d\mathbf{p}/dt = -\nabla V$$

for the motion of a classical particle in a potential V and then we have;

$$dE/dt = \partial V/\partial t$$

for the change in energy. The classical Newtonian equations have been recovered from the Schrödinger equation in the limit $\hbar \to 0$ where Hamilton's principal function was taken as the phase of a travelling wave. The motivation for this simple example is by analogy with optics, but many other illustrations of apparent classical limits could be given. Of course, not all solutions of the Schrödinger equation have classical analogues and it cannot be said that the above demonstrates a correspondence principle in the sense that classical theory can be deduced from quantum theory. We examine this further.

There is a famous theorem in elementary quantum mechanics that enables the expectation values (or mean values) of quantum mechanical operators to replace those of classical variables and with the quantum commutators replacing the Poisson brackets, resulting in analogous equations, which sheds considerable light on the quantum-classical correspondence in relation to the domain of applicability. For this reason we briefly discuss this apparent reduction from the quantum to the classical realm.

Recall that the *expectation* value of a quantum mechanical operator Ω is given by;

$$\langle\Omega\rangle = \int \psi^*(\mathbf{r},t)\Omega\psi(\mathbf{r},t)d\mathbf{r}$$

On differentiating this with respect to time;

$$\frac{d}{dt}\langle\Omega\rangle = \int \left(\frac{\partial\psi^*}{\partial t}\Omega\psi + \psi^*\frac{\partial\Omega}{\partial t}\psi^* + \Omega\psi^*\frac{\partial\psi}{\partial t} \right) d\mathbf{r}$$

Now if we substitute the Schrödinger equation and its complex conjugate;

$$i\hbar \frac{\partial}{\partial t}\psi = H\psi$$

$$-i\hbar \frac{\partial}{\partial t}\psi^* = H\psi^*$$

into this expression and, noting that the operator Ω has no explicit time dependence, we obtain;

$$\frac{d}{dt}\langle\Omega\rangle = \frac{i}{\hbar}\langle[H,\Omega]\rangle$$

where $[H,\Omega] = H\Omega - \Omega H$ is the *commutator*. Comparing this with the classical equations of motion in generalised momentum and position co-ordinates p_j and q_j (see §3.9);

$$\frac{d}{dt}f(p_j,q_j,t) = \{H,f\}$$

for some arbitrary function f and where;

$$\{f,g\} = \frac{\partial f}{\partial q_j}\frac{\partial g}{\partial p_j} - \frac{\partial f}{\partial p_j}\frac{\partial g}{\partial q_j}$$

is the *Poisson* bracket of the two function f and g. This suggests that to obtain the quantum analogues of the classical equations of motion, one should replace the classical variables with their mean values and Poisson brackets with the appropriate commutator multiplied by $i/\hbar$.

If we now set the Hamiltonian operator as $H = \mathbf{p}^2/2m + V(\mathbf{r})$ then first with $\Omega = \mathbf{r}$ and then $\Omega = \mathbf{p}$ we obtain, respectively, the two equations;

$$\frac{d}{dt}\langle\mathbf{r}\rangle = \frac{\langle\mathbf{p}\rangle}{m}$$

$$\frac{d}{dt}\langle\mathbf{p}\rangle = -\langle\nabla V\rangle = \langle\mathbf{F}(\mathbf{r})\rangle$$

which assert *Ehrenfest's Theorem* for this particular case, the more general form being the inclusion of the time dependence of the operator in our earlier expression. The last of these equations is Newton's second law of motion if $\langle \mathbf{p} \rangle = m\mathbf{v} = \mathbf{p}$ and $\langle \mathbf{F} \rangle = \mathbf{F}$, the classical momentum and disturbing force respectively. We may therefore be satisfied that quantum mechanics embraces the entire domain of classical theory, for all we have to do is replace operators and brackets in the manner mentioned above and the matter is resolved. But of course, we already knew this; it was in fact essentially where we began when identifying the momentum and energy operators with classical variables in our discussion of the Schrödinger equation. We realise now however, that this was a special case and that a direct identification can only be made under certain conditions. Essentially, we are attempting to identify the mean value of a quantum mechanical operator with the mean value of a classical variable. Thus, if A is a hermitian operator and a the corresponding classical variable, then we wish to write $\langle A \rangle = \langle a \rangle$. It should be obvious that there are difficulties here. If $H(q_1,....,q_n,\ p_1,.....,p_n)$ is the Hamiltonian in the usual Cartesian co-ordinates of position and conjugate momentum then we may write Ehrenfest's theorem as;

$$\frac{d}{dt}\langle q_i \rangle = \left\langle \frac{\partial H}{\partial p_i} \right\rangle$$

$$\frac{d}{dt}\langle p_j \rangle = -\left\langle \frac{\partial H}{\partial q_j} \right\rangle$$

as the quantum analogues of the canonical equations of Hamilton as discussed in §3.9. But there is no direct correspondence here in any general sense. In order for the mean values of the p_i and q_i in the above to obey the laws of classical mechanics, it is necessary for the variables in the Hamiltonian function on the right to also be replaced by their average values. This means that;

$$\left\langle \frac{\partial}{\partial p_i} H(q_1,.....,q_r, p_1,....., p_r) \right\rangle \rightarrow \frac{\partial}{\partial p_i}\left(H(\langle q_1 \rangle,....., H\langle q_r \rangle, \langle p_1 \rangle,.....,\langle p_r \rangle) \right)$$

$$\left\langle \frac{\partial}{\partial q_j} H(q_1,......,q_r,p_1,......,p_r) \right\rangle \rightarrow \frac{\partial}{\partial q_j} \left(H(\langle q_1 \rangle,......,H \langle q_r \rangle,\langle p_1 \rangle,......,\langle p_r \rangle) \right)$$

and this can happen only in a restricted number of cases when the Hamiltonian is of a certain form and specifically when the wave function is localised and any fluctuations of the variables about their mean values is negligible. This is hardly what we wanted with respect to the domain of applicability of quantum mechanics. To illustrate this with a specific example, let us combine the two equations for the Newtonian example given earlier into the more suggestive form;

$$m \frac{d^2}{dt^2} \langle \mathbf{r} \rangle = \langle \mathbf{F}(\mathbf{r}) \rangle$$

For this to hold classically, it is necessary that the mean value of the force;

$$\langle \mathbf{F}(\mathbf{r}) \rangle = \int \psi^*(\mathbf{r},t)\mathbf{F}(\mathbf{r})\psi(\mathbf{r},t)d\mathbf{r}$$

be replaced by its value $\mathbf{F}(\langle \mathbf{r} \rangle)$ at the point $\mathbf{r}$. In such case, Ehrenfest's theorem says that the classical equations hold for the mean values and we can have $\mathbf{F}(\mathbf{r}) = \mathbf{F}(\langle \mathbf{r} \rangle)$ only in the case of a free particle when the force vanishes, or when it depends linearly on the co-ordinates $\mathbf{r}$ as in the case of the harmonic oscillator. Otherwise, as mentioned above, one requires that the wave-function be highly localised so that the force maintains a virtually constant value in a sufficiently small region. Again, this is hardly satisfactory from our viewpoint and we must regard the Ehrenfest theorem as more of an interesting and remarkable result than a demonstration of correspondence.

We mentioned earlier that classical and quantum mechanics are very different creatures, to continue that biological simile. That the variables of each are of a different pedigree is already evident from the disparate nature of the wave function and the abundant clarity of Newtonian dynamical precision. Essentially, if two objects are to be equated then they must be entities of the same type, which means that they are required to either exist within a similar theoretical framework that has a common axiomatic foundation, or that there is a clear and general method for transforming, in a strict mathematical sense, between one framework and the other, in other words, the two structures must be isomorphic. No such correlation exists between quantum and classical mechanics.

Nevertheless, we require that the domain of applicability of quantum mechanics be greater than and yet still encompass those of classical systems, otherwise the quantum revolution is not a major one of the first kind. But, from the preceding discussion it would appear that we have no definite or conclusive way of deriving Newtonian mechanics from the quantum formalism. Nor does it seem that such may even be possible considering the extraordinary differences that clearly obtain between these two theories. But we are missing the point and in fact all of the above remarks are quite irrelevant to our concerns. If there is a correspondence principle that clearly demonstrates domain absorption, as in the case of Special Relativity and other physical theories, then that is all well and good and furthermore may save a great deal of time and unnecessary speculation such as we have become embroiled within here, but it is not anything other than a convenience in this respect, for Definition 7.4.1 does not require a correspondence principle in any of the possible forms in which the latter has been given in the philosophical literature. All that our definition requires is that the domain of applicability *increases* and that a minimal number of assumptions are replaced by ones that are inconsistent with those that they supersede. We have shown that the latter condition is satisfied and we have discussed various attempts to satisfy the former, but none of these have been convincing in the way that we would wish and now it is clear why this is the case. Essentially, correspondence principles may, if they are successful, allow one to derive one theory from the scientific model that supersedes it by some sort of limiting process, thereby establishing the appropriate increase in the domain of applicability of the new theory over that of the old and in this sense may serve as a sufficient condition for domain enlargement, but it is not a necessary one. It is quite possible that the new scientific theory *already* embraces the domain of the old theory irrespectively of whether the latter can be derived from the former and is contained therein, or indeed, might be subsumed thereby in some vague sense. Evidently, it is important to make a clear distinction between the scientific theories themselves and their domains of applicability, for confusion will otherwise doubtless ensue.

This is the situation with quantum mechanics, at least in its present form. There is no question that quantum mechanical laws cannot be applied to Newtonian macroscopic phenomena, for they are obviously applicable within the classical regime. It is just that from a practical rather that a theoretical perspective, it would be rather pointless to use the Schrödinger equation to compute, for example, the orbits and wave functions of planetary bodies in orbital motion or that of large galactic stellar systems (with consequent very large quantum numbers) when Newtonian and relativistic mechanics is already available for and wholly adequate for that purpose, but of course one is free to do so even if one finds that one is really doing a classical calculation anyway. Thus, quantum mechanics actually already embraces classical mechanics and encompasses its domain – such is the scope of this extraordinary empirical axiom system. It now becomes apparent why we failed to include any requirement of a correspondence principle in Definition 7.4.1. Then we can say that the domain of applicability of quantum mechanics includes that of Newtonian theory in addition to its own special domain in the micro-world of atomic and molecular phenomena where the classical theory fails. It is therefore probably better to think of quantum theory as unrelated to Newtonian mechanics in the sense of a correspondence principle, but that it nevertheless covers the domain of the latter without specifically 'containing' classical physics as such, for the mathematical structure of the two theories are too dissimilar. However, this has no bearing upon our definition. Hence, we may conclude that quantum mechanics is a particularly unique example of a major revolution of the first kind.

It is not often in the history of science that the work of a single individual precipitates more than one scientific revolution. Yet remarkably, this has been observed in the early part of the twentieth century. We have accepted Special Relativity as a revolution of this category, yet only some ten years after the presentation of the latter, we find that Einstein developed a far more revolutionary theory that must also constitute a major revolution of the first kind. If this is so, then we need to consider whether General Relativity (GR) satisfies the requirements of Definition 7.4.1.

Just as in Special Relativity (SR), the foundations and assumptions of General Relativity (GR) also have a strong geometrical flavour and are similarly iconoclastic in that regard. Yet, this was not the primary motivation for the development of GR, for as it stands now, the extant theory in its modern formalism would probably not be recognised by its creator, for much time has passed and in the intervening years many insights into the structure of GR have been painfully elucidated by many investigators. However, although new revelations have gradually come to light, the foundations of General Relativity remain the same, or *nearly* the same. Let us briefly review the basic tenets of the theory and try to isolate some of the fundamental assumptions that are of particular significance in our context.

General Relativity is all about geometry and the 'geometrization' of physics. Special Relativity was also of this character but that was not realised until it was cast into a more rigorously geometrical formalism by Minkowski, as we have noted earlier with regard to the space-time metric. By contrast, GR is at the outset at once necessarily geometrical in its mathematical structure. Special Relativity dealt with uniform motion whereas GR extends this to accelerating reference frames, hence the name. But GR is also a theory of gravitation and it is so because of a well established empirical fact that was known to Newton and familiar to all subsequent Classical Dynamicists, but realised by Einstein to hold the fundamental key to not only extending SR, as was his wont, but requiring that any such generalisation must simultaneously be a theory of the gravitational field.

There are three concepts of mass in Newtonian mechanics, between which it is not usually necessary to differentiate except in special circumstances such as consideration of the foundations of General Relativity. Thus, the quantity m_I that appears as the constant of proportionality in Newton's second law;

$$F = m_I \frac{dv}{dt} = m_I a$$

(assuming it is constant) is the *inertial* mass and is a measure of the body's resistance to a change in motion caused by the impressed force F, which is the inertia of that body. Then there are two types of *gravitational* mass, the *passive* and the *active* gravitational mass. The former occurs when a body is placed in a gravitational field of potential Φ;

$$F = -m_p \nabla \Phi$$

where it experiences the force F. Hence, it is a measure of the response of the body to a gravitational field. On the other hand, the mass occurring in the expression for the Newtonian gravitational potential at a distance r;

$$\Phi = -\frac{Gm_A}{r}$$

is the active gravitational mass m_A and is a measure of the strength of the gravitational field produced by a body, here located at the origin. From Newton's Law of Gravitation $m(dv/dt) = GMm/r^2 = ma$ for a body of mass m falling freely in the field of a much larger mass M, we see that the acceleration $a = g$ due to gravity is independent of the mass of the body, a fact that has been well tested experimentally. Using this fact we may consider two bodies falling freely with inertial and passive gravitational masses m_{I1}, m_{I2} and m_{P1}, m_{P2} respectively, then the above expressions give for the accelerations a_1 and a_2;

$$F_1 = m_{I1}a_1 = -m_{P1}\nabla\Phi$$
$$F_2 = m_{I2}a_2 = -m_{P2}\nabla\Phi$$

and since $a_1 = a_2$ we have $m_{I1}/m_{P1} = m_{I2}/m_{P2}$ and we see that the ratio m_I/m_P is constant. Choosing suitable units we can therefore set $m_I = m_P$. Similarly, if we consider two bodies interacting gravitationally with respective potentials $\Phi_1 = -Gm_{A1}/r$ and $\Phi_2 = -Gm_{A2}/r$ we have;

$$F_1 = -m_{P1}\nabla\Phi_2$$
$$F_2 = -m_{P2}\nabla\Phi_1$$

from which, since $\nabla\Phi_2 = -\nabla\Phi_1$ we obtain;

$$F_1 = \frac{Gm_{P1}m_{A2}}{r} \qquad F_2 = -\frac{Gm_{P2}m_{A1}}{r^2}$$

but since from Newton's third law $F_1 = -F_2$ we get $m_{P1}/m_{A1} = m_{P2}/m_{A2}$ and as before we conclude that for any body the passive and active gravitational masses are equal. Hence we need speak only of the gravitational mass m_G which is then equal to the inertial mass. It is this result of classical mechanics that Einstein considered as fundamental in formulating General Relativity and is the experimental fact known to Newton alluded to above.

The equality (or proportionality) of inertial and gravitational mass is essentially the *equivalence principle* upon which the General Theory of Relativity may be built. In Newtonian mechanics this equality appears to be only a remarkable co-incidence; a lucky accident for which there is no explanation in classical theory but has nevertheless been verified experimentally to a very high degree of accuracy. Even so, Einstein was able to see that such equivalence between apparently different natural variables could provide a vital tool in the generalisation of SR to the accelerated frames of reference that thereby would encompass a less restricted domain. He therefore interpreted it in different terms and in such a suggestive way as to make it apparent that motion in a gravitational field and accelerated motion were, under certain conditions, essentially indistinguishable in a practical manner from the vantage point of the physical body concerned. Thus we recall Einstein's famous *gedänken* experiment concerning an observer inside an elevator that is freely falling under the influence of a (uniform) gravitational field.

Hence, we envisage an unfortunate (but dedicated) experimental physicist inside an elevator of which some malicious and homicidal individual has cut the cable (not Einstein) and where all safety measures have been similarly sabotaged. Yet, our courageous physicist, now realising his impending demise and being at once an objective and intrepid investigator, attempts to make the best of his situation in the only way that he can. In true scientific spirit, he has with him sufficient measuring apparatus and scientific instrumentation to conduct instantaneous experiments, which with great foresight, he carries with him for just such an occasion and this is as well considering the very finite time now available to him. Since the elevator is in free fall, any object within it will accelerate at the same rate and appear to be weightless. Our physicist attributes this to the acceleration due to some external gravitational field, but realises that neither he himself nor any object in the elevator can 'feel' any gravitational influence. Hence, for all practical purposes, there may as well *not be* an external gravitational attracting body causing the acceleration of the elevator. Gravity has apparently disappeared for our physicist *because* of the acceleration and hence, the two phenomena are in some sense equivalent.

Conversely, let us now suppose that our physicist, having survived this ordeal, is again in the elevator but this time the latter is accelerating vertically upwards. Objects in the elevator therefore fall to the floor as it accelerates upwards. This looks like there is some massive external body that is causing bodies to fall within the elevator via a powerful gravitational field. Then the elevator may just well be stationary and not moving at all from this point of view. Our physicist concludes that he cannot tell whether he is experiencing a gravitational field or whether the elevator is accelerating and there is no experiment he can perform that will answer this question. Therefore, acceleration is equivalent to gravity.

Of course we have made certain assumptions in the above scenario. Firstly, the elevator must not be rotating, for then it would be possible to isolate the cause of the acceleration. Secondly, the field must be uniform, which is rarely the case in practice where real fields vary in strength from point to point. It is also the case that since the gravitational field of a massive body is concentrated at its centre, the radial motion of bodies falling towards that point would be measurable since trajectories would converge thereto. Clearly, these facts of reality are sufficient for our physicist to distinguish between a real and fictitious field, but that is hardly the point. The elevator experiment must be considered in an 'ideal' sense much as a perfect gas or molecular kinematical model is ideal; meaning that the brute facts of reality that are not important for the purposes of the construction of the model or the argument concerned, may be removed. Einstein's *gedänken* experiment is significant in that it reveals clearly and actually *explains* the equivalence between gravitational and inertial mass. They are the same because the inertial mass occurring in Newton's second law is the resistance to the force causing the acceleration in that equation and the gravitational mass is a measure of the strength of the gravitational field. Then, if acceleration and gravitation are indistinguishable, so is inertial and gravitational mass. It is now easy to see why GR is also a theory of gravity, for the latter now becomes a fictitious force, at least locally.

But we have also noted above that this equivalence cannot be realised in a real, non-uniform gravitational field except at a given set of points that cover a region of 'sufficiently small' extent such that any inhomogeneities of the field are negligible. Mathematically, this is a limiting process and things must be referred to each point in the manifold, but from a physical perspective we may consider regions of space-time that are 'local' in the sense that the equivalence principle holds within a 'small region', when technically we should say 'at a given point'. We shall return to this matter in more detail later, but since our primary concerns are the fundamental assumptions upon which GR is founded, we first need to consider the appropriate formulation of the equivalence principle from which it may be built.

Then, we may state in the first instance that the mere equality or more precisely the proportionality of inertial and gravitational mass may be expressed thus;

> *At each point in the space-time manifold in the presence of a gravitational field we may select an inertial frame such that the laws of motion of a freely falling particle are the same as in a (non-rotating) unaccelerated inertial frame in the absence of a gravitational field.*

which is one form of the *weak* equivalence principle known to Galileo and Newton and of course to all subsequent physicists prior to GR, including Mach and Einstein. There is nothing new here, however, it is the manner in which Einstein's interpretation of the above strengthens it in such a way that it leads directly to General Relativity which is important in our context. From our earlier discussion, we see that that interpretation essentially amounts to a new principle that is evidently related to the above, but with far-reaching connotations that could not hitherto have been seen without the extra component that stemmed from the *gedänken* experiment of the elevator described above.

The weak equivalence principle is a Newtonian concept. Special Relativity supersedes Newtonian mechanics as we saw earlier and therefore we would certainly expect SR to be applicable in the absence of a gravitational field. The 'non-rotating unaccelerated inertial frames' in the above definition are just those of Special Relativity. We may therefore strengthen the above as follows;

> *At each point in the space-time manifold in the presence of a gravitational field we may select an inertial frame such that the laws of physics are the same as in a (non-rotating) unaccelerated inertial frame in the absence of a gravitational field.*

and in view of the above remarks we are merely saying that at a given space-time point Special Relativity applies. This is often called the *strong* equivalence principle. Essentially, it states that 'locally', within a sufficiently small region of space-time, we can 'transform away' the gravitational field and replace it with a reference frame that is accelerating with respect to an inertial frame. Another way to state this idea is that in a locally freely falling frame the laws of Special Relativity hold. But obviously this cannot be done globally since; as we have seen, real gravitational fields are not uniform, not least because a body under the gravitational influence of another body will itself produce a gravitational field. We are forced to the conclusion that gravitational fields are incompatible with Special Relativity in a global sense, i.e. there can be no global inertial frames. This means that, although, locally, space-time is Minkowskian and flat, it cannot be so globally. Hence, General Relativity must, of necessity, be formulated in a curved space-time or a non-euclidean manifold - which eventually became clear to Einstein after many years of deep thought.

By the 'Laws of Physics' with respect to the strong equivalence principle, we really mean Special Relativity which, of course, includes Lorentz invariant theories such as Maxwellian electrodynamics. This is quite a generalisation of the weak equivalence principle and although it has its origins in the latter classical theories, it effectively embodies an entirely new principle rather than just mere reformulation and, as such, becomes one of the fundamental assumption upon which GR is founded. Since SR obtains locally, then the assumptions upon which that theory is based and which we discussed earlier, are also fundamental assumptions of GR. We have already considered these and found them to be in accordance with Definition 7.4.1. However, as far as GR is concerned they are not *new* fundamental assumptions in the sense of our definition. But the equivalence principle requires that physical laws take the same form in all frames and the statement of this is often called the *Principle of Covariance*. We do not regard this as a fundamental assumption of GR *per se*, since it is a consequence of the strong equivalence principle. However, it does imply a mathematical formulation of GR where physical laws can be expressed without regard to the co-ordinate systems chosen. Traditionally, this has been the tensor calculus but GR can also be constructed equally well in the spinor formalism, which is arguably more appropriate for its needs.

The space-time of General Relativity is curved as we have mentioned. Recall that from SR the Minkowski metric;

$$ds^2 = c^2 dt^2 - dx^2 - dy^2 - dz^2$$

can be written in the form;

$$ds^2 = (dx^0)^2 - (dx^1)^2 - (dx^2)^2 - (dx^3)^2$$

where we have made the identifications $x^0 = ct, x^1 = x, x^2 = y, x^3 = z$ with indexed co-ordinates x^μ. This can then be written more compactly as;

$$ds^2 = \eta_{\mu\nu} dx^\mu dx^\nu$$

where the indexed quantity;

$$\eta_{\mu\nu} = \begin{pmatrix} 1 & 0 & 0 & 0 \\ 0 & -1 & 0 & 0 \\ 0 & 0 & -1 & 0 \\ 0 & 0 & 0 & -1 \end{pmatrix}$$

is the usual covariant metric tensor in Minkowski space-time (Minkowski metric tensor). This metric holds only locally in GR and we need something more general. This leads to;

$$ds^2 = g_{\mu\nu} dx^\mu dx^\nu$$

which is the metric of a *pseudo-Riemannian* manifold and $g_{\mu\nu}$ is the *metric tensor*. It is the same as the Minkowski tensor locally at a given point by the equivalence principle. This metric describes a generally non-Euclidean curved space-time where in the case of GR the $g_{\mu\nu}$ are to be considered as 'gravitational potentials' by analogy with Newtonian theory. They are therefore expected to similarly satisfy second order differential equations. Thus, in GR the gravitational field is to be a manifestation of the curvature of space-time with the components of the metric tensor playing the role of the potentials. But since the source of the gravitational field is mass, then the source of space-time curvature is also mass (or energy since from SR mass and energy are equivalent). Hence, an expression is needed that connects the curvature of the manifold to the mass-energy content of the system under consideration, which may be the entire universe. By the principle of covariance, this will be a tensor equation.[*]

---

[*] The question of whether the equations of physics need to be expressed in covariant form is largely a matter of choice. Einstein insisted on covariance as a criterion, but later came to realise that it makes little difference from a mathematical perspective if one's equations are covariant or contravariant because these are, geometrically, dual concepts. Thus, in a geometrically formulated physical theory, the physics is the same anyway.

We therefore expect that the mass-energy will be described by a tensor and that the latter must satisfy the all important conservation of energy condition - and whether the total mass-energy is positive or negative should not matter because a negative pressure component is perfectly admissible in whatever cosmological model we obtain from the eventual equations. It is easily found that a second-rank tensor can therefore be taken as the source for the gravitational field (and thus a source of curvature) and that, by conservation of energy, we demand;

$$\nabla_v T^{\mu v} \equiv 0$$

where the covariant derivative is indicated. This expresses the conservation of energy and the tensor $T^{\mu v}$, which is the *energy momentum tensor* describing the entire matter and energy content of the physical system under consideration and, in the cosmological scenario, equates to the entire universe. This tensor must therefore be related to the space-time curvature, which again must also be described tensorially and satisfy an identical conservation property. We are therefore seeking an equation of the form;

$$G_{\mu v} = \kappa T_{\mu v}$$

where the second rank tensor $G^{\mu v}$ describes the curvature of the space-time. In Riemannian geometry the curvature is described by the *Riemann-Christoffel Tensor* or *Curvature Tensor*;

$$R^{\lambda}_{\mu v \kappa} = \frac{\partial \Gamma^{\lambda}_{\mu v}}{\partial x^{\kappa}} - \frac{\partial \Gamma^{\lambda}_{\mu \kappa}}{\partial x^{v}} + \Gamma^{\eta}_{\mu v} \Gamma^{\lambda}_{\kappa \eta} - \Gamma^{\eta}_{\mu \kappa} \Gamma^{\lambda}_{v \eta}$$

in terms of the affine connections or *Christoffel symbols of the second kind*. These latter are expressed in terms of the metric tensor as in, for example;

$$\Gamma^{\lambda}_{\mu v} = \tfrac{1}{2} g^{\lambda \kappa} \left(\frac{\partial g_{\kappa v}}{\partial x^{\mu}} + \frac{\partial g_{\mu \kappa}}{\partial x^{v}} - \frac{\partial g_{\mu v}}{\partial x^{\kappa}} \right)$$

Now, the $G^{\mu v}$ of our proposed tensor equation should be constructed from the curvature tensor and the metric tensor and it should be linear in the partial derivatives of the metric tensor no higher than the second order. Since the curvature tensor contains second order partial derivatives of the metric tensor, we may expect that these might be involved in the process of constructing a suitable form for $G^{\mu\nu}$. However, the only second rank tensor obtainable from the curvature tensor that would be a possible candidate is the *Ricci Tensor*;

$$R_{\mu\nu} = R^{\lambda}_{\mu\lambda\nu} = g^{\lambda\kappa}R_{\kappa\mu\lambda\nu}$$

which is obtained by contraction with the metric tensor. A further contraction gives the *Ricci Scalar*;

$$R = g^{\mu\nu}R_{\mu\nu}$$

often called the *curvature* scalar. From these one may now construct the tensor;

$$G_{\mu\nu} = R_{\mu\nu} - \tfrac{1}{2}g_{\mu\nu}R$$

which is the *Einstein Tensor* and has the conservation property that we require;

$$\nabla_{\nu}G^{\mu\nu} \equiv 0$$

and we therefore take this as the left hand side of our proposed equation;

$$G_{\mu\nu} = R_{\mu\nu} - \tfrac{1}{2}g_{\mu\nu}R = \kappa T_{\mu\nu}$$

We can identify the constant κ through comparison with Poisson's equation in Newtonian theory and the weak field approximation;

$$\nabla^2\Phi = 4\pi G$$

which results in $\kappa = 8\pi G/c^4$. Hence we have found an appropriate tensor equation for the description of the gravitational field as space-time curvature with the mass-energy as the source of that curvature. Then, we find that the appropriate equations for expressing the dynamical influence of matter and energy upon the geometry of the space-time manifold are just the Einstein *Field Equations*;

$$R_{\mu\nu} - \tfrac{1}{2} g_{\mu\nu} R = \tfrac{8\pi G}{c^4} T_{\mu\nu}$$

These result in ten (non-linear) partial differential equations for the ten unknown components of the metric tensor. They are a remarkable achievement and show that the space-time of General Relativity is itself a dynamic entity, quite different from that of Newton's world or even SR.

We mentioned in §3.6 that Einstein later modified his equations by the inclusion of an extra term, the famous and notorious Λ-term, or cosmological constant;

$$R_{\mu\nu} - \tfrac{1}{2} g_{\mu\nu} R - \Lambda g_{\mu\nu} = \tfrac{8\pi G}{c^4} T_{\mu\nu}$$

where it was taken as an illustration of theory modification. These may be called the *full* Einstein field equations and although the Λ-term was dropped; it has now been re-instated and may be of great significance in evolutionary cosmology.

Before returning to the question of the fundamental assumptions of the theory, it is useful to consider how its formulation differs from previous ideas, for we would expect that a revolutionary theory in this category would exhibit, *ipso facto*, a more novel presentation of its subject matter. This is true of GR and it was also true of SR. The preceding adumbration is intended only to isolate the salient points, which for us are the fundamental assumptions that are pertinent to our context[*]. Yet, the historically intuitive development of the theory although mathematically unsophisticated is sufficient to isolate the essential principles, but it must be said that the modern presentation using the full power of differential geometry and which now has largely superseded the historical approach is essential for a complete understanding of the *structure* of General Relativity. It is by such methods that the contrast with classical concepts becomes most illuminating.

There are some principles that we have not discussed either because we do not regard them as sufficiently fundamental or they follow from one or

---

[*] This is not a physics book and our treatment of GR and SR as well as QM is only cursory and omits many details which a standard textbook would include.

the other of the equivalence principles anyway. Thus, the fact that the influence of gravity on a physical body is independent of its mass or composition obviously follows from the weak form of the equivalence principle since all bodies fall at the same rate in a gravitational field and this is no more than a statement that inertial and gravitational mass are proportional. From this one concludes that no body can escape the effects of gravity or can be 'shielded' from that influence as can charged particles from electromagnetic forces for example. This is sometimes stated as 'gravity couples to everything' and by SR this also includes energy. However, as noted, it is not independent of the weak equivalence principle and again is merely a re-formulation of the latter. Even if either of these two 'principles' were of independent importance in the foundations of GR they would be of little concern as fundamental principles in relation to Definition 7.4.1 since, as with the assumptions of SR, they may well be incorporated into GR, but the definition allows for that circumstance and we need only consider new fundamental assumptions; a point we have stressed several times.

The Principle of General Covariance is related to the Principle of General Relativity. The latter states that all observers are equivalent in relation to the laws of physics, i.e. there are no preferred co-ordinate systems. In Special Relativity all inertial observers were considered to be equivalent and now this is generalised to the extent that, whatever the co-ordinate frame, it is possible ascertain the same physical laws as in any other co-ordinate frame. The Principle of General Covariance says that since the laws of physics are the same in all co-ordinate systems, whether inertial or not, they must take the same form in all such frames. In other words, the laws of physics are invariant under a co-ordinate transformation in the sense that they will be the same observed physical laws irrespective of the observer's frame of reference. For this to be the case the equations of physics must be expressible in a 'co-ordinate free' manner and the way to do this is to write them in tensorial form. This then, is the statement of the Principle of General Covariance: The Laws of Physics should be expressed as tensor equations. Einstein considered this to be an important fundamental principle, but on reflection it is seen to be just the *strong* equivalence principle in disguise. If a law of physics holds in an arbitrary co-ordinate system in a gravitational field then it holds in all such co-ordinate systems, but by the strong equivalence principle there is a local inertial frame in which the law also holds. Hence, this is just a version of the Principle of General Relativity, which in turn is the strong equivalence principle. It is therefore the latter that we regard as the fundamental assumption of GR. We note that the Principle of General Covariance doesn't actually have any physical content and that in fact *all* physical theories can be formulated with the use of tensors and differential forms. Indeed, today they invariably are.

Then there is Mach's principle, also important to Einstein to the extent that he considered it to be, at least in his early years, at the heart of General Relativity and a primary foundational tenet of the theory. Mach's principle is rather vague. It asserts that the inertia of a body is determined by the distant masses of the universe, usually taken to be the 'fixed stars'. Thus, when a body is accelerated the source of the resulting inertial forces is the distant material content of the universe and therefore in an empty universe there would be no inertial forces. Mach's principle has a vast literature. It cannot be disproved because it is too vague. No mechanism is specified by which inertial forces are related to the overall mass-density of the universe, but it certainly played a significant role in the initial development of GR, even though it is not really incorporated into the theory and in fact is not necessary for the formulation of the latter. Furthermore, although the geometry of the universe is governed by the energy-momentum tensor in accordance with the field equations which may thus seem to confirm Mach's principle, this ignores the initial conditions on the geometry which should also be specified through the metric tensor and its time derivative. This simply reduces Mach's principle to the notion that matter determines geometry, which we already know.

We need to show that in accordance with Definition 7.4.1, the strong equivalence principle, taken as a fundamental assumption of GR is inconsistent with the *status quo* and that the domain of applicability of GR is greater than that of the latter. But what do we mean here by the *status quo*? Do we mean Special Relativity or Newtonian mechanics? It may seem that we mean both. Special Relativity supersedes Newtonian mechanics but deals only with inertial frames at rest or in uniform motion. It does not deal with gravitational fields or non-inertial frames. On the other hand, Newtonian theory does just that via the inverse square law of universal gravitation. The domain of applicability of General Relativity must therefore include both the domain of SR and the domain of the Newtonian theory of gravitation.

The strong equivalence principle requires that the space-time of GR be generally pseudo-Riemannian, as we have already noted. It is a curved space-time and only flat on the tangent space at each point of the manifold, where it corresponds, locally, to SR. The geometry of SR however, is flat everywhere because, as we have seen, its metric is pseudo-euclidean and linear, this applies throughout the SR manifold in contradistinction to the space-time of GR which is globally noneuclidean. But the space-time of SR is fixed and static; it is, in fact, 'absolute' in the Newtonian sense. There is no relation between mass-energy and geometry in SR and the Minkowski metric of that theory cannot encompass any formulation of the gravitational field without extra assumptions. Additionally, there are no global inertial frames in GR. Since the strong equivalence principle is a fundamental assumption of GR and leads directly to these differences and incompatibilities, it replaces that of SR with which it is inconsistent. We must conclude that in this respect Definition 7.4.1 is satisfied.

There remains the question of the domain of applicability of GR. First of all, we already know that GR includes SR as a special case, namely at each point in the space-time, so there is no problem with the domain of applicability in that respect. But GR is a theory of gravitation and already must include the domain of Newtonian Gravitation as well as that of SR. Hence, the domain of applicability is increased.

This latter can be confirmed directly through a correspondence principle in the weak-field limit of GR. It would be out of place to reproduce a complete and rigorous proof of this here, but we shall briefly indicate the general idea since it is important for our purposes. Thus, we consider the metric tensor in a weak gravitational field which is 'nearly' Minkowskian. In this case, by the equivalence principle it is possible to choose co-ordinates x^μ so that;

$$g_{uv} = \eta_{\mu v} + h_{\mu v}$$

where $|h_{\mu v}| \ll 1$. Let us take all time derivatives of the metric to be zero, i.e. $\partial g_{\mu v}/\partial x_0 = 0$, then, the metric is *stationary*. We recall that the world-line of a particle freely falling under the influence of a gravitational field is given by the *geodesic* equation;

$$\frac{d^2 x^\mu}{d\tau^2} + \Gamma^\mu_{v\sigma} \frac{dx^v}{d\tau} \frac{dx^\sigma}{d\tau}$$

where the *proper time* interval $d\tau$ is defined by $d\tau^2 = ds^2/c^2$. This is the analogue of Newton's law of motion. Now in this approximation we are taking the Newtonian limit of General Relativity so that the velocity of the particle $v = dx^i/dt \ll c$, where $i = 1, 2, 3$ so that this is the ordinary three-velocity and the time $t = x^0$. This means that $dx^i/d\tau \ll dx^0/d\tau$ and we need only consider the time component t. Under these conditions the geodesic equation reads;

$$\frac{d^2 x^\mu}{d\tau^2} + \Gamma^\mu_{00} c^2 \left(\frac{dt}{d\tau} \right)^2 = 0$$

and using the expression given earlier for the Christofell symbols with this metric we find to first order in $h_{\mu\nu}$;

$$\Gamma^\mu_{00} = \tfrac{1}{2} g^{\kappa\mu} \left(\frac{\partial g_{0\kappa}}{\partial x_0} + \frac{\partial g_{0\kappa}}{\partial x_0} - \frac{\partial g_{00}}{\partial x_\kappa} \right)$$

$$= -\tfrac{1}{2} g^{\kappa\mu} \frac{\partial g_{00}}{\partial x^\kappa}$$

$$= -\tfrac{1}{2} \eta^{\kappa\mu} \frac{\partial h_{00}}{\partial x^\kappa}$$

then because the metric is static this gives;

$$\Gamma^0_{00} = 0, \quad \Gamma^i_{00} = \tfrac{1}{2} \delta^{ij} \frac{\partial h_{00}}{\partial x^j}$$

and with the above geodesic equation we obtain;

$$\frac{d^2 t}{d\tau^2} = 0, \quad \frac{d^2 \mathbf{r}}{d\tau^2} = -\tfrac{1}{2} c^2 \left(\frac{dt}{d\tau} \right)^2 \nabla h_{00}$$

where we have reverted to standard vector notation in 3-space. We see that since $dt/d\tau$ is obviously constant we can set $dt = d\tau$ and obtain;

$$\frac{d^2\mathbf{r}}{dt^2} = -\tfrac{1}{2}c^2\nabla h_{00}$$

describing the motion of our particle. If we now set $h_{00} = 2\Phi/c^2$ and multiply by the mass m of the particle we can write this as;

$$m\frac{d^2r}{dt^2} = -m\nabla\Phi$$

which is of course, the Newtonian equation of motion for a particle moving in a field of potential Φ. The left hand side of the equation is simply the force $\mathbf{F}$ and for a Newtonian gravitational field we have for the potential; $\Phi = -GM/r$. If we insert this into the above equation we obtain;

$$F = \frac{GmM}{r^2}$$

(in scalar form). Hence we have recovered Newton's law of gravitation in the classical limit of General Relativity. Then we see that GR does indeed include the domain of applicability of Newtonian gravitation via a correspondence principle and this fact, together with the above observations concerning SR and the equivalence principle shows that Definition 7.4.1 is fully satisfied. Hence, General Relativity is a major revolution of the first kind.

The examples given thus far of revolutions in this category have all been confined to empirically well-attested areas of theoretical physics. This is inevitable, for in order to determine the fundamental assumptions of a scientific theory it is usually necessary that the theory be presented in such a precise form that those assumptions, by mere examination of the structure of the theory, become immediately evident. Whilst this is certainly the case in physics where the required precision and perspicuity obtains, it is less so in other scientific disciplines. The reason for this is of course the inexactitude of the language used to describe the foundational aspects of a given hypothesis. If natural language is used, then at some stage of the development it will be found to be inadequate and a technical vocabulary will have to be introduced. But, this is not to say that any notion, idea or concept cannot be clearly expressed in natural language, for we have noted earlier on many occasions that all reasonable languages are equally suitable for such descriptions when appropriately employed. Hence, when a scientific theory is expounded in natural language, it should in principle be possible to isolate the fundamental assumptions involved, but it is the complexity of natural language with its accompanying ambiguities, as opposed to the precision of formal languages that makes this process difficult.

There is another reason why major revolutions of the first kind occur primarily in the well-developed physical sciences and that is again a result of the clear formulation of theories in such disciplines. Evidently, in order to have a revolution at all, the theory which is being modified must be capable of modification by the standard techniques and therefore appeals to the empirical data which can thus be expressed in a quantitative manner. This circumstance clearly favours the physical sciences, which because of the structural form of the latter, make them far more accessible to empirical methodology as well as theoretical development. However, this is not always so with major revolutions of the second kind.

Let us therefore, straying initially not too far from full precision, consider an example from Geophysics. We ask whether the modern theory of Plate Tectonics constitutes a major revolution of the first or the second kind. Let us first outline the main ideas in order to gain some sort of perspective on this theory.

The idea of *continental drift* in its modern form dates back to Frank Taylor in 1910 and was elaborated upon in 1912 by Alfred Wegener. The motivation for such a notion was inspired by the observation that the coastlines of the continents suggest that they would fit together much like pieces of a jig-saw, a point noted by Francis Bacon in relation to Africa and South America as long ago as 1620 when cartography had made such a comparison possible. Thus, Wegener imagined that at some stage in geological history all the continents were conjoined in a large 'supercontinent' called Pangaea that subsequently broke apart. Wegener proposed that the mechanism driving the movements of the continents was due to the tidal forces of the Sun and the Moon or the centrifugal force due to the Earth's rotation. However, this was quickly shown to be untenable because such forces would be inadequate for the purpose. This, together with the prevailing view that the mantle was too rigid to admit of such motion led to the idea of continental drift being abandoned, although it occasionally surfaced again at various points during the first half of the twentieth century.

However, we are not concerned with continental drift *per se*, but with plate tectonics. The former is merely a consequence of the latter. Let us therefore briefly discuss the principle notions of that model. Firstly, it is observed that the lithosphere, which is that part of the upper mantle immediately below the oceanic and continental crust, is comprised of numerous *plates* which are slowly moving about the globe and carrying the continents with them. These plates are being continuously re-generated by material in the Earth's interior rising from below in the *mid-ocean ridges* which mark the boundaries between the plates. There must also be regions where the material of the plates is returned to the interior, if the surface area of the Earth is to remain constant. This process is called *subduction* and occurs at *ocean trenches*. Immediately below the lithosphere is the *asthenosphere* which is the layer of rock upon which the plates move. The rocks in the asthenosphere therefore behave like a fluid, something called *solid-state creep* and is a very contrary notion to that held prior to plate tectonics. In fact, only the lithosphere is rigid over long time scales, although of course it must be able to bend at subduction zones so that material can pass into the mantle. This movement of the plates leads to all sorts of geological phenomena from mountain-building (*orogenesis*) to volcanism and other seismic phenomena.

One still needs a mechanism for plate movement to complete this picture. The modern view is similar to that first proposed by Arthur Holmes in 1929. The explanation is thermal convection in the mantle. Hot rocks from below are less dense than the cooler rocks in the lithosphere and therefore ascend through the mantle where they cool and accrete to the ocean floor at the mid-ocean ridges. This is *sea-floor spreading*. They eventually descend back into the mantle by the same principle of buoyancy and gravitational instability according to the equation of hydrostatic equilibrium. This is a continuous cycle and as the plates collide and one passes below the other (albeit extremely slowly) the geological process mentioned result. But what heats the rocks deep in the mantle so that they may ascend to the surface? The source of the heat is the radioactive decay of isotopes of uranium, thorium and potassium, specifically ^{235}U, ^{238}U, ^{232}Th and ^{40}K. This convection results in various body forces acting upon the plates, forcing them towards ocean trenches where they descend.

The situation is rather more complicated than we have described but we have the essential details that we need to answer our original question. We summarise the fundamental assumptions of plate tectonics as follows:

(a) The surface of the Earth is divided into a number of crustal plates, (b) the plates are continually being regenerated and subducted, (c) convection currents in the mantle moves the plates in different directions and (d) radioactivity deep in the mantle is the source of the heat that drives the convection currents.

The first thing to note is that each of these hypotheses satisfies Definition 7.4.3, for the negation of any one of them would negate the theory. Secondly, neither (a), (b), (c) or (d) is consistent with the prevailing view before the advent of plate tectonics. More specifically, (a) is contrary to the prevailing view that the Earth's crust was solid and (c) is contrary to the view that the continents cannot move and that the Earth is composed of solid rock. Finally, (d) is contrary to the pre-tectonic view that the interior of the Earth does not generate heat but is merely hot and cooling down. In fact the description of the Earth according to plate tectonics is very different from the preceding model. Then it seems that all of the assumptions of the pre-tectonic picture of the Earth have been replaced or modified.

What about the natural variables of the theory? If we had presented plate tectonics more fully we would have drawn upon fluid mechanics and rheology, the mechanics of rigid bodies and gravitational physics, thermodynamics and nuclear physics, etc. Essentially, large areas of theoretical physics are employed in this subject, as would be expected. Thus the natural variables are just those that are used in these physical theories when applied to geophysical problems and suitably defined in relation to observable quantities appropriate for that discipline under the natural interpretation, which was also the case in pre-tectonic models, but less so.

This leaves the domain of applicability to consider and whether it has increased or decreased. In pre-tectonic models the domain of applicability was the entire physics of the Earth, i.e. geophysics. In plate tectonics the domain is also the entire physics of the Earth, namely geophysics but is simply developed in more detail than was hitherto possible. Hence, the domain of applicability is unchanged. Therefore, together with the above remarks we find that this revolution accords precisely with Definition 7.4.2 and Plate Tectonics is a major revolution of the *second* kind and indeed, is our first example of such.

We have mentioned the Copernican revolution several times in the course of our discussions concerning theory-change and modification. What we shall mean by this is the replacement of the geocentric Ptolemaic system of planetary motion by the heliocentric model of Copernicus. We shall see that this also corresponds to a major revolution of the second kind and a particularly simple one to describe. We first state the principle characteristics of the Ptolemaic system.

Recall that in the Ptolemaic model of the universe, the Earth is at the centre of a number concentric spheres of varying but unspecified radii. Note that the Earth is also spherical in this scenario; a fact that was known to the Greeks, but is stationary and non-rotating. There were eight spheres that rotated about the central Earth each day. The first sphere was the sphere of the Moon, the next was that of Mercury and then respectively Venus, the Sun, Mars, Jupiter and the seventh sphere for the known planets was that of Saturn. The final eighth sphere was the sphere of the 'fixed stars'. With the exception of the latter, the motions of these celestial bodies are not accounted for in this simple system. The observed movements of the planets are sometimes prograde, sometimes retrograde and even appear to be stationery during the period of transition between these two directions of motion. This of course, is due to the Earth's motion around the Sun, but that is not relevant here. Therefore, theory modification was employed, which in spite of our musings here was thus clearly invented in the Hellenic era. The axiom modification required was that each of the planetary bodies would also perform an additional circular motion about the points of the primary circular orbit as the latter rotated about the Earth. These are the *epicycles* of the Ptolemaic model. An epicycle therefore represents circular motion about a point upon the primary orbit. This was done to varying degrees depending upon the celestial body involved but for some bodies there needed to be further circular motion so that one had epicycles within epicycles. As well as all this, the Earth was allowed to be slightly off-centre in relation to the diurnal planetary spheres. These additional degrees of freedom are similar to a perturbation series but they accounted for celestial movements to sufficient accuracy for the time and indeed for a considerable time afterwards. It is surprising that the Greeks did not invent the Standard Model of Particle Physics, or at least QED! There were also some Aristotelian mythological attributes attached to these ideas that we need not be concerned with. The point is that this was a workable scientific model; it was a scientific theory that correctly described the known universe. Using this system with different velocities for each planet, considerable accuracy could be achieved. Let us therefore identify the fundamental assumptions of the Ptolemaic system.

The first of these is that (a) the Earth is the centre of the universe. Secondly (b) the Earth is stationary and does not rotate. Then, (c) each planet, including the fixed stars, inhabit an 'orb' or 'sphere' that rotates diurnally in a circular fashion about the Earth and (d) along its primary 'trajectory' a planet performs an additional circular motion about its trajectory and possibly further circular motion about that additional circular motion, etc. Thus, the movements of celestial bodies are accounted for through a succession of epicycles that, in the science of the time, could be taken to any required degree of accuracy, even though the Greeks had difficulty with the concept of infinity. Clearly, this is a scientific approach intended to account for the available empirical observations of the bodies involved.

Now, let us look at the heliocentric view in contrast to the above. It is obvious at the outset that the ever increasing number of epicycles in the Ptolemaic system could result in considerable geometric complexity from a computational standpoint. But the Greeks were rather adept at geometry, at least the Euclidean geometry that we have all learned in our childhood. Nevertheless, it may be possible to simplify the above description of the operations or mechanics of the known universe by relaxing one or more of the fundamental assumptions of the Ptolemaic *theory*, as it must now be called.

The simplification that Copernicus proposed was just the heliocentric hypothesis, which had been considered already by Aristarchus in Hellenistic times. The essential characteristics, now well-known are these: (e) The Sun is at the centre of the universe. (f) The Earth is not stationary but rotates diurnally upon its axis. (g) The planets, including the Earth, orbit the central Sun in *circular* orbits. (h) The fixed stars do not rotate about the Earth, for their diurnal motion is due to the rotation of the Earth itself, as indicated by (f).

This results in a computational as well as a conceptual simplification but it is no more accurate than the Ptolemaic model which accorded very satisfactorily with observation. However, let us examine the logical differences between the hypotheses of each model in the context of Definitions 7.4.2 and 7.4.3. In the first instance, it is clear that the negation of any of the hypotheses (a) to (d) of the Ptolemaic theory would render that theory untenable. Hence Definition 7.4.3 is easily satisfied. It is obvious also that (e) $\wedge$ (a), (f) $\wedge$ (b), (c) $\wedge$ (g) and (c) $\wedge$ (h) are quite inconsistent with one another. But this exhausts all of the hypotheses of the Ptolemaic model. Then all of these fundamental assumptions are replaced by their Copernican counterparts. Hence Definition 7.4.3 is satisfied in this respect. Notice also that the natural variables involved are also the same, namely the positions and velocities of each celestial body at any given time.

The domain of applicability of each theory should be clear. They both describe solar system phenomena or more specifically the movements of the planets, which as far as the protagonists of these models were concerned is the entire universe. Then the domains of applicability are the same for the Copernican and Ptolemaic systems. We see that all the conditions of Definition 7.4.2 are satisfied and therefore the heliocentric model of the world is indeed a major revolution of the second kind, as we knew intuitively from the outset.

What we have described is the logical nature of the transition from a geocentric to a heliocentric picture of the world. We have used the adjectival term 'Copernican' in this context and this is quite correct since it was Copernicus who first proposed, or rather revived the heliocentric model. However, what is usually referred to as the 'Copernican Revolution' is actually more than that described above. The three Keplerian laws of planetary motion are generally included in this latter description. In fact, that the planets actually orbit the sun in bound elliptical orbits with the Sun occupying one foci rather than moving in perfect circles is nothing more than theory modification. Therefore, if there is to be a Copernican Revolution it is merely the transition from geocentrism to heliocentrism with the Keplerian adjunct being a familiar instance of theory modification in the way that we have described in earlier chapters. If the notion of elliptical orbits had been proposed by Copernicus from the beginning, then the situation would of course, have been very different.

Major revolutions of the second kind are relatively infrequent in the developed sciences and for a very good reason. If one scientific theory is to replace another, then there must in the first instance be a scientific theory to replace. However, any theory that may properly be called scientific has been to some extent empirically verified. Thus, it is established in the sense that it accords with the empirical data, which by definition it is required to do and no empirical or observational data is contrary to the axioms of the theory. Hence, the complete removal of a theory could only arise from new empirical information that is sufficiently extensive so as to negate all of the fundamental assumptions of the model and this is unlikely to be the case in highly sophisticated physical theories which have already been continually refined and subjected to experimental and observational scrutiny over a long period of time. Clearly, theory modification is far more probable in such instances and therefore we would expect major revolutions of the first kind to preponderate in the more theoretically mature sciences. This is indeed seen to be the case as the examples given above of major revolutions of the second kind adequately demonstrate.

It may seem surprising that we have not yet considered Newtonian mechanics in revolutionary terms. We may ask: In which category of revolutions does the Newtonian Revolution fall? The answer of course is that it is unlike the revolutions discussed thus far and does not easily fit into such categories. We began this work by quoting Newton's laws of motion as an illustration of axioms for a scientific theory and we further remarked upon some of the assumptions and variable definitions relating thereto. But there is more to the Newtonian Revolution than just Newton's laws and it must be considered as an integral whole which includes the work of Galileo and Descartes in addition to other more or less contemporaneous figures. This illustrates a fundamental difficulty in the classification of scientific revolutions, for if we were merely to regard Newton's laws of motion as a modification of Galilean principles for example, then there would be no category into which the former could possibly belong, for no assumptions of the latter are replaced in the Newtonian conception, on the contrary, they are incorporated within the Newtonian scheme and extended and expanded upon. Nor are any axioms replaced or modified, but added to. Indeed, there is an entire theoretical edifice that is constructed from a motley collection of preceding ideas and empirical observations that subsumes and integrates the latter into a consistent whole, yet to which the Newtonian model need not pay homage. Rather, such a theory must be seen as an *ab initio* formulation that is largely independent of any preceding ideas. If one is required to place new formulations of the Newtonian sort into a special category that embraces all former suppositions, however vague or ill-conceived, and also incorporate new fundamental assumptions and variable definitions one would find oneself faced with an infinity of such categories. It would seem that a more perspicuous and accurate view of revolutions of this kind may be realised in the manner suggested, that they are part of the continuing process of the scientific accumulation of empirical knowledge and in fact represent the culmination of that endeavour within the appropriate context and domain of applicability. Hence, scientific revolutions of this type must be thought of as the beginning of new science which does not however build upon, but may possibly incorporate former achievements, for there is no well-defined theoretical structure upon which to erect the new edifice. Then, we would expect that Newtonian mechanics, or more properly the Newtonian Revolution should satisfy Definition 7.4.4. Let us see if this is so.

Whatever Isaac Newton's view of Aristotelian physics may have been, he was certainly well-versed in that subject since to be so would have been a pre-requisite for all *Natural Philosophers* of the time. Although the beginnings of what would now be seen as the scientific method - the empirical determination of the facts of Nature by means of observation and experimentation – had been made by Galileo and others, including Newton, there still remained a strong adherence to Greek science and philosophy as well as Christian Biblical teachings. The study of the Natural World was essentially therefore a philosophical and religious discipline and, with the exception of Galileo who attempted to quantify motion, would not constitute science in our sense of that term and certainly not a scientific theory. However, Galileo's researches on the motion of material bodies were incorporated into the Newtonian system and his 'Principle of Inertia' is essentially Newton's first law of motion. Then as noted above, we must regard the Newtonian Revolution as not the work of a single individual, although of course the primary role was played by Newton himself. We would observe also that we do not equate the Newtonian Revolution with the so-called 'Scientific Revolution'. The latter has a much wider and more general scope whilst by the former we really mean only Newtonian mechanics.

Then, since Aristotelian ideas concerning the motions and interactions of physical bodies cannot for us constitute a scientific theory, there was no prior model upon which a scientific mechanical theory could be built. Hence, Newtonian mechanics is an *ab initio* formulation of a scientific theory, although this is *not* a requirement of Definition 7.4.4. Furthermore, it had a strong empirical basis through the work of Galileo and indeed Newton's own experimentation, as well as that of others. It is clear also that Newtonian mechanics involves new fundamental assumptions, some of which we discussed earlier in connection with Special Relativity and need not repeat here. That new natural variables are also introduced such as mass and momentum as the product of mass and velocity is also obvious from our previous remarks and certainly new axioms are obtained that express relations between these variables in the form of the second and third laws of motion. All of this is seen to accord precisely with Definition 7.4.4 and we find that Newtonian mechanics is indeed a major revolution of the third kind.

There is however, another way in which we may perceive Newtonian mechanics in terms of a kind of unification of two theories. In this picture, Newtonian mechanics is not created *ab initio*, for we may regard the first law of motion as a pre-existing scientific theory in the form of Galileo's principle of inertia and then, by axiom addition, we include Newton's second and third laws into the theory, thereby extending it. Definition 7.4.4 does not preclude this, it would be too for it restrictive to do so, but it is of little consequence since the two processes are entirely equivalent. We can then alternatively view the *ab initio* description as theory replacement over the 'null' scientific theory, i.e. the theory with no axioms. Evidently, there are other possibilities also but they are all equivalent, which beautifully illustrates the flexibility of the theory-change operators acting upon theories as sets of axioms. Note however, that the revolution categories themselves are not affected by such manipulations.

Darwinian evolution provides a similar example of a major revolution of the third kind, but in an entirely verbal yet well-defined way and which certainly exhibits on the part of its main protagonist the most detailed and meticulous observational and taxonomic adherence to the empirical technique of the scientific method characteristic of the emergence of the latter as a legitimate investigative tool. Then, we need to examine the Theory of Evolution in relation to Definition 7.4.4 if we are to establish our claim that it falls into this category. First however, we should briefly examine the situation before the publication of Darwin's *On the Origin of Species* that we may discover thereby the historical background of the pre-Dawinian conceptual framework.

In the first instance, the Biblical creationist account of the origin of species as described in the first book of the Tora, *Genesis*, is not a scientific one on any understanding of that term and whatever beliefs may be irrationally held by its adherents, they play no part in our present discussion and are thus summarily dismissed as mere mythology. Then we are left with various prevailing views relating to the development of species that may or may not lay claim to scientific status. Yet, the idea of actual change in the ontogeny and phylogeny of biological organisms is not new and once again first emerges in early Greek philosophy. Thus, more generally, morphological variation is a concept that has appeared and re-appeared throughout history, although perhaps sometimes in forms that would now be seen as veiled or imbued with religious or mythological and certainly in the modern sense, non-scientific and non-empirical content. It is therefore difficult to find a precursor to Darwin whose ideas and notions are not merely speculative in relation to the evolution of species or even individual organisms that can be conceived as scientific in our context, but most prominent amongst those views that deserve description, even if they might not be worthy of such attention, as some may have it, are those of the French naturalist Jean-Baptiste Lamarck.

For Lamarck as well as other evolutionists, the phenotypical aspects of an organism are determined entirely through interaction with the environment on the part of that organism acting individually of its own volition, together with the transmission of those characteristics to its offspring. The famous example of the Giraffe is often quoted; by stretching its neck in order to reach higher and higher leaves the Giraffe is able to modify its morphology, perhaps only slightly, but this modification is inherited by its offspring and over many generations the result is what we see today. Thus, species change in Lamarckism is confined to the particular lineage of the species concerned and extinction is not possible since appropriate adaptation to the environment by this 'striving' of the organism will always be possible. No specific biological or biochemical mechanism is given in explanation for this process and it is therefore generally dismissed as being too teleological and vague. However, the important point is that the morphology is altered to accommodate the needs of the organism, a position also adopted by Darwin and which is in direct contradistinction to the Biblical account. Thus, Lamarck sees the development of a species as being driven by slow environmental change, which ultimately and inevitably leads to perfection, culminating in the modern *homo sapien*. Whether this can be regarded as a scientific theory is quite another question and one which is significant in the context of scientific revolutions. We shall return to this point shortly, but first let us contrast the Darwinian approach with the foregoing.

The Darwinian portrayal of species development is of a quite different character insofar as all teleological aspects are immediately removed from serious consideration in relation to the adaptive evolution of either individual organisms or species as a whole, if such can be properly defined. The essential points of the Darwinian picture are (a) no individual is identical in all aspects to any other of the species concerned, (b) there is hereditary transmission of the characteristics from a given organism to its offspring and (c) any such characteristics so transmitted will be either advantageous to the organism within its environment or not. Finally, (d) if it is advantageous, then those individuals endowed with such ontogeny will be more likely to propagate, i.e. the probability of the survival of the organism (and hence its traits) that is in possession of such characteristics is thereby increased. Thus, as opposed to the Lamarckian scenario, extinction is highly possible and indeed, expected to occur. The fruitful point here then, is that all teleological notions are instantly irrelevant by the conjunction of (c) and (d). Additionally, (a) and (b) allow for an actual mechanism for species change.

If Darwinian evolution is a scientific theory then it needs to be empirically verified or at least have a strong empirical basis. Whilst there is now overwhelming empirical evidence for the model, it was not at first seen in such scientific terms. Partly, this was due to religious objections which seem to have obfuscated the very extensive empirical observations and experimentation by which Darwin was led to the theory and like Faraday, must be considered to be an exemplar of the scientific and empirical method. Another difficulty was that although a hereditary mechanism for the transmission of characteristics is proposed, it is not described in anything like the detail that would be necessary in order to escape criticism and Darwin was unaware of the work of Mendel, the adoption of which would have in large measure resolved this problem. Nevertheless, this simple model of species development has an empirical foundation through Darwin's own researches and largely explains the fossil record, albeit that the latter was incomplete at the time (and still remains so). We must therefore accept Darwinian evolution as a scientific theory.

Then, how does the Darwinian picture fare with respect to Definition 7.4.4? Recall that Darwin was well aware that he had not specified a precise physical mechanism for the hereditary transmission of biological traits and for this reason he was prepared to allow for other additional means by which characteristics of individuals may be passed to their offspring, including the Lamarckian conception outlined above. Of course, this is no longer necessary with our present understanding of genetics and we have already mentioned that *neo*-Dawinism or the Modern Synthesis is actually theory unification (§3.10) and is therefore an intermediate revolution of the third kind. But, whether or not Lamarckian or other ideas are incorporated into Darwinism as it stood in 1859 is of no consequence to Definition 7.4.4 which allows for this provided that no inconsistency results. There is then the question of the determination of new variables and axioms. As far as the latter are concerned (c) and (d) above are clearly in accordance with the definition and (a) allows identification of new variables in which these may be expressed. Thus, if x is some trait of one individual then another of the same species possesses the trait $x + \Delta x$, exhibiting a slight variation of the (but not infinitesimally different) characteristic in question, which will now be transmitted hereditarily by (b). Then we must conclude that the Darwinian Theory of Evolution does indeed satisfy Definition 7.4.4 and is therefore an excellent example of a major revolution of the third kind.

There are of course many other illustrations of revolutions in this category throughout the history of science and they tend to be first attempts at codification and empirical theory formulation as indicated earlier. We may mention the Chemical Revolution of Lavoisier *et al* that took place in the late nineteenth century. This is the oft quoted removal of the 'phlogiston' theory as the primary agent of combustion and the substitution of oxygen in that role, but the chemical revolution is much more than that and involves many individuals over an extended period of time that finally results in the determination of the major constituents of the air and the identification of the chemical elements and the periodic table. Much of this work is of an empirical nature and new variables and axiom schemes are obtained in a far more scientific way than before, which was still essentially the four-element notion of Greek science. We are oversimplifying here, but a closer examination will show that this is also a major revolution of the third kind, followed by many minor and intermediate revolutions. We see now in this example that revolutions are not entirely independent.

We have discussed plate tectonics earlier in the context of a different revolution category, but there was also the 'geological revolution' beginning around 1780 in which the new empirical methods were extensively employed and many geological features and mineralogical characteristics determined, leading finally to a scientific view of the Earth that may have been incomplete, but at least had an empirical basis. This too was a major revolution of the third kind and as with the chemical revolution and Darwinian evolution demonstrates one of the special attributes of such revolutions, namely that they tend to be (but not always) the first scientific and properly empirical formulation of a theory within a previously ill-defined and mystically imbued discipline. We shall return to this point later, first let us take stock of what we have attempted to describe so far in this chapter.

7.5 Revolution taxonomy

We have now illustrated three categories of scientific revolution and in each category there are three distinct types. It cannot be denied that this classification is not free from arbitrariness and many other schemes are possible. Indeed, we could easily extend these categories to cover any possible eventuality that may be identified in the development of a scientific theory or perhaps even reduce the number that we have. But this does not matter, for our purpose is to examine how scientific theories evolve and to some extent the specific taxonomical structure is of secondary importance provided that the main features and characteristics of theory-change have been captured. Our classification scheme is intended to be reasonably representative of the most salient points in theory evolution. We may summarise our discussions so far in the following table:

REVOLUTION TYPE	CONSTANT SYMBOLS	VARIABLE SYMBOLS	AXIOM FORM	DOMAIN OF APPLICABILITY	FUNDAMENTAL ASSUMPTIONS	AXIOM COMBINATION
Minor I	change	no change	no change	no change	no change	inapplicable
Minor II	change	change	no change	no change	no change	inapplicable
Minor III	no change	no change	change	no change	no change	inapplicable
Intermediate I	no change	no change	change	increased	no change	inapplicable
Intermediate II	change	change	change	increased	no change	inapplicable
Intermediate III	no change	no change	change	increased	no change	yes
Major I	change	change	change	increased	change	inapplicable
Major II	no change	no change	change	non-decreasing	change	inapplicable
Major III	new	new	new	inapplicable	change	inapplicable

It is also useful to have a single symbol for each type of revolution. Hence, we shall write R_n^T for a revolution in category $T = \{m, I, M\}$ of type $n = \{1, 2, 3\}$ where m = minor, I = Intermediate and M = Major. In this notation an intermediate revolution of the second kind for example would be written R_2^I and similarly for the others. We note that axiom combination occurs only in the R_3^I case, which is to be expected since it is only necessary when theory unification is dominant.

As mentioned, the classification given here is not the only one possible, but any reasonably comprehensive alternative system must obviously include the essentials of our description. Clearly, we could refine the above by the introduction of many more categories, where each additional type or sub-type of revolution would describe just one small facet of theory-change on a finer and finer scale. This would, of course, require an equal number of definitions of scientific revolutions since the taxonomy is determined by that definitional choice and each one of these new definitions would relate to an alteration of some individual characteristic of a theory such as a constant or variable symbol, numerous types of axiom change, axiom combination and fundamental assumption change, etc. The resulting list would certainly be a long, but because one is dealing with discrete entities would at least be finite.

However, to proceed along such lines would render the meaning of 'scientific revolution' so trivial as to make that designation seem to be quite inappropriate. In essence, we expect a revolution to be a significant event, indeed, even a milestone or turning point in the attainment of scientific knowledge in relation to the history of that endeavour. Nevertheless, the very fact that we may 'fine grain' scientific revolutions through the inclusion of additional categories tells us a great deal about the acquisition and growth of science in connection with the empirical method as it is employed by particular biological or mechanistic investigators and the dependency of the technique with regard to the limitations of such researchers as placed upon them by their own constitution and the nature of their interaction with the external world. This would appear to suggest that scientific revolutions are somehow species dependent or even that the scientific enterprise harbours en element of subjectivity. However, we see that this is not entirely true, for we expect that any sufficiently sentient organism endowed with the appropriate level of intellect and curiosity will adopt the scientific method in the pursuit of understanding the Natural World and these are the only criteria necessary for so doing. It is rather that technological progress enables each species to improve on the empirical data available and it is the speed at which this may be achieved that results in a revolution in the first place, as we have frequently observed in the foregoing. Therefore, any species dependency on scientific revolutions will be due to this limitation, itself variable from organism to organism and unrelated to the empirical method *per se* except insofar as technological enhancements and methodological refinements facilitate a more expeditious or efficient employment of the scientific method.

Accordingly, the taxonomy of revolutions will be highly dependent upon the technology available for the investigation of the physical world at any given time and one may ask whether scientific revolutions may therefore vary in both frequency and magnitude, or indeed not occur at all. Such questions are not merely philosophical, for there is a definite pragmatic component here that is important from a scientific as well as an epistemic viewpoint insofar as the latter has practical applications in the search for knowledge. These and similar issues are discussed in greater detail in the next chapter.

Chapter 8

Revolution Evolution

The relationship between different types of revolutions is examined and the significance of each considered in the development of a particular scientific (sub)discipline. A hierarchical structure of revolutions is posited and specific sequences of revolutionary progress imagined. The ontology of scientific revolutions is discussed in relation to the mechanisms of theory change. The frequency of revolutions within a given revolution sequence is examined. We next consider a graphical representation of sequences of scientific revolutions and envisage socio-economic factors relating thereto.

8.1 Revolution hierarchies

We have observed that scientific revolutions fall into various categories and may be further subdivided into numerous types or *kinds*, as we have called them, within a given such category. Clearly, there is some sort of hierarchical structure intended by the construction of this particular taxonomy which has presumably been devised in order to describe in a reasonably representative manner, the processes and progress of science as it is actually experienced and practiced from a pragmatic viewpoint. This is indeed the case, but as indicated in §7.5 the system of classification given is designed to reflect the general historical development of science as it has been experienced by human kind, with all the empirical restrictions and psychological baggage which may accompany that condition. Such limitations in no way imply a lack of objectivity in the methods of science, or any sociological or psychological content in theory development whatsoever, but they do suggest the present or a similar taxonomy for scientific revolutions. It is therefore appropriate to consider the relative significance and order of precedence for the revolutions described thus far in hierarchical terms.

The first thing to note is that an R_3^M revolution in a particular scientific area must precede all others, for it does not modify or replace an existing scientific theory. Such a revolution presages the first (empirical) scientific theory in the discipline concerned and if it is to supersede anything, that which it replaces is merely speculative, mythological or religious and certainly not scientific. Then, in terms of precedence, R_3^M comes first and can only happen once and not on subsequent occasions, for if it did then the theory would have to be completely abandoned, in which case there could hardly have been an empirical foundation in the first instance. Hence, once R_3^M has occurred, it is at some stage followed by one of the eight remaining types of revolution, viz;

$$\{R_1^m, R_2^m, R_3^m, R_1^I, R_2^I, R_3^I, R_1^M, R_2^M\}$$

But of course, R_3^I could only follow if there is another discipline with a scientific theory that can be unified with the now existing R_3^M theory and usually only minor revolutions are likely at this stage. In fact, as discussed earlier, we expect major revolutions to occur less frequently than intermediate revolutions, which in turn are less frequent than minor revolutions. The reason of course is the empirical process which drives such revolutions, requiring greater and greater sophistication as we progress from minor to major. This then, is the hierarchy that we envisage, at least in human scientific history. The rarest event is R_3^M and the most frequent is R_1^m. However, we do not insist upon a strict ordering in terms of the frequency of the other six revolution types because a great deal depends upon the empirical data available and the technical development of the scientific apparatus and data acquisition methods extant at any given time as well as the number of practising investigators, which is also, at least for organised biological entities such as *homo sapiens*, a function of the prevailing economic and political climate. These are further restrictions placed upon the speed of scientific progress that are specific to the human endeavour and possibly applicable to other social species also, but have nothing to do with the scientific method or how the latter works in practice, for only the *rate* of development is curtailed by such limiting factors. Thus, there are no sociological contributors that could be said to influence the growth and development of scientific knowledge *per se*, for the latter proceeds through the empirical method via the natural interpretation which is entirely objective and independent of such mundane considerations. Consequently, one may imagine a scenario in which virtually infinite resources are available. In such case scientific revolutions would be commonplace since the rate of progress would be proportionately increased. It would then be difficult to even define what a revolution would be, for now the term loses its essential meaning. We shall meet this situation later.

Suppose that a scientific theory S_1 is formulated in the appropriate empirical manner. This is a revolution of type R_3^M and we may denote this symbolically in the context of S_1 by $R_3^M [S_1]$. There will then be a sequence of further revolutions $R_n^T [S_1]$ that lead to the current form of the theory. Throughout this process one is of course employing the methods of theory-change and as always, the natural interpretation obtains. We may illustrate the hypothetical development of S_1 schematically as follows;

$$R_3^M [S_1] \xrightarrow{t_1} R_1^m [S_1] \xrightarrow{t_2} R_3^m [S_1] \xrightarrow{t_3} R_1^I [S_1] \xrightarrow{t_4} R_2^m [S_1] \xrightarrow{t_5} R_2^I [S_1]$$

as a possible sequence of revolutions, with no pretensions to reality at present. Let us briefly describe this scenario. Obviously, we begin at the initial time t_0 with a major revolution of the third kind which results in the determination and formulation of the theory S_1. Axioms and variables have been empirically obtained via the natural interpretation and are therefore in accordance with the phenomena described by the theory. Then, after a period of time t_1 during which scientific investigation has ascertained more precise values of the constants of the theory, a minor revolution of the first kind results. Nothing pertaining to the theoretical structure of S_1 is yet altered, for only constants are re-determined thus far. Hence the theory is merely empirically refined only with regard to the values assigned to the latter. However, after a time t_2 has elapsed a minor revolution of the third kind transpires. Now, at least one of the constants of S_1 has become a variable and additionally, one or more of the axioms are altered, but the domain of S_1 is as yet unchanged. That situation does not persist, for after a further time t_3 the theory undergoes an intermediate revolution of the first kind in which the axioms are further altered or replaced and now the domain of applicability is increased. Next, in time t_4 comes a minor revolution of the second kind with the re-interpretation of one or more of the constants of S_1 as variables but the axioms and domain remain unchanged. Finally, this is followed after time t_5 by another intermediate revolution, now of the second kind. New variables and axioms enter S_1 and again the domain is further enlarged.

Clearly, the theory S_1 has radically changed since its first conception, as it must if there are to be scientific revolutions. It began of course with a major revolution of the third kind and subsequently underwent all three types of minor revolutions and the first two types of intermediate revolution. However, no intermediate revolution of the third kind R_3^I is possible without another theory in a related discipline (but different domain) with which unification can take place. If S_2 is such a theory with a revolution sequence;

$$R_3^M[S_2] \xrightarrow{t_1} R_2^m[S_2] \xrightarrow{t_2} R_1^I[S_2] \xrightarrow{t_3} R_1^m[S_2]$$

we may imagine the two sequences for S_1 and S_2 converging with the unification $S_3 = S_1 \sqcup S_2$ so that the last term in both sequences is;

$$\mapsto R_3^I[S_3]$$

where the final unified theory S_3 results from a revolution R_3^I. Of course this is not necessarily the end of the story, for the new unified and presumably by now highly developed theory may and generally will continue to undergo subsequent revolutions. Obviously, many other variations are possible and the route to S_3 can be achieved in several ways. Hence, this example merely suggests one particular sequence of revolutions and is therefore arbitrary in that sense. The sequence that will actually obtain is dependent on the factors mentioned above and especially the rate of technological development of the necessary measuring apparatus so essential to the very practice of empirical science. Even so, that in itself does not exclusively determine the precise sequence of revolutions that may occur in a given situation or domain of enquiry, for there must also be a serendipitous element involved, as is made abundantly evident in the case of the human experience in the field of scientific discovery. With regard to the latter point, we note that accidental or fortuitous scientific progress can only be a consequence of imperfections in the execution of the scientific method, for in principle they should not otherwise occur.

With regard to the above schemata of a sequence of scientific revolutions, a more perspicuous and picturesque representation of this special case may be presented as in the following diagram where now we specifically show two revolution sequences with two different theories converging to form a new, unified theory;

$$R_3^M [S_1]$$

$$\downarrow$$

$$R_1^m [S_1]$$

$$\downarrow$$

$$R_3^m [S_1] \qquad R_3^M [S_2]$$

$$\downarrow \qquad \qquad \downarrow$$

$$R_1^I [S_1] \qquad R_2^m [S_2]$$

$$\downarrow \qquad \qquad \downarrow$$

$$R_2^m [S_1] \qquad R_1^I [S_2]$$

$$\downarrow \qquad \qquad \downarrow$$

$$R_2^I [S_1] \qquad R_1^m [S_2]$$

$$\searrow \qquad \swarrow$$

$$R_3^I [S_{12}]$$

$$\downarrow$$

$$R_2^m [S_{12}]$$

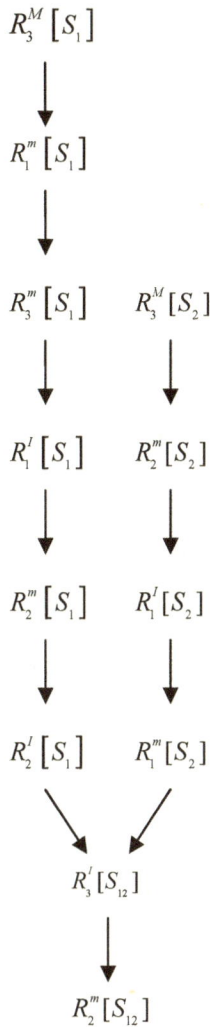

where now a minor revolution of the second kind on the theory $S_{12} = S_1 \sqcup S_2$ follows the unification of S_1 and S_2 We shall generally adhere to this notation and denote a unified theory by a subscript which is the concatenation of the subscripts of the component theories. Any number of further unifications and revolutions may subsequently take place until an entire scientific discipline is brought together into one single theory, but of course, although such a situation may be highly desirable, the matter may not end there and revolutions can continue indefinitely.

In the general case a sequence of revolutions in a given category T and of types $i, j, k, \ldots$, may be written as;

$$R_i^T \xrightarrow{t_1} R_j^T \xrightarrow{t_2} R_k^T \xrightarrow{t_3}, \ldots\ldots, \xrightarrow{t_N} R_n^T$$

Henceforth, we omit reference to the particular theory S where possible. Here, the time intervals t_x between each revolution vary in the manner that we have discussed briefly above and are dependent upon the factors relating to species capability in the technological and empirical sense as well as the political and sociological aspects described. However, for the reasons also mentioned in many places formerly, there is no psychological component involved in the scientific process *per se*. But now it may seem that there is a difficulty in relation to the objectivity of science, for how can there be objectivity when there is species dependency? In fact no problems of this kind actually arise, for we are not talking about the scientific method as such, but just the *rate* of scientific revolutions which is clearly a different issue and *cannot* be divorced from the technology employed by the given species of empirical investigators. This raises some interesting issues in the development of science as it has sometimes been portrayed in the literature, especially in relation to the classification and nature of scientific revolutions and we must now address this question lest any misunderstandings or ambiguities remain unresolved.

8.2 Revolution ontology

The traditional view of the development of science has generally consisted of two primary but apparently disparate and conflicting scenarios. The first of these is that scientific theories gradually unfold slowly and cumulatively via the application and benefit of the empirical method, at each stage further refining the existing model in an almost but not quite, continuous fashion. In this picture, if there is a radical change in the theory or departure from the current scientific outlook, as for example with relativistic or quantum physics, then that is simply an inevitable consequence of the gradual accumulation of empirical knowledge, which is thus seen to merely exemplify and confirm the scientific method. True, there *are* revolutions here, but when considered in their proper historical scientific context and in the light of the cumulative acquisition of empirical data as an ongoing process, such revolutions whilst representing significant historical 'turning points' in the growth of science are depicted not as events that are unrelated to the *status quo*, but rather as an integral part of the entire scientific endeavour. In this way revolutions remain revolutions but are not *revolutionary* insofar as a connection with earlier notions, ideas and scientific theories is lost. Nor are they to be considered as exceptional from an ontological perspective, but just simply the natural outcome of the empirical method. Evidence for this is our previous discussion of correspondence principles, which must therefore be of paramount importance on this view of science. Hence, whilst revolutions do indeed take place and are acknowledged as such, they are nothing more than a requirement of the scientific method and furthermore could have easily been anticipated if only one had closely observed the direction in which the given scientific field had been progressing. In this sense revolutions are not surprising, but may well seem to be so to practitioners at the time.

In apparent contrast to this picture an alternative understanding of scientific progress may be envisaged in which revolutions are *really* revolutionary. They are momentous events whereby old ideas are discarded with the development of new, very different notions. But this does not transpire within an entirely empirical context, for sociological and psychological elements also play a part. On this view, it is not that scientific revolutions occur because of non-empirical reasons, for that would be absurd. Rather, the onset of a revolution is hindered by the prevailing consensus of the *status quo* which resists radical revision of existing scientific theories that have clearly passed extensive empirical tests. It is the latter that constitutes the sociological and psychological influences in this scenario. Yet, the method of science is not thereby denigrated or derogated by this circumstance and indeed, the scientific method is still of fundamental importance. It is the *rate* of development that is affected by such non-empirical factors, as we have already indicated. However, revolutions so described are only revolutionary in the sociological context, from which they are inseparable and a purely logical viewpoint would clearly demand a different empirical status and ontology for scientific revolutions, as we have seen.

We can now more perspicuously obtain a realistic picture of scientific revolutions and their ontology. From the last section we have noted a species dependency on the frequency of revolutions which would seem to suggest that there is some element of truth in the latter scenario presented above. On the other hand, we have also emphasised the value of the empirical process itself as the cause of scientific revolutions. Pains have been taken to observe that it is the limitations of extant scientific apparatus and data acquisition methods for a given species that places such restrictions on knowledge development in relation to the physical world. Additionally, we have realised that the frequency of revolutions is a function of both these factors. We therefore may be inclined to think that each of the above scenarios might simultaneously pertain and although seemingly disparate and irreconcilable on a first encounter, each might contain, at least in part, a representation of the actual state of affairs. We might therefore seek a compromise in which we can combine elements from both pictures, however insufficient or unsatisfying such a prospect may seem to be.

In fact, it is not necessary to attempt a reconciliation of the two pictures of scientific revolutionary development presented above for the following reasons. We have seen that the frequency of scientific revolutions is species dependent and therefore already contains non-empirical factors in a similar way to the second scenario but with the psychological component replaced by political and economic modifiers that effect a similar decrease in the revolution rate. Thus, it may well *appear* that scientific revolutions occur less frequently than would be expected and are therefore attributed greater significance than would otherwise be the case, thus artificially presenting a picture of such revolutions that gives the impression of periodic radical reform in the sciences. Without these elements we would expect scientific knowledge to gradually and cumulatively grow as in the first scenario until such time as the empirical data reveals new and deeper structure into the workings of Nature. These two descriptions of scientific revolutions are not of course the only ones possible, or even particularly accurate portrayals of current thinking in the philosophy of science. Our purpose is merely to capture the essential ideas of two extreme possibilities, of which there are many variants. The basic tenets of the one, namely cumulative progress is then contrasted with those of the other, which is sudden revolutionary change. For us, neither and both of these positions are true, for revolutions are relative first to the particular area of the scientific discipline concerned and then to the specific species and their individual (but not empirical) *modus operandi*.

If we examine the structure of a particular revolution we find that since it consists of nothing more than applications of the mechanisms of theory-change (Chapter 3), or the various theory-change operators (Chapter 4) and since also these are incremental processes, then in such terms a scientific revolution is also an incremental process, which seems to suggest that a cumulative and gradual modification of scientific theories is all that is possible. But evidently, such a view would be incorrect, for the action of a theory-change operator can be quite devastating in its own right and although incremental in the sense that an axiom is removed, added, replaced or modified, just as with revolutions, the magnitude of the change to a theory or its domain of applicability may be very great indeed. Then, incremental theory-change this may be, but it is often far from trivial and it would be more appropriate use the term *discrete* rather than incremental in this context.

In fact, all minor and intermediate revolutions can be expressed in terms of the theory-change operators or the theory evolution equation of §4.5, but the situation is rather different for major revolutions because now it is the fundamental assumptions of a theory that become subject to modification and change. Then, for major revolutions one would need a theory to be completely abandoned and then re-formulated anew since the definitions involved, as we have seen, make no reference to the axioms themselves. One may therefore be inclined to think of major revolutions as the only 'true' revolutions in the sense of the common meaning of that term, but to do so would force the exclusion of too many historical instances that at least are generally agreed to be revolutionary in their effect. We have thus felt that it is more representative and realistic to construct the hierarchy that we have, since considerable flexibility thereby ensues, but we again emphasise that nothing is written in stone. Hence, scientific revolutions have a separate ontology from scientific theories. In order to describe the situation more explicitly we need to consider the frequency of revolutions in relation to revolution type and category.

8.3 Revolution frequency

In principle, with the possible exclusion of R_3^M any revolution may be followed by any other in a revolution sequence of the type described in §8.1. Thus, suppose we consider the unlikely but perfectly possible sequence;

$$R_1^m \xrightarrow{t_1} R_1^m \xrightarrow{t_2}, \ldots\ldots\ldots, \xrightarrow{t_N} R_1^m$$

Then all that has happened is a continuous empirical re-evaluation of one or more constants of the theory by a sequence of minor revolutions of the first kind. This would seem to be nothing more than the mundane everyday activity of empirical science, hardly meriting the special status afforded to it by the designation we have ascribed. Nevertheless, the measurement of some constant c may well be the first step towards more recognisable revolutions and in fact, if the value of an important constant is sensitive to a particular model or choice of models then it can certainly attain significance as a revolution of this minor type. But there is another reason that makes R_1^m interesting and that is that it is to be considered the most *frequent* of scientific revolutions, as we have pointed out before. We must ask why this should be so and how such an assertion may be justified.

The measurement of a well-defined physical constant is clearly a fairly simple operation from an empirical standpoint and one may expect such measurements to be performed routinely as part of the investigative process, perhaps on a monthly or even daily basis if the matter of the value of the constant is of particular concern. Since measurement and data acquisition techniques can often be further refined themselves independently of any improvements in the technology available, it is to be expected that new values for physical constants will frequently emerge. It is also the case that only a minor improvement or suitable adaptation of the experimental or observational measuring apparatus will similarly yield better values of the constants concerned and hence result in more accurate data. Then, this being so for revolutions of the type R_1^m more than any other, it would be surprising if the frequency of minor revolutions of the first kind were not the greatest within our hierarchy.

How *quickly* the revolutions R_1^m occur is therefore a function of the time intervals t_i between successive revolutions of this type, or more specifically the average value of these times, which we now expect to be less than that of other revolution types. In the above implausible revolution sequence the total time elapsed between the initial revolution at t_{0_n} and the final revolution at t_N is the sum of each of these intervals, i.e. $\sum t_k$ and we may denote the mean value of this by $\tau = \langle t \rangle = \frac{1}{N} \sum_{i}^{N} t_k$. This is therefore a measure of the (average) *speed* of revolutions of type R_1^m ; or how quickly they occur. In the general case of a revolution of type R_n^T we may denote this measure by τ_n^T. Thus τ_n^T is the average value of the time intervals for a revolution of type R_n^T for a given revolution sequence. In accordance with the foregoing remarks we anticipate that the smaller the value of τ_n^T, the greater will be the likelihood of occurrences of a revolution of type R_n^T within a given sequence of revolutions, or indeed any set of such converging sequences. Hence, the probability that a revolution of type R_n^T will occur within a given revolution sequence is some function of $1/\tau_n^T$, which if $P(R_n^T)$ denotes such a probability will be of the form $P(R_n^T) = Q(1/\tau_n^T)$ for some function Q. Of course the occurrence of R_3^M is certain since no revolution sequence can begin without it and R_3^M is always the first term. If we are given some arbitrary revolution sequence;

$$R_3^M \xrightarrow{t_1} R_1^m \xrightarrow{t_2} R_1^I \xrightarrow{t_3} R_2^I \xrightarrow{t_4} R_2^m \xrightarrow{t_5} R_1^m \xrightarrow{t_6} R_2^m \xrightarrow{t_7} R_3^m, \text{etc.}$$

we may be interested in the number of occurrences of each particular revolution type or the revolution *frequency* f_n^T within this sequence. These may be conveniently displayed in matrix form as;

$$\begin{pmatrix} f_1^m & f_2^m & f_3^m \\ f_1^I & f_2^I & f_3^I \\ f_1^M & f_2^M & f_3^M \end{pmatrix}$$

and for the above example this reads;

$$\begin{pmatrix} 2 & 2 & 1 \\ 1 & 1 & 0 \\ 0 & 0 & 1 \end{pmatrix}$$

for this particularly simple case. However, merely counting the number of times a given revolution appears within a sequence is not a very useful measure, for it is dependent upon the length of the sequence and if the latter is sufficiently long, we may expect the number of occurrences of many revolution types to attain large values, from which we may learn little in relation to the specific scientific society that is represented by the magnitude of such frequencies within the sequence peculiar to that culture. We therefore take the revolution frequency to be the ratio of the number of occurrences of a given revolution to the number of terms in the sequence, thus making it independent of the length of the latter and more specific to the species concerned. If we wish to refer to the former number of terms of a given type of revolution within a revolution sequence, then we shall denote that number by N_n^T, which is therefore just a natural number. Then the total number of terms R_N in a sequence is given by the double sum;

$$R_N = \sum_{T=1}^{3} \sum_{n=1}^{3} N_n^T$$

Thus, the definition we require is;

<u>Definition 8.3.1</u>
The *revolution frequency* f_n^T of a revolution of type R_n^T for a given revolution sequence is the ratio of the number of occurrences of R_n^T within the sequence to the total number of terms of the revolution sequence, i.e. $f_n^T = N_n^T / R_n$.

Then f_n^T takes on the values $0 \leq f_n^T \leq 1$. Note that if $f_n^T = 1$ then all the terms of the revolution sequence are the same and the number of terms can be greater than or equal to unity. However, since R_3^M must always be the first term of a sequence this is the only case when $f_3^M = 1$. With this more appropriate measure the above example now takes on the values;

$$\begin{pmatrix} \tfrac{1}{4} & \tfrac{1}{4} & \tfrac{1}{8} \\ \tfrac{1}{8} & \tfrac{1}{8} & 0 \\ 0 & 0 & \tfrac{1}{8} \end{pmatrix}$$

Each such representation illustrates just one way in which scientific revolutions may evolve from the first, guaranteed term, to the last current term which thus becomes the current state of knowledge or latest scientific theory and is species dependent, for there are many legitimate possible sequences by which the final term may be obtained. The specific revolution sequence that actually does obtain reflects the socio-political, economic and empirical status of the observers involved. Thus the species dependency is expressed by these matrices and the function Q.

Thus, these *frequency matrices* are specific to the development of scientific knowledge as it pertains to an individual society, whether that is a sub-group of the same species or an entirely separate culture. They provide an insight into the rate of development and current scientific status of the given society and the scientific field concerned and this may be seen at a glance in the following way. First, one would note that if f_1^m has a relatively high numerical value, then this would seem to indicate that there has been a sustained effort on the part of the investigators to ascertain certain empirical constants, which then suggests that the technology needed to accomplish such tasks has been quickly developed. One would conclude for this scientific society that there is some urgency or strong motivation for proceeding in such an investigative and empirical fashion with the vigour indicated by the value of this frequency and also that economic funding is readily available for such research and relatively few political restrictions apply. The value of f_2^m complements that of f_1^m, for successive measurements of the constants of Nature may well result in the realisation that some of these constants are actually variable quantities, thus further emphasising the diligence of this hypothetical society if f_2^m is greater than zero. The value of f_3^m suggests analytical ability on the part of the observers along with a facility to reform current ideas according to the empirical data available. Thus, an objective, empirical and scientific approach combined with intellectual and theoretical reasoning is thereby exhibited if this frequency attains a significant value.

Values for f_1^l further extend the implications of f_3^m for now the structure of the theory is altered from an axiomatic viewpoint with the express purpose, empirically motivated of course, of expanding the domain of applicability, thus indicating a determination on the part of our investigators to carry the empirical process through to its logical conclusions and embrace new natural phenomena by the incorporation of acquired data and the axiom modification that is necessary in order to accommodate the latter. Obviously, f_2^l similarly represents the continued and serious ambition of our hypothetical society to pursue knowledge of the external world, where it is now understood that new natural variables may be identified and are necessary for the description of the physical phenomena being studied. At this stage we may expect that such a society has evolved economically to the extent that dedicated teams of scientific researchers may be sustained. Thus, we clearly have here a civilised and scientifically orientated society and it is probably the case also that the pursuit of empirical knowledge has already been of practical benefit.

When we come to intermediate revolutions of the third kind, the socio-economic picture suggested by the value of the frequency f_3^l can only be extremely informative. If $f_3^l = 0$ then we are surely led to the view that the society in question, whilst empirically active and possibly having achieved much with regard to the elucidation and determination of physical laws and axioms, has nevertheless not yet reached the stage of maturity within the disciplines concerned such that the latter may be seen as a unified whole, or at least as a single scientific discipline. The contrary is the case if $f_3^l > 0$, for now it is clear that considerable sophistication obtains in that scientific advancement has allowed once separate theories to be unified into a single entity, with all that that entails. If $N_3^I = 1$ then the first steps in this process have been taken and we may imagine that our investigators are delighted to have perceived new insights into the physical world that the unification of scientific theories can provide. Then, in cases where $f_3^I > 1$ there has evidently been a great deal of further progress in this direction. Now, the process of unification of scientific disciplines is seen to present a far more perspicuous view of both the empirical world and the scientific enterprise itself than was available hitherto. Should this frequency attain a relatively high value and even approach unity, as it may well do for human-kind, then an immediate state of theoretical understanding, empirical technical expertise and an appreciation of the fundamental structure of Nature Herself is at once implied. Given that a species has been able to ascend to such a condition and given also that a fuller and more enhanced knowledge of the Natural World would necessarily imply the unification of scientific disciplines or sub-disciplines, it is then to be expected that revolutions of type R_3^I would become increasingly predominant. Hence, the higher the value of f_3^I, the greater the number of scientific disciplines that have been brought together and unified within an ever larger domain of applicability and therefore the more scientifically sophisticated and empirically astute will be any society that can lay claim to such a significant measure of this frequency. The question of just how large N_3^I can become is therefore related to the famous quest for a 'Theory of Everything', or an ultimate scientific theory or model of all the phenomena of Nature. This is related also to the issue of when a revolution sequence terminates, if at all and is a matter that we will address in the sequel.

The value of the frequencies f_i^M for $i = 1,2,3$, are of a different kind with respect to the scientific status of a given society and we have already remarked briefly upon their ontology in §8.2. Let us first consider f_1^M in the present socio-economic-empirical context, hitherto only slightly touched upon, but now given a specific value by its entry in the frequency matrix. If $f_1^M > 0$ then obviously a conceptual change has obtained for our hypothetical investigators, but as mentioned with alarming frequency in the prequel, such changes must always have an empirical basis. The significance and nature of major revolutions of the first kind has already been discussed, but how such epistemological transitions and their frequency within a revolution sequence relate to the characteristics of the particular society concerned has not. We therefore first observe that if

$f_1^M = 0$ then no serious epistemological transition can have ensued, for fundamental assumptions have not been modified. Hence, either this society has attained a sufficient degree of knowledge of the empirical world and a suitably rich theoretical apparatus to so describe that understanding, or there is an impasse within the developmental scientific enterprise that can only be of empirical origin insofar as no envisaged observational or experimental measurements could possibly alter or improve the current state of knowledge of the community in question within the prevailing societal bounds and limitations and we have already observed that these may be for empirical reasons relating to the accuracy of available scientific apparatus, or yet may also be due to a sociological or economic cause. Hence, stagnation would therefore result and an appropriate time period is required to pass before further progress can be made. The null value of f_1^M thus implies either complete knowledge or the failure of the scientific method, neither of which is probable and particularly the latter in view of the nature of the mechanisms of theory-change, but the former can never of course, be discounted on any reasonable grounds.

Therefore, when $f_1^M > 0$ we imagine that the scientific society has, within its empirical limitations and its sociological structure, found resources to pursue, envelop and indeed, fully incorporate the fundamental philosophical principles and epistemological assumptions of science within its domain of enquiry, for otherwise no such value of this frequency would be possible. It is certainly the case that a positive and non-zero value of f_1^M results in a scientific revolution merely by definition. Such a society, now aware of the principles upon which the scientific foundations of its empirical theories are based, may at once appreciate its own condition and status with regard to the possible progress that may now be made in the scientific domain and such self-awareness could well be a pre-requisite for further revolutions, first initiated by a major revolution of the first kind which then becomes the primary deterministic element for the comprehension of such. Hence, a realisation in this way of foundational aspects concerning the structure of the Natural World is thus seen to be merely a manifestation of the empirical process, evidently guided by the natural interpretation, as must always be so. Then, the value of the frequency f_1^M represents an understanding of the workings of the scientific method that has been achieved after much deliberation over a relatively long period of time, for clearly such revolutions are not possible in any instantaneous way.

The above remarks apply with even greater emphasis should the frequency f_2^M be non-zero, for now all of the fundamental assumptions of previously cherished theories are replaced or modified. This suggests *a fortiori* that reliance upon the empirical scientific method to correctly reveal the workings of the physical world is virtually total and supersedes all other possible modes of enquiry. Any society exhibiting such confidence can only have reached that state through the previous success of the empirical method. It is therefore to be expected that even in the minimal case when $N_2^M = 1$ there has been considerable endeavour on the part of researchers in such a culture to seek an understanding of the external world over a long period of time via the empirical technique and that a number of lesser revolutions have already taken place prior to f_2^M attaining its present value. Thus, the commitment to established, tried and tested methods together with an assurance in the natural interpretation is unquestioned in this society, which is then further demonstrated if $N_2^M > 1$. However, whilst this may well be so, a high value of f_2^M would seem to imply some difficulties in the development of scientific knowledge within a given culture. Clearly, continued revision of scientific theories over a relatively short time-period suggests that suitable scientific apparatus for the measurement of natural variables and the determination of empirical axioms is insufficiently decisive for the establishment of a firm foundation for the theories in question at a fundamental level. Sociological reasons may be discounted for they are inapplicable here, since the context is empirical and methodological only. By a *short* time-period we mean that in the revolution sequence of the discipline concerned, revolutions of type R_2^M occur either adjacently within the sequence or are separated by very few revolutions of a different kind. Thus, if we have a revolution sequence;

$$R_3^M \xrightarrow{t_1} R_2^M \xrightarrow{t_n}, \ldots\ldots\ldots, \xrightarrow{t_n} R_2^M$$

consisting of constant major revolutions of the second kind, we would suspect that some problem exists with the empirical techniques employed that have led to such a sequence. Even if there are intervening terms between the successive R_2^M the same conclusion may be drawn and a high value of f_2^M is suspect in this sense. This also applies, but to a lesser extent with f_1^M and we conclude that it is the *proximity* of revolutions of these types within a given sequence that is of importance in this respect.

Each revolution type therefore has a different status and we may envisage a 'revolution topology' in which the 'closeness' of revolutions of the same kind is described (to discuss this would take us too far afield into the homological realms of algebraic topology). Before proceeding further, we need to examine the final situation with regard to the frequency of major revolutions of the third kind.

We have mentioned that revolutions of type R_3^M can only occur once and this is hardly surprising since they are the first cogent and empirically scientific formulations or replacements of hitherto non-scientific and non-empirically based ideas, possibly of a mythological or religious kind. If a civilisation has undergone such a revolution then it has taken the first tentative steps along the empirical road towards an understanding of the physical world through the natural interpretation and may thus be expected, if such a society is so inclined, to continue to refine and develop its scientific theories by the utilisation of the appropriate empirical methods and full application of the theory-change mechanisms.

In view of this, we would therefore, anticipate that the frequency is $f_3^M = N_3^M / R_n$ at all times. However, this is true only for a single linear revolution sequence. In the example given earlier we have $f_3^M = 1/8$, but when f_3^I has a non-zero value, f_3^M is increased accordingly. Thus, in the last illustration of §8.1 where unification of two sequences has taken place we see that $f_3^M = 1/6$. Of course f_3^M can never be zero, for the first term of every revolution sequence must be a major revolution of the third kind, but for every instance of R_3^I there will be an additional instance of R_3^M. Then $N_3^I = N_3^M - 1$ because the unification of one or more sequences has been realised. Hence, in a given frequency matrix, if $f_3^M > N_3^I / R_n$ we know immediately that two or more revolution sequences have resulted in a unified theory and conversely with the value of the frequency f_3^I. Thus, if $f_3^I = 0$, then $N_3^I / R_n = 0$ and since R_n can never be zero we have $N_3^I = 0$, hence $N_3^M = 1$ and we are dealing with a single revolution sequence without unification. The frequencies f_3^M and f_3^I are therefore not independent and knowledge of one entails that of the other through the above relationship. This is hardly surprising considering the definitional schemata of scientific revolutions that we have constructed and many similar results follow for this particular scenario, which in the general case nevertheless also apply. Evidently, sequences of scientific revolutions can become quite complex when unification is involved as the following figure illustrates, where we have enclosed the unified theories in boxes;

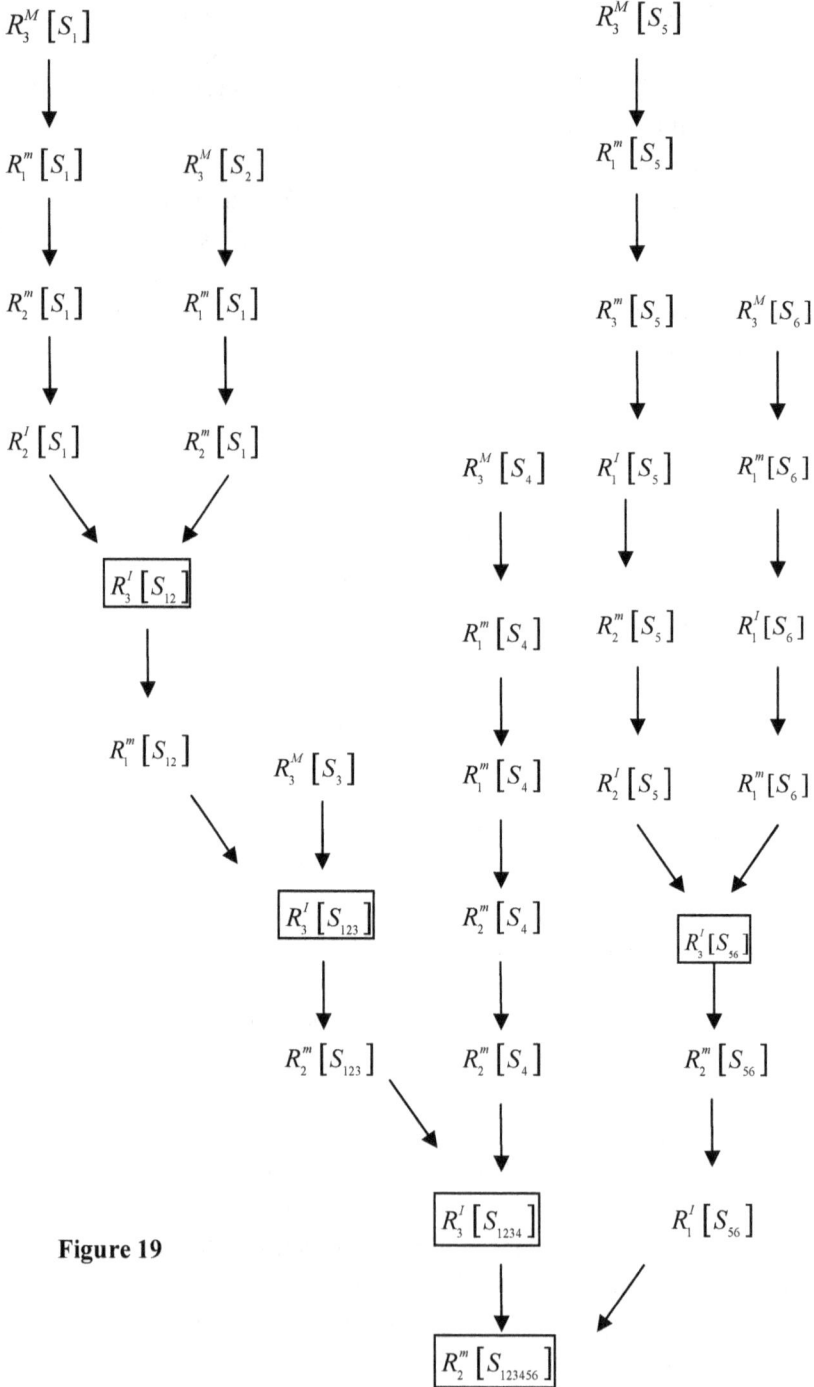

Figure 19

In Figure 19 we depict five unifications of six theories; $S_1, ..., S_6$. Each of these theories will generally come from the same scientific discipline such as physics or chemistry. The reason for this is clear. It is impossible to unify directly theories from completely separate scientific disciplines without considerable modification of one or the other, for they are unrelated either mathematically or linguistically. By this we mean that disciplines separated in this sense do not enjoy the same terms of expression even though they may seem to deal with the same broad range of phenomena. Essentially, they are conceptually disparate and have no common grounds or fundamental assumptions that can furnish a mutual foundation. Therefore, before unification of such theories can transpire, one of the disparate theories must be expressed axiomatically in terms of the other and this is just reductionism and the reason for revolution sequences in the first place and indeed, the whole idea of theory-change. Thus, reduction of theories in one discipline to those of a more fundamental kind, including the identification of similar natural variables is already incorporated into the notion of revolution sequences with theory-change and is not therefore a distinct epistemological concept. This is quite clear from even a cursory glance at Figure 19. But Figure 19, although apparently intricate, is far less complex than might be realised in a real-world situation where many extra terms and sub-sequences would be necessary. If we imagine actually drawing a diagram of the progress of physics since say, the time of Aristotle, we would certainly require a very large canvas on which to adequately depict it and would be exceedingly complicated, possible resembling a family tree in which much interbreeding has transpired[*].

Figure 19 exhibits thirty-two revolutions and five unifications resulting in the final unified theory S_{123456}. The frequency matrix for all this is;

$$\begin{pmatrix} \frac{1}{4} & \frac{7}{32} & \frac{1}{32} \\ \frac{3}{32} & \frac{1}{16} & \frac{5}{32} \\ 0 & 0 & \frac{3}{16} \end{pmatrix}$$

and as expected $3/16 > 5/32$, i.e. $f_3^M > f_3^I$ or $N_3^I = N_3^M - 1 \equiv 5 = 6 - 1$ as required. There is clearly much that can be discerned from a revolution

---

[*] It would surely be most inappropriate to liken such a diagram to the family tree of the British Royal Family and we certainly refrain from doing so.

frequency matrix with regard to the sociological and scientific proliferations of a given society, but complete knowledge is unattainable thereby. To see this, let us examine Figure 19 in greater detail. We first note that $f_1^m > f_3^M$. This is to be expected, as we have mentioned and in fact f_1^m is the highest frequency in the matrix with the value of ¼. The reason for this, we must assume, is that it is relatively easy for any culture or civilisation employing the empirical method in a deterministic way to ascertain the values of the constants within the axiomatic expressions of the theory concerned. It is therefore to some extent natural, though by no means required, that a predominance of minor revolutions of the first kind should obtain. This is certainly exhibited in Figure 19 and its frequency matrix and it may be thought that this is generally the case. However, we are hardly entitled to draw such a conclusion and may suggest only that the empirical technique merely favours a higher probability of this type of revolution. It is reasonable to expect that a revolution of type R_1^m will most likely follow one of another type, particularly a R_3^M revolution for these reasons, but only *probably* so, for as we have repeatedly emphasised, there exists a connection and dependency upon the political, sociological and economic as well as the scientific aspect of the society that contributes to the values entering the frequency matrix. There is however, much more that may be learned from a study of revolution sequences and their frequency matrices. Let us first develop these ideas in a more precise manner.

8.4 Revolution sets and diagrams

A complex diagram such as Figure 19 is clearly built from several, initially disconnected revolution sequences. We immediately know exactly how many such individual linear sequences there are from the values of f_3^I and f_3^M in the frequency matrix. Any such sequence that is contained within a diagram like Figure 19 is a *sub-sequence* or *sub-diagram* in the sense that it has the properties of a revolution sequence and the set of terms in the sub-sequence that obey those properties is a subset of the set of terms within the whole diagram. We need therefore to define these properties, if they are not already abundantly clear.

Given a set of revolutions of various kinds;

$$\mathbb{R} = \{R_1, \ldots\ldots\ldots, R_n\}$$

where each R_i is one of the nine revolution types for a given theory or set of theories, then the elements $R_i \in \mathbb{R}$ of this *revolution set* satisfy the following relational properties in a revolution diagram:

1) for every $R_i \in \mathbb{R}$, $R_i \overset{t}{\longrightarrow} R_i$ (*reflexivity*)
2) if $R_i \overset{t}{\longrightarrow} R_j$ and $R_j \overset{t}{\longrightarrow} R_i$ then $R_i = R_j$ (*anti-symmetry*)
3) if $R_i \overset{t}{\longrightarrow} R_j$ and $R_j \overset{t}{\longrightarrow} R_k$ then $R_i \overset{t}{\longrightarrow} R_k$ (*transitivity*)

which simply defines a partial order relation on the revolution set $\mathbb{R}$. Thus a revolution diagram is nothing more than a partial order over the set of revolutions which is in fact, temporal according to the relation $\overset{t}{\longrightarrow}$. Hence, a revolution diagram $\mathbb{D}$ is simply a *poset* $\mathbb{D} = (\mathbb{R}, \overset{t}{\longrightarrow})$ under this order relation. In fact, revolution diagrams are a little more than this.

Many elements of Figure 19 are not comparable, but the figure is composed of several sub-diagrams each of which satisfy the stronger condition that for any pair R_i, R_j within such a sub-sequence we have either $R_i \overset{t}{\longrightarrow} R_j$ or $R_j \overset{t}{\longrightarrow} R_i$. Thus, these sub-diagrams are *totally* ordered. In other words, although any subset of $\mathbb{D} = (\mathbb{R}, \overset{t}{\longrightarrow})$ will naturally inherit the partial order of $\mathbb{D}$, some of them are totally ordered. Actually of course, revolution diagrams are *graphs* of a special sort. Each particular revolution is then a vertex in such graphs and since there is a specified direction according to the partial ordering, they are directed graphs or *digraphs* and the directional arrows $\overset{t}{\longrightarrow}$ are then referred to as the *edges* of the graph. Thus the vertices are connected by the edges. Furthermore, there are no 'loops' or cycles in revolution diagrams (or graphs); i.e. no vertex is connected to itself. Additionally, such graphs are *simple* for there are no multi-arcs connecting the vertices. Thus revolution diagrams are acyclic and all vertices (of the underlying graph – i.e. ignoring direction) are connected, which is the definition of a (directed) *tree**. There is a profusion of terminology within Graph Theory, but only a few concepts and properties need concern us here. We shall therefore continue to refer to these pictorial depictions as revolution diagrams or revolution sequences, but occasionally referring to *revolution trees* when it seems more appropriate to do so, especially when dealing with sub-diagrams, introducing additional terms only when needed or concision so requires.

---

* Thus a tree is just a one dimensional simplicial complex which is both connected and simply connected.

We have noted above that a complex revolution diagram such as portrayed in Figure 19 contains within it totally ordered sub-diagrams such as appear *prior* to unification, as for example each of the sequences describing the revolutions for the theories $S_1, ..., S_6$. These sub-diagrams are trees in themselves and hence the building blocks of the full diagram, which as already indicated, is also a tree. Then, the full tree diagram can also be construed as part of a larger, as yet unrealised diagram which is also a tree. Revolution diagrams are always tree diagrams just because they are connected and acyclic. The latter property follows automatically since a loop or cycle in a revolution diagram for some revolution R_n^T of a scientific theory S such as;

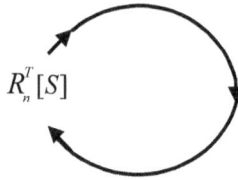

$$R_n^T[S]$$

would clearly be an absurdity in the context of scientific revolutions since theory modification or replacement is impossible if it is the same theory that results thereby.

There are of course many possible sub-graphs in a revolution diagram but of special interest to us are those that precede unification, for these represent the progress of a scientific discipline before it lost its independence by that process. Hence, the identity of a theory is changed by unification which is thus the amalgamation of two or more graphs or revolution trees with the addition of a vertex and an appropriate number of edges to incorporate that union. With revolution diagrams or trees each vertex except those that indicate unification have just one edge, which are just the arrows of the partial order relation as depicted in our previous drawings. The *degree* of a vertex is the number of such arrows or more properly edges, that emanate from that vertex. Therefore, the degree of a vertex will represent the number of scientific sub-trees that have thus combined under the unification process and is indicative of the evolution of the individual theories concerned up to the point of unification, but it is important that the entirety of any revolution sub-tree or sub-graph be presented prior to the vertex of unification so that the whole revolution sequence may be seen before unification becomes operative. This is just a useful pictorial way of representing the evolution of scientific theories, but by the use of Graph Theory all the theorems and established facts relating thereto become at once available. This then, opens up an entire area in which not only the structure of scientific revolutions may be more properly explored but as alluded to earlier, the important aspect of the topology may also be elucidated. Returning to Figure 19, we may list those revolution sequences for each of the theories S_i prior to the first instances of unification:

$$R_3^M[S_1] \xrightarrow{t_1} R_1^m[S_1] \xrightarrow{t_2} R_2^m[S_1] \xrightarrow{t_3} R_2^l[S_1] \xrightarrow{t_4}$$
$$R_3^M[S_2] \xrightarrow{t_1} R[S_2]_1^m \xrightarrow{t_3} R_2^m[S_2] \xrightarrow{t_4}$$
$$R_3^M[S_3] \xrightarrow{t_1}$$
$$R_3^M[S_4] \xrightarrow{t_1} R_1^m[S_4] \xrightarrow{t_3} R_1^m[S_4] \xrightarrow{t_4} R_2^m[S_4] \xrightarrow{t_5} R_2^l[S_4] \xrightarrow{t_6}$$
$$R_3^M[S_5] \xrightarrow{t_1} R_1^m[S_5] \xrightarrow{t_2} R_3^m[S_5] \xrightarrow{t_3} R_1^l[S_5] \xrightarrow{t_4} R_2^m[S_5] \xrightarrow{t_5} R_2^l[S_5] \xrightarrow{t_6}$$
$$R_3^M[S_6] \xrightarrow{t_3} R[S_6]_1^m \xrightarrow{t_5} R_1^l[S_6] \xrightarrow{t_6} R_1^m[S_6] \xrightarrow{t_7}$$

where now we have indicated the specific time sequence $t_1 < t_2 <, \ldots, < t_8$ for the progress of each theory. Thus the unified theory S_{12} appears at time t_4 whilst those of S_{123} and S_{56} both occur at t_6. Next, the theory S_{1234} appears at the late stage of t_8. For the second stage of unification we have the revolution trees:

$$R_3^l[S_{12}] \xrightarrow{t_5} R_1^m[S_{12}] \xrightarrow{t_6}$$
$$R_3^l[S_{123}] \xrightarrow{t_7} R_2^m[S_{123}] \xrightarrow{t_8}$$
$$R_3^l[S_{1234}] \xrightarrow{t_9}$$
$$R_3^l[S_{56}] \xrightarrow{t_7} R_2^m[S_{56}] \xrightarrow{t_8} R_1^l[S_{56}] \xrightarrow{t_9}$$

which finally results in the theory S_{123456} at time t_9. The length of these sub-trees tells us something about how quickly a theory has been developed in relation to the accompanying theories. Thus, theory S_5 underwent six revolutions before its unification with S_6, which needed only four revolutions. But S_5 begins at t_0 with a major revolution of the third kind whereas S_6 starts off later at t_2, suggesting that either progress on S_6 was an easier matter than with S_5 or that greater effort or funding was available for work on S_6 and similar remarks apply to the other sequences. Note that the time intervals between revolutions are specific to the theory in question and we do not expect that in general t_i for S_i is equal to t_i for S_k, for that would imply that each of the theories $S_1, \ldots, S_n$ are undergoing scientific revolutions simultaneously. However, we *do* suggest that the initial times t_0 of the first major revolutions R_3^M for each theory take place with respect to those of the others at the times indicated both by their position in Figure 19 and the above revolution sub-trees of that graph. It is therefore the number;

$$R_N[S_i] = \sum_{T=1}^{3} \sum_{n=1}^{3} N_n^T$$

in a *pre-unification sub-tree* of the theory S_i that is significant when considering the relation of one theory to another and their relative development. We have thus singled out these particular sub-trees for special consideration.

It is quite possible that several theories may be unified all at once, but in the example of Figure 19 no more than two theories are ever unified simultaneously and hence each vertex representing theory unification or more properly in our context, intermediate revolutions of the third kind, is of degree two. Then, since from the frequency matrix we see that there are five such revolutions, there must be in the figure $5 \times 2 = 10$ pre-unification sub-trees and in general when only revolutions of two theories are allowed at any given time there will be $2N_3^I$ pre-unification sub-trees. We also know from the frequency matrix the number of theories involved since this is just N_3^M. The question therefore arises as to whether it is ever possible to unify more than two theories at the same time for surely, given say three theories S_1, S_2, S_3, one could obviously form the theory unifications $S_1 \sqcup S_2$, $S_1 \sqcup S_3$, or $S_2 \sqcup S_3$, which could then each be unified with the remaining theory. This, of course, is just a combinatorial process and given n theories one has $_nC_2 = n!/2!(n-2)!$ possibilities of such pairwise unifications. Thus, in the case of the six theories of Figure 19, we have $_6C_2 = 15$. In the more general case of unifying n theories r at a time there are clearly $_nC_r = n!/r!(n-r)!$ ways in which one may do this. However, it is not so immediately clear what is meant by this simultaneous unification of theories when $n > 2$, for surely such a procedure is a multi-step process in which one would first unify two theories and then adjoin additional theories thereafter. Then, unification would never involve more than two theories and every revolution tree would be binary – each unification vertex being at most of degree two.

However, it would be unreasonable to assume that theory unification must always be a binary operation, for that is merely the human experience and other intelligent beings of a different species may well be capable of multiple simultaneous unifications of their scientific theories. There is also the time element between successive unifications and herein lies the difficulty mentioned above. At what point should we consider the unification of more than two theories to be simultaneous? Significant scientific revolutions are often thought to be at least a generational and often a centurial phenomenon and certainly revolutions of type R_3^I would normally be accorded such rarity. Yet, we have seen that the rate of scientific revolutions is not only species dependent, but also bears a functional relationship to the socio-economic factors that may pertain at any given time within the culture of that species, which in practical terms demands a species dependency of the interval between revolutions. In the special case of R_3^I when n theories are unified those factors are particularly pertinent, for if n scientific theories are unified at a given instant, we would need to define what is meant by 'instant' in this context. Consider the simple case of three theories and the following sub-graphs of a possible revolution diagram for unification;

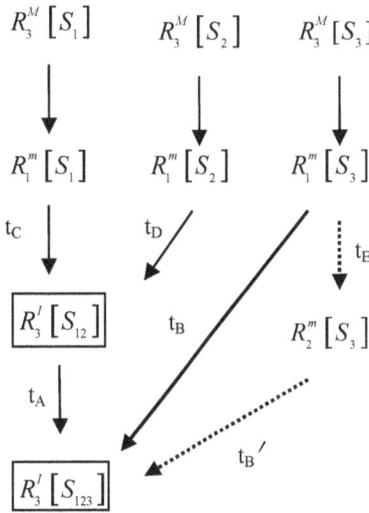 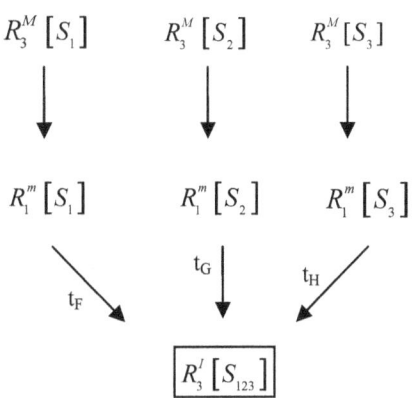

Figure 20   **Figure 21**

In Figure 20 (ignoring the dashed edges for the moment) the three theories are unified in a two step process, first S_1 and S_2 to produce S_{12} which is subsequently unified with S_3 to produce the final theory S_{123}. This is just the normal procedure that we have discussed above and each of the unification vertices is of degree two. Quite the contrary is the case with

Figure 21 where the three theories are unified directly with no intermediate stage and hence there is just one unification vertex of degree three. In Figure 20 we have labelled the edges of the final unification resulting in the theory S_{123} by the time intervals t_A and t_B which are supposed to have elapsed prior to the concluding revolution. Since theory unification occurs at some given instant of time we must have $t_A = t_B$ for if $t_A > t_B$ or $t_B > t_A$ then one would be unifying two theories at different times, which is absurd. Thus $t_C = t_D$ and in Figure 21, $t_F = t_G = t_H$. We may say that whenever an intermediate revolution of the third kind takes place, then the edges in the directed revolution diagram leading to the vertex of that revolution represent equal times. All this of course, is rather obvious since if theories are to be unified then such unification could hardly occur at different times, for then it would not be unification, as already noted. However, there may well be a time delay before unification becomes possible. Effectively, this means that a theory is not yet sufficiently developed, or has not been cast into an appropriate form for theory unification, which may be due to a lack of suitable empirical information or axioms, or difficulties with the axiom combination process. Then in such case further development of a theory becomes necessary before a revolution of type R_3^l can ensue. This is illustrated specifically by the dashed edges in Figure 20, where the theory S_3 undertakes an alternative path by undergoing a revolution of type R_2^m before finally being unified with the already unified theory S_{12}. Then, now we have instead that $t_A = t_B'$ and the delay in this revolution is represented primarily by t_E if $t_E > t_C$ as we have assumed and inversely will be caused by a lack of progress in the theory S_1 if $t_C > t_E$. Evidently, such a situation obtains in every complex revolution diagram and glancing again at Figure 20 we observe two more such instances. This accords particular significance to intermediate revolutions of the third kind and in the sequel we discuss this point more fully.

Given a revolution diagram, the cardinality of its revolution set $\mathbb{R}_1$ is just the number of vertices in that diagram and if we have another such diagram $\mathbb{R}_2$ with the same number of vertices, i.e. $card(\mathbb{R}_1) = card(\mathbb{R}_2)$, then we may enquire as to whether there is any sort of structural relationship between them. Clearly, in this case it is possible to define a map that takes each vertex of $\mathbb{R}_1$ into a corresponding vertex of $\mathbb{R}_2$ and *vice versa*, so that there is a bijection between the two diagrams. If two diagrams are to be in some sense structurally equivalent then evidently they need first to be vertex bijective in this way but of course, whilst such a condition is necessary it is hardly sufficient for we must also consider the other structural properties of a revolution diagram and in particular the edges or 'arrows' as we have sometimes called them. Recall from Graph Theory that the *degree* (or valence) of a vertex is the number of (proper) edges incident upon that vertex. Since revolution diagrams are simple graphs, the edges are always *proper* because they connect distinct vertices, which is the meaning of that term and with which we may now dispense. It would therefore certainly be expected that any structure preserving mapping should maintain the degree of a vertex. Now, *adjacent* edges in a graph are two edges that have a common endpoint, otherwise they are *non-adjacent*. Hence, we anticipate structural equivalence between two graphs when there is a bijective mapping that preserves vertex degree and adjacency and non-adjacency of the edges. This is to say that given two graphs G, H and a vertex bijection $f : G \to H$ then vertices v_1, v_2 are adjacent in G if and only if $f(v_1), f(v_2)$ are adjacent in H and similarly for non-adjacency. Then in such case the *simple* graphs G and H are *isomorphic* and the vertex bijection *f* is *structure preserving*. As usual one writes $G \cong H$ to indicate such an isomorphism. Additionally, if we drop the requirement of bijectivity, then we may define a *graph homomorphism* as a pair of mappings, one for the vertices f_V which preserves adjacency and another f_E for an *induced* mapping of the edges such that $f_E(v_1 v_2) = f_V(v_1) f_V(v_2)$, but, for the present, we are concerned primarily with graph isomorphisms.

However, this is still not sufficient to fully describe an isomorphism of a revolution diagram, for recall that the latter are digraphs and therefore we would require that the direction of each edge should also be preserved. Thus, given an isomorphism f between the underlying graphs of two revolution diagrams we expect additionally that if for any edge e from say v_1 to v_1, then $f(e)$ is directed from $f(v_1)$ to $f(v_2)$ and conversely.

We have observed that revolution diagrams are more than simple digraphs; they are also (inverted) trees and furthermore, *rooted* trees. A tree is rooted if there exists a unique vertex, the *root*, such that all edges are directed away from the root. Thus a rooted tree may be pictured as;

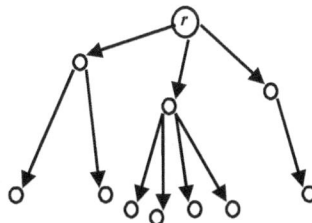

and our abuse of terminology becomes clear, for the vertex **r** is the root of the tree and all the edges are directed *away* from the root. However, for us a revolution diagram is an inversion of this;

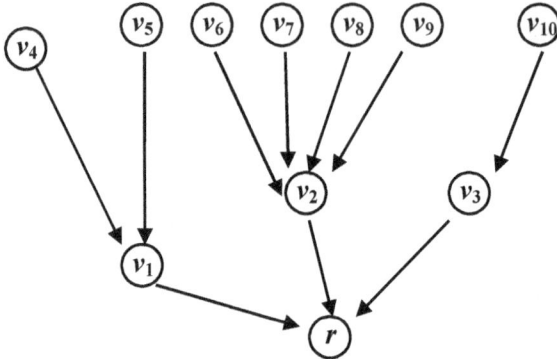

Figure 22

where now the edges are instead directed *towards* the root **r** and we have labelled the other vertices appropriately. None of this makes a great deal of difference of course, but accounts for our use of the term *revolution* diagram in our context rather than a rooted *tree*. The number of vertices and their degree obviously remains unchanged and the *vertex set* $\{r, v_1, \ldots, v_{10}\}$ is just the revolution set as discussed earlier, with the root **r** being an intermediate revolution of the third kind.

Then, given an isomorphism with all the above requirements, we also expect that the root of the tree will be preserved *as the root*, i.e. roots map to roots, although obviously the root is maintained injectively anyway. Thus, in isomorphic revolution diagrams we always have $f : r \rightarrow f(r)$ and since the vertex degree is preserved we automatically map intermediate revolutions of the third kind into their corresponding images of the same type. Then, isomorphic revolution diagrams are of identical structure and preserve both revolutions of type R_3^M and R_3^I but may represent many different revolution sequences with regard to the remaining categories of scientific revolutions. Nevertheless, these conditions certainly place restrictions upon the nature of the vertices if the revolution diagram is not too complex. Hence, in Figure 22 the vertex subset $\{v_4, v_5, v_6, v_7, v_8, v_9, v_{10}\}$ are all major revolutions of the third kind and the subset $\{r, v_1, v_2\}$ are intermediate revolutions of the third kind. Only v_3 is undetermined in this example and left open, but it cannot be mapped into either of these revolution types. Therefore, with any such mapping there are limitations in the degree of freedom with regard to the kind of revolutions that may appear in the image set and to a greater or lesser extent depending upon the complexity of the diagram. Thus, if we removed v_3 from Figure 22, then any diagram to which it is isomorphic is not only of the same structure but also of exactly the same revolution type for each corresponding vertex. In such case the two diagrams G and H are identical and we will say that they are *equivalent* and write $G \equiv H$. It is convenient to invent some further terminology here. Suppose we are given two isomorphic diagrams with revolution sets $\mathbb{R}_1$ and $\mathbb{R}_2$. They have the same cardinality, i.e. $|\mathbb{R}_1| = |\mathbb{R}_2|$ but there are three special subsets that are also of equal cardinality, namely, the subsets of each which consists of revolutions of type R_3^M, the subsets of each which consists of revolutions of type R_3^I and the subsets of each which consists of the remaining revolution types which must have the same cardinality because of the one-to-one correspondence carried by the bijection. Hence, any revolution set can more informatively be written as the union of these three subsets;

$$\mathbb{R} = \{R_{3_1}^M, \ldots\ldots, R_{3_i}^M\} \cup \{R_{3_1}^I, \ldots\ldots, R_{3_j}^I\} \cup \{R_{n_1}^T, \ldots\ldots, R_{n_k}^T\}$$

and $|\mathbb{R}| = i + j + k$. It is the latter subset of cardinality $|k|$ that is of interest here for it represents the freedom of variability of choice for isomorphic revolution diagrams. In view of the relevance of these subsets of the revolution set, it is expeditious to give them a name. We shall call the first such subset consisting of all the revolutions of type R_3^M the *formulation* (sub)set and denote it by $\mathbb{R}_F$ and the second consisting or all the revolutions of type R_3^I the *unification* (sub)set and denote it by $\mathbb{R}_U$. The third (sub)set we shall name the *equivalence* (sub)set and denote it by $\mathbb{R}_E$ so that $\mathbb{R} = \mathbb{R}_F \cup \mathbb{R}_U \cup \mathbb{R}_E$. Let us denote these sets for the graph G by ${}^G\mathbb{R}_F, {}^G\mathbb{R}_U$ and ${}^G\mathbb{R}_E$ with a similar convention for H. Then, given an isomorphisim f between two revolution diagrams G, H, if ${}^G\mathbb{R}_E = {}^H\mathbb{R}_E$ then G is equivalent to H, i.e. $G \equiv H$ and revolutions are mapped into the same revolution type, but in general this will not be the case. Thus, the equivalence set is a measure of how 'identical' two revolution diagrams are with regard to revolutions of the same type and its cardinality is also an indicator of the complexity of a revolution diagram, amongst other things which we shall discuss later. It is convenient also to write the relative complements of the two equivalence sets. Thus, the difference $^{G}\mathbb{R}_E - {}^{H}\mathbb{R}_E$ consists of those revolutions (elements) of $^{G}\mathbb{R}_E$ which are changed under the graph isomorphism $f : G \to H$ and $^{H}\mathbb{R}_E - {}^{G}\mathbb{R}_E$ are those revolutions of $^{H}\mathbb{R}_E$ which have been changed under the isomorphism $f : G \to H$, or which are changed under the isomorphism $f^{-1} : H \to G$. The formulation set measures the number of scientific disciplines, possibly within the same general field, that have contributed to the evolution of those areas of scientific investigation and the unification set indicates the sophistication level that theories have attained in so far as unification cannot occur unless there exists a common language and understanding of the Natural World that permits such a process as R_3^I. One of course always has $\left|{}^{G}\mathbb{R}_F\right| = \left|{}^{H}\mathbb{R}_F\right|$ and $\left|{}^{G}\mathbb{R}_U\right| = \left|{}^{H}\mathbb{R}_U\right|$. Obviously, $\left|\mathbb{R}_F\right| > \left|\mathbb{R}_U\right|$ but $\left|\mathbb{R}_E\right|$ can take any integer value greater than or equal to zero.

It really doesn't matter how isomorphic revolution diagrams are drawn, for that is merely a question of personal taste, as long as the above conditions are satisfied. For example, the two diagrams;

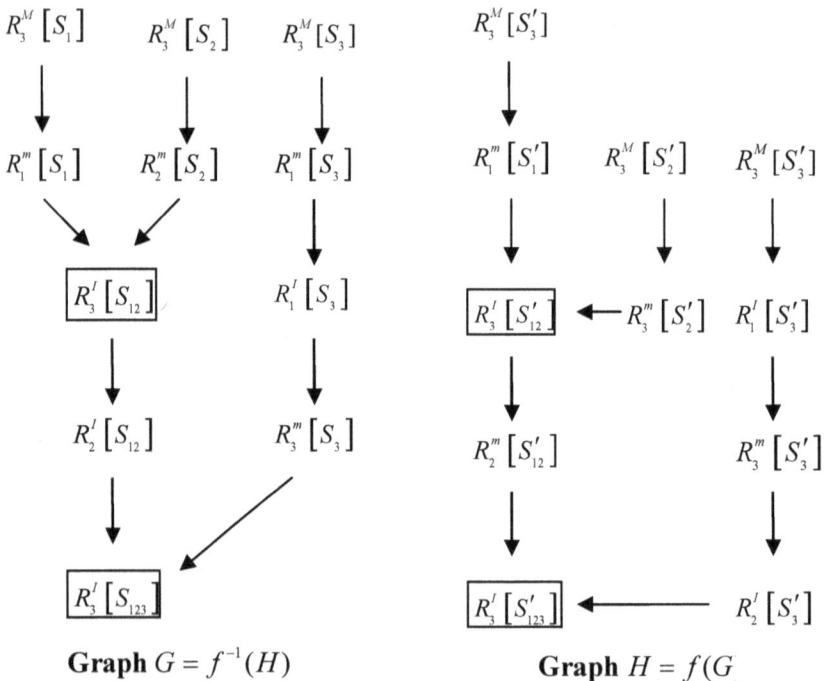

Graph $G = f^{-1}(H)$ **Graph** $H = f(G$

are isomorphic, but diagram G is somewhat easier to visually assimilate (at least for humans) than is H since the path of unification is more explicit in the former. For this example the revolution sets $\mathbb{R}_G$ and $\mathbb{R}_H$ of the graphs G and H respectively are;

$$\mathbb{R}_G = \{R_3^M[S_1], R_3^M[S_2], R_3^M[S_3], R_3^I[S_{12}], R_3^I[S_{123}]$$
$$R_1^m[S_1], R_2^m[S_2], R_1^m[S_3], R_1^I[S_3], R_2^I[S_{12}], R_3^m[S_3]\}$$

$$\mathbb{R}_H = \{R_3^M[S_1'], R_3^M[S_2'], R_3^M[S_3'], R_3^I[S_{12}'], R_3^I[S_{123}']$$
$$R_1^m[S_1'], R_3^m[S_2'], R_1^I[S_3'], R_3^m[S_3'], R_2^m[S_3'], R_2^I[S_3']\}$$

The formulation sets for G and H are thus;

$$^G\mathbb{R}_F = \{R_3^M[S_1], R_3^M[S_2], R_3^M[S_3]\}$$

$$^H\mathbb{R}_F = \{R_3^M[S_1'], R_3^M[S_2'], R_3^M[S_3']\}$$

and the unification sets are just;

$$^G\mathbb{R}_U = \{R_3^I[S_{12}], R_3^I[S_{123}]\}$$

$$^H\mathbb{R}_U = \{R_3^I[S_{12}'], R_3^I[S_{123}']\}$$

and finally the equivalence sets for this particular isomorphism are;

$$^G\mathbb{R}_E = \{R_1^m[S_1], R_2^m[S_2], R_1^m[S_3], R_1^I[S_3], R_2^I[S_{12}], R_3^m[S_3]\}$$

$$^H\mathbb{R}_E = \{R_1^m[S_1'], R_3^m[S_2'], R_1^I[S_3'], R_3^m[S_3'], R_2^m[S_{12}'], R_2^I[S_3']\}$$

and we may also write the relative complements;

$$^G\mathbb{R}_E - {}^H\mathbb{R}_E = \{R_2^m[S_2], R_1^m[S_3], R_1^I[S_3], R_2^I[S_{12}], R_3^m[S_3]\}$$

$$^H\mathbb{R}_E - {}^G\mathbb{R}_E = \{R_3^m[S_2'], R_1^I[S_3'], R_3^m[S_3'], R_2^m[S_{12}'], R_2^I[S_3']\}$$

We see that the mapping from G into H preserves adjacency and non-adjacency and from these set theoretic differences we have the following correspondence between the vertices of the latter;

$$R_1^m\left[S_1\right] \xrightarrow{\ f\ } R_1^m\left[S_1'\right]$$

$$R_2^m\left[S_2\right] \xrightarrow{\ f\ } R_3^m\left[S_2'\right]$$

$$R_1^m\left[S_3\right] \xrightarrow{\ f\ } R_1^l\left[S_3'\right]$$

$$R_1^l\left[S_3\right] \xrightarrow{\ f\ } R_3^m\left[S_3'\right]$$

$$R_2^l\left[S_{12}\right] \xrightarrow{\ f\ } R_2^m\left[S_{12}'\right]$$

$$R_3^m\left[S_3\right] \xrightarrow{\ f\ } R_2^l\left[S_3'\right]$$

$$G \qquad\qquad H$$

and in only one case is a revolution in G mapped into a revolution of the same type in H and as noted above, major and intermediate revolutions of the third kind *must* be mapped into the same revolution type. There are of course many other ways in which two graphs can be isomorphic. The above example shows one particular isomorphism.

Note that it is not necessary for the theories pertaining to a particular revolution to be of the same type; we are concerned only with the structural aspects of revolution diagrams. The frequency matrices for the graphs G and H in the above illustration are;

$$\begin{pmatrix} \frac{2}{11} & \frac{1}{11} & \frac{1}{11} \\ \frac{1}{11} & \frac{1}{11} & \frac{2}{11} \\ 0 & 0 & \frac{3}{11} \end{pmatrix} \qquad \begin{pmatrix} \frac{1}{11} & \frac{1}{11} & \frac{2}{11} \\ \frac{1}{11} & \frac{1}{11} & \frac{2}{11} \\ 0 & 0 & \frac{3}{11} \end{pmatrix}$$

$$G \qquad\qquad\qquad H$$

from which we see that G and H differ only in the values of the frequencies f_1^m and f_3^m and from this it might be thought that we can make a number of observations concerning the nature of the scientific theories involved, but several circumstances are possible and need to be considered before any assumptions are made or conclusions drawn.

Each vertex of G is mapped to the corresponding vertex of H and in such case the equivalence set completely determines the differences between the two revolution diagrams. The structure of each is the same but it is only for the elements of $\mathbb{R}_E$ that revolutions can be mapped to different revolution types and this reveals much about the nature and development of the theories in question. We consider the particular isomorphism for the above two graphs in connexion with the types of revolutions given and what may be said about the developmental process as a result. We note however, that there are a number of possibilities depending upon the theories S_i and S_i'.

First, we suppose that S_i and S_i' are completely unrelated theories that pertain to different scientific disciplines. Thus, $\{S_1, S_2, S_3\}$ may be theories in physics for example, whereas $\{S_1', S_2', S_3'\}$ may be theories in theoretical biology or some other unconnected discipline and we assume that both theories have been developed within the same scientific culture. Then a comparison of each of the revolution vertices in the revolution set for G and H should tell us something about the sophistication of each of these scientific areas and the cultures involved. Hence, looking at the two diagrams we see that the theories S_1 and S_1' undergo the same revolutions up to the point of unification with S_2 or S_2', but that the minor revolution $R_2^m[S_2]$ for S_2 in G is replaced by the minor revolution $R_3^m[S_2']$ in graph H.

Since a minor revolution of the third kind is a little more iconoclastic than one of the second kind the implication is that the theory S_2' was not properly formulated in the first instance, for it has been necessary to alter the axioms of that theory in a non-trivial manner before the unification with S_1' could be achieved. With the theory S_{12} in G the contrary situation applies. After the first unification, an intermediate revolution of the second kind is required before final unification with S_3 becomes possible, but in H only a revolution of type R_2^m is needed. Doubtless, this is due to the fact that considerable modification of the theory S_3 obtains in G, but to an even greater extent with S_3' in H. We may therefore conclude that the theory depicted by the revolution diagram H is generally less developed than that of G. There may be several reasons for this, but they cannot be empirical or technical since the same scientific method and technique is equally available for the investigation of both areas. Hence, we are led to the view that theory S_3' is either a new scientific domain or that it has been particularly difficult to isolate appropriate natural variables in which to precisely formulate the axioms.

Secondly, we imagine now that the two sets of theories belong to the same scientific discipline. Then, in such case attempts are being made to unify one branch of the scientific field in question with another and therefore diagram G must represent the progress of a different culture than that of diagram H.

Much of the preceding discussion applies verbatim in this case except that if the realm of scientific enquiry is in the well-defined physical domain where the natural interpretation readily supplies appropriate variables, then we must assume that the particular fields of S_3 and S_3' are relatively new pursuits for their respective cultures since S_1, S_2 and S_1', S_2' do not suffer from so much revolutionary development. If this is not the case however, we would be inclined towards the view that some extraneous influence is hindering the progress of S_3 and even more so for S_3'. Such constraints can only be political or economical and only empirical or technological in the above case of a new domain of enquiry. Thus, we now envisage that the culture relating to diagram H experiences the greater measure of these difficulties.

It is possible to imagine many other scenarios in this sort of socio-political context. We may have it, for example, that the revolution diagrams G and H refer to different sets of scientific theories within different cultures, or even that each of the elements of the sets $\{S_1, S_2, S_3\}$ and $\{S_1', S_2', S_3'\}$ are separate individual axiomatic theories of disconnected scientific disciplines either within the same culture or otherwise. In this latter instance, the revolution diagrams would be describing the reduction of one or more scientific disciplines to that of another and therefore conjoining them in a way that looks very much like the familiar philosophical problem of *reductionism,* or the reduction of scientific theories in disparate areas, one to the other.

Clearly, there are myriad possibilities that may be considered with isomorphic revolution diagrams and one may speculate endlessly upon the cultural and scientific implications, but that such philosophical and sociological musings are even admissible in serious discourse is a direct consequence of the nature of the isomorphism between two diagrams and indeed, the equivalence class of that relation. In particular, it is the equivalence sets $^G\mathbb{R}_E$ and $^H\mathbb{R}_E$ or more directly their relative complements that permit such analysis and the cardinality of the latter, $\left| {}^G\mathbb{R}_E - {}^H\mathbb{R}_E \right| = \left| {}^G\mathbb{R}_E - {}^H\mathbb{R}_E \right|$ immediately enumerates the diversity of similar revolution diagrams and the many paths that may be followed in the quest for empirical knowledge.

It might be a little surprising to find that the nature of a scientific culture can be discerned by the way in which its scientific theories evolve, but that should not be so, for evidently the two must be related at the sociological level simply because the practical pursuit of science cannot proceed independently of the circumstances of the society in question and indeed, we have encountered this circumstance before. We have noted several factors that may be of relevance in this connection ranging from the technological to the economic, but it may also be the case that theory formulation and hence scientific revolutions cannot be achieved for reasons inherent in the subject matter of the theory itself or even the nature of the observing entity. Thus, in the human case we might find that fields such as psychology have not even developed beyond the purely speculative stage and no revolutions in this area have ever been realised, not even of type R_3^M. If the same applies to a different human culture with an isomorphic diagram, or even a different species, then a direct comparison can be made and conclusions relating thereto drawn from this equivalence as above.

On the other hand, for non-human societies considerable progress may have been made in these fields with respect to understanding the psychological dispositions, motivations and inclinations of a given species, possibly involving many revolutions and this discipline may then be properly afforded scientific status with a full set of axioms and natural variables. This would seem to imply that humans have some difficulty in isolating the appropriate natural variables and axioms which would allow the formulation of scientific theories in this area and there may be good reasons for thinking that this is so. We shall discuss this topic in more detail in Chapter 11. Note that it is only in the isomorphic case that such comparisons and cultural analysis can be made.

We enjoy no such privilege in the non-isomorphic case, for without a relation that can facilitate equivalence of some kind, there are no properties between the objects of interest that can be precisely discerned. Yet, it is also true that an order relation can always be imposed upon that which at first seems to be disordered.

We mentioned reductionism in a previous paragraph and suggested that, at least by implication, a revolution diagram is equally capable of describing such a process in so far as it doesn't matter to the diagram where the scientific theories come from since it is just an abstract structure, and yet we have given no substance to this notion.

The reduction of one scientific domain to that of another cannot be ignored or dismissed in any enterprise that purports to elucidate the structure and evolution of scientific theories in a systematic fashion and is connected with theory unification in certain circumstances. This has traditionally been a philosophical topic, but we must now consider it from the point of view of theory-change and revolution diagrams. We should therefore examine the philosophical content of this problem first and then connect it with theory-change and scientific revolutions in general. This we shall attempt to do in the next chapter.

Chapter 9

The Reduction of Scientific Theories

The concept of the reduction of scientific theories in one field to that of another is considered. We discuss the nature and mechanism of such a process and its philosophical import. We liken inverse reduction to correspondence between a theory and its successor and conclude that the whole idea of inter-theoretic reduction is not a particularly useful or viable concept. Theory reduction is then seen to be nothing more than theory change. The Doctrine of Emergence and its relation to reduction is then briefly discussed within the context of theory change. Finally, the relevance of both these philosophical notions with regard to theory change and scientific revolutions is considered.

9.1 The concept of reductionism

The notion that one scientific theory may be subsumed by another is an old one in the Philosophy of Science, originating, as with so many other concepts, in ancient Greek Philosophy, but has attracted attention since the scientific explosion following Newton through to Maxwell, and became of particular and special interest in the mid-twentieth century, presumably because of the advances made by science during that period when relativistic and quantum formulations of theories seemed to demand such re-examination. In its modern form, we envisage a scientific theory S being 'embraced' by a more 'fundamental' theory S' which in some sense 'explains' the theory S in that S may be deduced from S', i.e. $S' \vdash S$ in some clearly definable fashion, and we say that S has been *reduced* to S'. Then, S' is the *reducing* theory and S is the *reduced* theory. The whole process is referred to as *inter-theoretic reduction*, or just theory reduction.

From the purely pragmatic standpoint of the practising scientist, this idea is stimulating and may well provide strong motivation for future research, but from the philosophical perspective it is immediately seen to be fraught with difficulties. There are questions relating to the precise meanings of all these terms and more questions concerning the mechanism of reduction and the respective domains of applicability of the theories involved. There is yet the further issue of the ontological status of one such theory after it has been reduced to another and the semantic connection of the variables or terms employed in either theory when the latter are dissimilar or the domains are disjoint. Most importantly of all, we would wish to know exactly what it means to say that one theory is *deducible* from another, for this implies a logical or computational and algorithmic methodological process which is hard to sustain when such ill-defined concepts obtain. Then, let us try to describe this situation more succinctly and in linguistic terms that are more amenable to analysis in our present context.

Traditionally, if the reduced theory S contains a subset T of the terms and variables T' of S' then the reduction of S to S' is said to be *homogeneous* and if this is not the case, then the reduction is *inhomogeneous* (or sometimes *heterogeneous*). Thus, an inter-theoretic reduction is homogeneous when there are no terms or variables of the reduced theory that do not also occur with exactly the same meaning or interpretation within the reducing theory and inhomegeneity obtains when one or more terms of the reduced theory are *not* terms of the reducing theory, or have no meaningful counterpart within the latter. We see that homogeneity of reduction does not require that the reducing theory be restricted only to the terms of the reduced theory and therefore a distinction, often ignored, may be made here. If the terms of the reducing theory and the terms of the reduced theory are the same, then we would prefer to say that they are *exactly* homogeneous. This distinction is important because it clearly relates to the domain of applicability of the reduced and reducing theories which must thereby be constrained by this criterion. We have encountered a number of illustrations of homogeneous reduction already, for example the reduction of Classical Mechanics to Relativistic Mechanics, where however it is sometimes argued that homogeneity does not obtain here since the definition of *mass* is different in the two theories. However, this would not be the case if we allow relativistic mechanics to contain several kinds of mass; rest mass, relativistic mass, etc. Observe the similarity with correspondence principles here.

The principal idea then, is that a new theory supersedes the old theory and that the latter is a logical consequence of the former. It is assumed therefore, that given a reduced theory S and a reducing theory S', and that having empirically established S' as a theory that adequately describes all the physical phenomena formerly described by S, with possibly a greater range of such phenomena, then S *must* be deducible from S' because S was already a successfully established scientific theory, for if that were not the case then the two theories would be describing different phenomena. Yet, herein resides the problem with terminology. If any of the terms of S are not also terms of S', then in what sense may it be said that S is deducible from S'? Surely, if S' is to be meaningful and indeed explanatory with regard to S then both theories should at least employ the same terminology with identical linguistic meaning. If there is a term in S that is not a term of S' then no possible notion of logical consequence may be envisaged between these two theories, for uniformity and consistency of variables and terms is a necessary condition of that relation. In other words, within both theories we expect to be talking about the same things and it is hard to see how this would be possible without a common terminology.

Two frequently quoted examples of the semantic difficulties that can arise in inhomogeneous reduction can be illustrated in the relation of Classical Thermodynamics to Statistical Mechanics on the one hand, and Maxwellian Electromagnetic Theory to Physical or Geometrical Optics on the other. In the former, as it is traditionally argued, the notion of temperature or heat is a macroscopic quantity and so has a familiar and quite distinct meaning from that ascribed to it in Statistical Mechanics or Kinetic Theory, where it is realised as the energetic molecular or atomistic motion of gaseous particles, specifically the mean kinetic energy $\frac{2}{3} E = kT$. Thus, the definition of temperature in thermodynamics is, it is alleged, entirely different to that given in the Kinetic Theory of Gases and therefore no reasonable correlation can be sustained between these two scientific fields in spite of the fact that they describe identical phenomena. How then, it is asked, may thermodynamics be a logical consequence of Statistical Mechanics when the terminology is not common to both theories?

The second example is broadly similar but with just a few differences. It is suggested that the electromagnetic theory of radiation subsumes and incorporates the wave nature of light as described in Physical Optics. However, as before one would be inclined to argue that the picture of visible light as portrayed in Physical Optics is quite different from that of undulating electric and magnetic fields as envisaged in Electromagnetic Theory and therefore that the ontology and meaning of these two descriptions are not in any way comparable, although both refer to the same physical phenomena.

There are many similar illustrations throughout the physical sciences and the same problem of how to account for the seemingly irresolvable difficulty of semantic compatibility between the reduced and reducing theories arises in each case and must, therefore, be properly addressed before these concepts can even begin to make sense. The usual 'solution' is to postulate a correspondence between the terms of one theory and those of the other such that those terms which denote the same thing are correlated by 'correspondence rules' or 'bridging laws'. Then, in the case of Classical Thermodynamics the appropriate bridging law is the rule that the temperature as defined in that theory is related to the temperature as defined in Statistical Mechanics by the equation above relating mean kinetic energy of molecular motion to the classical temperature. Note that 'temperature' in Statistical Mechanics is essentially undefined in the classical or 'everyday' use of that term and exists only in order to connect the two disciplines with each other that thereby a reduction may be realised, which to some extent may appear to be rather *ad hoc*. Bridging laws are thus a mechanism by which the terms of the reduced theory are related to those of the reducing theory so that all terms common to both can be construed to refer to the same phenomena by the correspondence of the bridging law that so relates each one to the other within the respective theories, and thereby provides some degree of semantic equivalence between such corresponding terms. Clearly, this is circular and may be applied to any situation at all where translation is necessary. Nevertheless, there is surely non-negligible content here, for it is doubtless the case that the reduced theory bears *some* kind of relation to the reducing theory that should be quantifiable with at least a small measure of precision in order for reduction to occur at all, and yet that most precious attribute seems not to obtain with this formulation of inter-theoretic reduction. Hence, the nature of the bridging law is crucial for theory reduction and requires explicit formulation. But, it is clear also, that an appropriate bridging law can always be found in the homogeneous case since the terms of both theories are identical and the bridging law is simply the identity mapping. Thus, for any term T of the reduced theory S and the corresponding term T' of the reducing theory S' we obviously just have $T = T'$ and therefore we need not be concerned with homogeneous inter-theoretic reduction.

In inhomogeneous reduction the bridging law maps the terms of S onto a subset of the terms of S'. Thus, if B is such a mapping we may write $B : T \rightarrow T'$. In the above example of thermodynamics and statistical mechanics this is just the expression $B : T \rightarrow T' = 2E/3k$, where T and T' are the temperatures in S and S' respectively. Of course, *some* of the terms of S' may also be terms of S, so that there may well be an exactly homogeneous subset within the two theories and this is another reason for considering only inhomogeneous reduction. However, we may ask whether the mapping B has an inverse $B^{-1} : T' \rightarrow T$, for surely this is possible. In fact, B^{-1} is just a correspondence principle of the type we have discussed earlier. Yet, we have seen that a strict correspondence principle of this type does not always obtain between a given theory and its successor. What is more, the very notion of reduction would not permit this inverse mapping, for we would then have to interchange the roles of the reduced and reducing theories so that the reduced theory becomes the reducing theory and *vice versa*. Evidently, the concept of inter-theoretic reduction is somewhat muddled and in the scientific context we would rather say that the superseding theory S' reduces in some limit to the theory S, as we have already seen in several cases in §7.4. We are inevitably led to the conclusion that the philosophical concept of a reducing theory is not the one that pertains within science as it is actually practised and we must wonder why such a notion is even necessary from a philosophical viewpoint and indeed, what it really means.

It seems that the ideal of inter-theoretic reduction is primarily motivated by theory-change, a topic that has been of central concern to us throughout the present musings. That scientific theories supersede others is observed to be an evolutionary process, as by now we well know. Then, since an already empirically established theory is superseded by a new theory that is subsequently established by empirical means also, it is thought that the new theory must not only 'explain' the former theory, but that the latter is a logical consequence thereof. Hence, the reduced theory must be derivable from the reducing theory purely on this basis. But we have noted that only in the homogeneous case may such a concept be entertained and even then the matter is doubtful. There is no reason to suppose that *any* scientific theory is logically deducible from another, for if this were the case then both theories would be identically correct or equivalent in their account of the physical world and supersedence could not thence obtain, rendering theory-change inoperable. It would be useless to argue that a subset of the superseding or reducing theory is equivalent to the reduced theory and that because the former incorporates the latter with a greater domain of applicability, then the reduced theory is a consequence of the reducing theory, for the domain expansion of the reducing theory requires that its axioms extend beyond the domain of the reduced theory, and therefore cannot be the same axioms of the reduced theory since theory-change by definition, forbids such a circumstance. Essentially, the domain of applicability of an axiom is the area within which it holds and the domain of a theory we recall, is the domain of the axiom of the theory with the least domain of applicability. Thus, any domain enlargement must entail axiom change and therefore no subset of the reducing theory can be equivalent to the reduced theory in any axiomatic sense, even though such an axiom subset may, with a suitable choice, have the same domain if only by axiom addition. Hence, any two such theories are different and one is not derivable from the other. More concisely, if the domain of one theory is not the same as another, then they are different theories and one cannot be derivable from the other. Only in the sense of a correspondence principle such as $B^{-1}: T' \rightarrow T$ may we say that one theory is reducible to another and as noted earlier, this does not always obtain, and when it does it is not logical deduction but the result of a limiting process.

So what do we want inter-theoretic reduction to mean? It is surely a simple idea and all we require is that one scientific theory be expressed in terms of another. Thus, if we can express biology in terms of biochemistry and the latter in terms of physical chemistry and then chemistry in terms of physics, then biology is expressible in the precise language of physics. It is clear that 'reduction' in the philosophical sense is nothing more than the replacement of one theory by another such that the replaced theory, or rather the description of the Natural World that it was able to so accurately depict at the moment of its formulation, is also replicated by the replacing theory, though with greater accuracy and perhaps more comprehensively. This has little to do with logical consequence and is essentially language translation. The terms of the reduced theory are expressed in the terms of the reducing theory in such a way that it is seen that the reduced theory was able to describe the same physical phenomena as the reducing theory simply by virtue of the successful translation of terms, but they are still not the same theory because the axiom sets are different even if the axioms of the reduced theory are contained within the reducing theory as mentioned above. It is this possibility of translation of terms that suggests that the reducing theory 'explains' the reduced theory because the former is more 'fundamental' or provides a deeper description than the latter. In fact, all that has happened is that new variables and axioms have been introduced in which the tenets of the old theory may be expressed. In this way it becomes immediately apparent that the use of the transitive verb 'reduced' and the gerund 'reducing' are too suggestive, as also is the noun 'reduction', for they do not correspond to their descriptive content. Such terms must therefore be considered to be ill-defined and *ipso facto* must lead to some conceptual confusion. It is evident also that the philosophical use of these terms neither accords with the practice of science nor do they satisfy the philosophical objectives that so optimistically signalled their creation. Yet, it is also clear that there is some merit in the basic notions underlying inter-theoretic reduction as we have (rather loosely) described it and perhaps a change of terminology would more appropriately capture the essential concepts. However, we do not need to construct such a new technical vocabulary since we already have within our account of the theory-change mechanisms everything that we need, for the natural interpretation operates as a kind of bridging law that effects translation directly in the empirical world. There is however, one other dominant philosophical topic closely related to reduction that needs to be discussed. This is the Doctrine of Emergence.

9.2 Emergence

As with the case of reduction there are several notions of emergence in philosophy, but we confine ourselves only to the most relevant for our context. In common parlance one often hears the expression 'The whole is more than the sum of its parts' and presumably this means that there is some extraneous attribute other than those exhibited by the individual parts that miraculously appears when the said parts combine to form the 'whole'. Then, it is this attribute that *emerges* or is stated to be *emergent* or an *emergent property*, which was hitherto unrecognised before combination of the individual constituents took place. So, a property is emergent if it cannot be inferred, or rather deduced from knowledge of the constituent parts that make up the whole. A property is not emergent if such a deduction is possible.

The usual example given to illustrate emergence relates to the transparency of water. The water molecule H_2O is composed of two atoms of hydrogen and one of oxygen and these are the constituent parts. But water is optically transparent, which is a property that could not be inferred from consideration of the individual atoms of hydrogen and oxygen when they are combined in this way. Therefore, the transparency of water is an emergent property that appears without explanation. Let us examine this example.

First of all, it is not true. Whilst in the normal course of events one would not take the trouble to compute the optical properties of H_2O, it is certainly possible (at least in principle) to do so. Performing such a calculation would then reveal the transparency of water and hence a supposedly emergent property *could* have been inferred and so cannot be emergent. Secondly, even if one could not carry out the calculation one may argue that this is simply because of incomplete knowledge, which was probably the case at the time this famous illustration was first presented.

However, this does not necessarily negate the essential idea of emergence, which in the manner in which it is used in the present context asserts that even with complete knowledge of the properties of all the constituents of a set of entities, that upon combination of those constituent parts, for each such entity there will in some cases arise a property for some entities so constructed that could not have been deduced by the given complete knowledge of their constituents. Expressed differently; the fact that complete knowledge of the properties of the constituent parts of one entity may entail similar complete knowledge of the properties of the entity after combination, does not imply that this is always the case or preclude the contrary situation, but merely avers that it was so for the particular entity in question. Then, one cannot categorically state that emergent properties do not arise following the combination of constituent parts, whether with complete knowledge or not, but one can say that if such an emergent property does present itself, then it cannot be deduced from that fact of emergence that the property is not deducible from knowledge of the properties of the constituent parts. Conversely, from full and complete knowledge of the constituent parts of an entity, it cannot be logically deduced that emergent properties of the whole entity will arise upon the combination of the constituent parts. Hence, it is not possible to settle the question of emergence by purely logical means. Clearly, there are ambiguities here. What do we mean for example by 'complete knowledge'? Surely, one's knowledge is *incomplete* if one cannot predict all the properties of something, including the fact that an emergent property will or will not arise on combination of the constituent parts of an entity. Obviously, if complete knowledge implies that one may predict the gross properties of some object formed by its constituent parts then the claims of emergence would be vacuous, but that is certainly not what is intended, for if that were that the case, then no emergence would be possible by definition. It is also the case that an emergent property is peculiar to the object under scrutiny, or at least the theoretical context within which the precise constituents of the properties of the entity or object in question are defined, for otherwise the concept of an emergent property would hold equal status with all things of all kinds, including those in which no property can be emergent. Thus, we must consider such terms to be relative to some particular context, since if they had meaning in an absolute sense then the predictability of emergent properties would not be specific to the particular entity in question, but apply equally to all things irrespective of type and category, which as we have just noted, is not the case.

Evidently, emergent properties, if they exist, must satisfy certain criteria. An emergent property of one class of entity would be distinct from that of another, simply because of the referential nature of the constituent parts. Hence, knowledge of the constituents is essential to the prediction of the attributes that may be exhibited upon combination, but unless such detailed knowledge is available for all such objects or entities, or indeed the entire universe, then the concept of emergence must be particular and applicable only to the specific entity in question, or the class of similar entities. We are saying therefore, that if one entity possesses an emergent property, then all entities of the same class must also possess emergent properties and if this is not the case for a given class of entities, then no member of that class can have an emergent property. In this way emergent properties are made relative with respect to an entity within a given category and even individualised to a special such entity, since no restriction is placed upon the number of members of a category or class which, if it is thought necessary, may have only a single member, although this is not generally expected to be the case.

It is important to note that the 'Doctrine of Emergence' requires that an emergent property of an object be either deducible or not deducible from the known properties of the constituents in the strictly logical sense of derivation, for clearly it has no value otherwise. The connection with theory reduction is just that, if one scientific field of enquiry is reduced to another in the aforementioned sense, then the reduced theory cannot possess emergent properties that are not already properties of the reducing theory since they would then be predictable from knowledge of the latter and hence, there can be no such emergent properties at all for this reason, because reduction would entail a bridging law that must characterise any such property in a precise manner, and therefore that property would, *ipso facto*, be deducible. Hence, inter-theoretic reduction and the notion of emergent properties are intimately related in this way and are therefore really two sides of the same coin. But we need to look at this in a little more detail in relation to the mechanisms of theory-change and the connection with scientific revolutions.

9.3 Reduction and emergence in theory change

The question of emergent properties acquires particular significance when the constituent parts are actually scientific theories and we have seen this already through the unification process. When two scientific theories are unified, new and interesting phenomenological results often follow or *emerge* as a consequence in the sense that such phenomena could not have been predicted from the knowledge of the constituent theories alone that are engaged in the unification process, i.e. such new phenomena could not have been foreseen prior to the combination of the two individual theories, or so it appears *prima facie*. We recall from §3.10 the MHD illustration of unification and find that Alfvén waves are one of the emergent properties of the unified theory, for they were not predicted from the constituent theories of hydrodynamics and electromagnetism prior to unification. Yet, no such affirmation can be made on the basis of the nature of the constituent theories since the issue of whether something or some property is deducible bears no relation to whether it is or is not actually deduced in practice. Magnetohydrodynamic waves appear to be a consequence of unification, but it may well have been *possible* to surmise their existence from the component theories through appropriate theoretical insight. However, this does not mean that phenomena of this kind are deducible from the component theories, for clearly they are not, since Alfvén waves arise only as a consequence of both theories when they are unified. Theoretical foresight plays no part in emergence and we must conclude that whenever theories are unified, emergent properties in the form of hitherto unknown physical phenomena can be expected to appear.

A moment's reflection suffices to reveal the inevitability of this conclusion. The purpose of the scientific enterprise is to uncover the workings of the Natural World and therefore, the discovery of new physical phenomena is part and parcel of the scientific endeavour. Then, since theories undergo change through the theory-change mechanisms that we have described, the latter must result in emergent phenomena in so far as the modified theories possess a new property not envisaged or hitherto deducible, and theory unification is the principle means by which such emergent properties appear. We have, of course, seen many examples of this during our investigations. Theory unification involves axiom combination which is a linguistic or mathematical process that results in a new theory that satisfies the axioms of each of the formally separate component theories. Since the physical processes within the domain of the component theories were of a different kind, but are now combined by unification and furthermore, the domain of applicability is increased to that of all components, it is hardly surprising that new phenomena emerge. Of course, this is intuitively understood in the practice of science, but nevertheless needs to be explicitly stated in the present philosophical context.

Then what of theory reduction? We have already noted in §8.3 and other places, that revolution diagrams involving theory unification where the component theories come from entirely unrelated scientific disciplines, is essentially theory reduction and, as such, emergent properties may be anticipated. There are though, some differences which will be dealt with shortly. First however, we may now ask if any of the other mechanisms of theory-change could also give rise to emergent properties. The answer to this question must be in the negative, for all other theory-change mechanisms have independent domains of applicability and it is only with theory unification and now inter-theoretic reduction, that a combination of domains occurs, and hence by the definition of emergence, it is this process alone that can result in emergent properties or new physical phenomena.

In effect, all other theory-change mechanisms consist of the empirical modification of theories in which the domain is either static or enlarged entirely within the context of the theory concerned, but it is still the same theory and domain of applicability without the individual components that arise in unification, where the requirements of emergence is met and the existence of components is a definitional prerequisite for emergence, as we have noted. This illustrates a trend in scientific progress that we have alluded to before, that unification of scientific theories is the primary means of subsequent development providing that the component theories have been sufficiently well-established or verified independently by standard empirical techniques. Thus, emergence entails new properties and phenomena, and this process serves only to emphasise the importance of theory unification or inter-theoretic reduction in that respect and it may be reasonably inferred that the human experience of theory and domain expansion may apply in a far more general arena, since methodological matters are of sole consideration here.

However, when one theory is reduced to another, it is postulated that bridging laws are the principle agents that effect reduction between the reduced and reducing theories, but in theory unification no such stipulation is necessary. Additionally, when two theories are unified one theory is not reduced directly to another in precisely the manner of conventional inter-theoretic reduction; but rather both theories are combined to form a new theory through the process of axiom combination.

This then, is how theory reduction must be seen within the context of theory-change and revolution diagrams, not as an identical procedure but as an equivalent one. In this way, axiom combination plays an analogous role in unification to those of bridging laws in theory reduction. Yet, inter-theoretic reduction does not *really* reduce a theory to an already existing theory just because of the necessity of bridging laws, for the introduction of the latter effectively modifies the reduced theory with respect to the definitions of natural variables and the axiomatic structure of both theories. Axiom combination is the same process in another guise and emergent properties arise in each case, but with unification this emergence is made much more explicit and amenable to analysis, enabling us to dispose of the philosophical issues associated with the Doctrine of Emergence and replace them with the precisely defined concepts of theory unification and axiom combination. This is why we regard these two ideas as essentially equivalent. Henceforth, we may entirely dispense with the notion of inter-theoretic reduction and consider only theory unification within the wider context of theory-change and theory-change operators. Similarly, the concept of emergent properties may now be perceived to be an inevitable consequence of unification with no philosophical baggage attached thereto, for these ideas are now well-defined and simply become deducible properties of the new, unified theory.

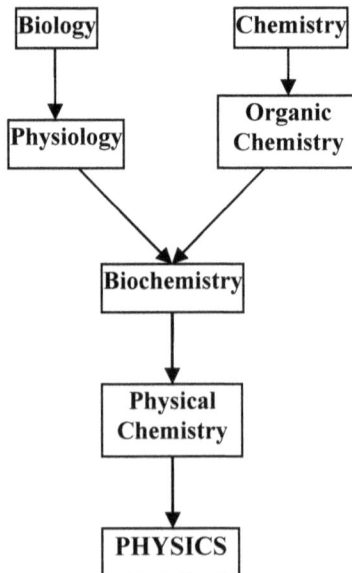

Figure 23

Theory reduction, when expressed in terms of theory unification, may be illustrated as before with revolution diagrams, which we might now be inclined to call *reduction diagrams*, as being perhaps a more appropriate term when they apply to whole and initially disconnected scientific domains. Consider for example the reduction of biology and chemistry to physics as it is naively pictured in Figure 23 above and which we now discuss.

This is of course, a very over-simplified portrayal of the vastly complicated reduction scenario that would actually obtain, but will serve to illustrate a number of points as well as the essential ideas. Figure 23 depicts a directed tree just as in our former revolution diagrams, but now each vertex is an entire domain of scientific enquiry, at least at the outset, meaning those furthest from the root. We have enclosed these separate and distinct disciplines in boxes in order to indicate that property. However, each of these boxes represents an entire revolution diagram or even a set of such diagrams and within every one of them there will be many unifications of sub-disciplines, with the consequent emergent phenomena or properties that result thereby. Then, each discipline is evolving towards unification with the others and it is doing so through the mechanisms of theory-change as usual, but theory unification is vital for such scientific progress and hence, emergent properties inevitably ensue as indeed they must if theories are required to embrace the domains of others. Figure 23 is only illustrative and could be written in many ways. If written out in full for all the pure sciences so that each vertex box contained the full sequence of revolutions for every discipline and sub-discipline concerned, then it would be a truly enormous graph; quite unwieldy and impractical. Nevertheless, in principle this can be done (when such reductions are possible) and all the emergent properties and new phenomena arising from unification would be evident; in effect one would have reduced all science to fundamental physics entirely by the simple application of the theory-change mechanisms.

In a reduction diagram, not only does theory unification take place within the sub-diagrams occupying each vertex (which are not shown), but also at the vertices of the scientific disciplines concerned. Thus, in the figure we see that physiology and organic chemistry are unified into biochemistry. However, this is not the same biochemistry that obtained prior to such unification because axiom combination has taken place and thus modified both theories, but we can still speak of reduction when we really mean unification, for they are essentially the same process, as we have discussed above. In fact, the merging of two entire areas of science is a sort of *super-unification*, since it involves unifications of unifications. The emergent phenomena in super-unification are of profound significance in the philosophy of science, for they represent a new understanding of the relationship between concepts in one field and those of another hitherto, distinct domain of enquiry. A revolution of this magnitude may therefore be thought of as a *super-revolution* and it is a matter of debate whether any scientific disciplines have really been fully reduced in this way, although chemistry or at least physical chemistry is often quoted as an example of complete reduction to physics, in principle if not in practice.

Yet, until every axiom of chemistry can be shown to follow from the axioms of physics, one cannot say that a super-revolution has occurred. Perhaps a better example is the reduction of astronomy to physics if one confines oneself to astrophysics, but then the latter may simply be regarded as physics anyway. Whatever view may be taken as to the current status of the merging of the sciences, it is clear that there is a long way to go and many interesting developments for both science and the philosophy thereof may be anticipated. We shall return to this issue later when the direction of science and final theories must be placed within the context of theory-change.

We see therefore that the traditional philosophical problems of reduction and emergence are interpreted differently within the scenario presented here. More pointedly, these philosophical notions must now be considered to be ill-conceived from our perspective and effectively disappear within the context of theory-change, where they at once become unnecessary.

Chapter 10

The Science Machine

We discuss the relation between scientific revolutions and
theory change mechanisms and find that they are essentially
the same kind of process. This tends to blur the distinction
between cumulative progress and sudden change in the
development of scientific theories and scientific revolutions in
general. The question of whether it is possible to construct a
machine to perform scientific investigations is considered and
the resulting philosophical consequences examined.

10.1 Revolutions and theory change

Scientific revolutions are often thought of as single momentous events that
occur occasionally during the history of science, by which one's
perspective of the Natural World is at once and forever changed, if only
temporarily, until the next revolution, or at least this is the historical view
generally propagated and the one that is responsible for the nature of
revolutions to have evaded full analysis and detailed scrutiny. Yet, there
are both truths and untruths in this perspective. We have described various
mechanisms of theory-change and for the most part these are seen to be
incremental and discrete operations that act upon the variables or axioms
of a theory such that the theory in question is altered in a precise manner
according to which operator is currently active. Revolutions, as we have
described them, should inherit this property since they are also a result of
the actions of theory-change operators and if this is the case, then
scientific revolutions must therefore be defined in terms of them. We have
seen this before, but now need to detail more explicitly the nature of the
mechanism that enables us to express these operator actions more
succinctly and in a more precise form. Let us therefore return to the
evolution equation that we discussed in §4.5 and consider its structure
with regard to the actions of the theory-change operators that it supposedly
encompasses in the context of scientific revolutions. We recall the
principal expression for the theory-change operator;

267

$$T_E S = T_\exists T_\sqcup S = S'$$

which takes the axioms of a theory S and turns it into the new theory S'. This will work for minor and intermediate revolutions, but for major revolutions there is a problem, for the latter involve changes to the fundamental assumptions of a theory and no reference to these is made in the theory evolution equations. We cannot say that a change in one or more axioms of a theory necessarily entails a change in the fundamental assumptions since there may be many different axioms that can satisfy those same assumptions. As we have discussed, the fundamental assumptions are *hypotheses* and not axioms, but from which the axioms of a theory may be formulated in conjunction with the natural interpretation and empirical methods. The fundamental assumptions are crucial to the axioms only in so far as the negation of them will negate the theory and not *vice versa*, i.e. the negation of an axiom may or may not negate a fundamental assumption. Then, if we are to express scientific revolutions in terms of theory-change operators, we must also incorporate the fundamental assumptions into expressions such as the above.

Now, the theory-change operators act upon the *axioms* of a theory, whereas revolutions operate with respect to entire theories. Furthermore, they are concerned not only with the axioms but with the variable and constant symbols contained therein in addition to the fundamental hypotheses, and hence their province is also the language $\mathcal{L}$ of the theory. Then, empirical processes are at work here, with particular emphasis upon the natural interpretation since the variable and constant symbols of the language $\mathcal{L}$ result under the natural interpretation in axioms of the form $f(x_i, c_j) = 0$ or $\varphi(x_i, c_j)$, which says nothing about the fundamental assumptions. Suppose then, that there is associated with a theory S a set of fundamental assumptions or hypotheses $\{h_i\}$ as well as the sets of variables and constants $\{x_j\}$ and $\{c_k\}$ that are specific to the axioms of S. Thus, for a particular axiom $\varphi \in S$ we may write $\varphi(h_i, x_j, c_k)$ where the sets $\{x_j\}$ and $\{c_j\}$ pertain only to the axiom φ and are subsets of the sets of variable and constant symbols of the language $\mathcal{L}$. However, $\{h_i\}$ is the *entire* set of hypotheses for S and is not specific only to φ. These hypotheses are therefore necessary and sufficient for all the axioms of the theory and the negation of any one of them will negate S, i.e. $\neg h_i \rightarrow \neg S$ according to Definition 7.4.3.

It would be nice therefore, by analogy with the evolution equation for the theory-change operators, to have a similar expression for scientific revolutions. Hence, we would try to envisage something of the form;

$$\hat{R}_n^T S = S'$$

where $\hat{R}_n^T$ is an operator that takes a theory S and transforms it into the new theory S'. The *type* of the resulting revolution in this scenario is as usual $T = \{$minor, intermediate, major$\}$ and the *kind* within the revolution category is $n = \{1, 2, 3\}$. As we have noted, this is in general a more complex process than simple theory-change because $\hat{R}_n^T$ must be regarded as acting upon the *whole* theory S and not just individual axioms, even though a theory consists only of its axioms. The difference here is that we must take account of the variables, constants and hypotheses of the theory.

This is of course, just a mapping from S to S'; $\hat{R}_n^T : S \to S'$. In fact, the operator $\hat{R}_n^T$ is a set of mappings $\{\mu_m\}$ from S to S', one mapping for each of the m axioms of the theory S. Let us try this for various revolutions where the axioms φ_m are now expressed by $\varphi_m(h_i, x_j, c_k)$. We first consider minor revolutions of the first kind $(T = m, n = 1)$ in which the constants of the language $\mathcal{L}$ are merely assigned different values within the axioms $\varphi_1, ..., \varphi_m$ of the theory S. This is just the set of mappings;

$$\mu_1 : \varphi_1(h_i, x_j, c_k) \to \varphi_1'(h_i, x_j, c_k')$$
$$\vdots \qquad \vdots \qquad \vdots$$
$$\mu_m : \varphi_m(h_i, x_j, c_k) \to \varphi_m'(h_i, x_j, c_k')$$

where all that has happened is that the constants c_k' have replaced the constants c_k *for the particular axiom φ_r only*, i.e. the c_k' are specific to the axiom in question. Of course, for a given axiom it may well be the case that $\forall k(c_k' = c_k)$ in which case $\varphi_r(h_i, x_j, c_k) = \varphi_r'(h_i, x_j, c_k')$, but there must be at least *one* axiom in which at least *one* constant is changed in order to produce the theory S'. Also, for this type of revolution we obviously have $|c_k| = |c_k'|$ so that we retain the same number of constants but neither the hypotheses nor the variables are altered and the form of the axioms remains the same. With these provisos we may represent this revolution as;

$$\hat{R}_1^m S = S'$$

Similarly, for minor revolutions of the second kind $(T = m, n = 2)$, the mappings μ_r of the operator $\hat{R}_2^m$ take the general form;

$$\mu_r : \varphi_r(h_i, x_j, c_k) \rightarrow \varphi_r'(h_i, y_{j+v}, c_{k-v})$$

and now it is the variables x_j that are replaced by variables y_v through the interpretation of v constants as variables. In this case we have $|c_k| = |c_{k-v}| + v$ and $|x_j| = |y_{j+v}| - v$ and again the *form* of the axioms φ_r are unchanged. Other than this, the same remarks concerning the constants in minor revolutions of the first kind apply *mutatis mutandis* to the variables in minor revolutions of the second kind and we may now write;

$$\hat{R}_2^m S = S'$$

But for minor revolutions of the third kind the operator $\hat{R}_3^m$ involves a change in the algebraic form of the *axioms* of the theory S in addition to the re-interpretation of at least one of the constant symbols as a variable symbol. The general expression for the μ_r is therefore the same as with revolutions of type R_2^m with the corresponding new form of the axiom(s). Thus;

$$\mu_r : \varphi_r(h_i, x_j, c_k) \rightarrow \psi_r(h_i, y_{j+v}, c_{k-v})$$

where the new axiom ψ replaces the axiom φ on at least one occasion. We then write;

$$\hat{R}_3^m S = S'$$

Let us now consider intermediate revolutions. For the latter type of the first kind we simply replace at least one of the axioms of the theory. Hence;

$$\mu_r : \varphi_r(h_i, x_j, c_k) \rightarrow \psi_r(h_i, x_j, c_k)$$

and the ψ_r are of a different algebraic or linguistic form to the φ_r such that the domain of applicability is increased. Nevertheless, the constants c_k and the variables x_j need not be altered. Again, we have;

$$\hat{R}_1^I S = S'$$

For revolutions of type R_2^I we replace axioms and variables, thus;

$$\mu_r : \varphi_r(h_i, x_j, c_k) \rightarrow \psi_r(h_i, y_j, c_k)$$

as was discussed in §7.3. This is then seen as the action of the operator;

$$\hat{R}_2^I S = S'$$

again with domain increase. Finally, for this category, we have intermediate revolutions of the third kind which by definition involves axiom change but no change of variable or constant symbols. However, unification is the principal operation here and the initial theory S consists of several theories $S = \{S_1, \ldots, S_p\}$ each with its own set of hypotheses and the final theory is the unification of these $S' = S_1 \sqcup, \ldots, \sqcup S_p$. The axioms of these theories are combined by axiom combination to produce new axioms but the hypotheses of the component theories are unchanged. Hence;

$$\mu_r : \varphi_r(h_i, x_j, c_k) \rightarrow \psi_s(h_i, x_j, c_k)$$

where it is understood that $h_i = \{h_{i_1} \cup h_{i_2} \cup, \ldots, \cup h_{i_p}\}$ is the union of each set of hypotheses for each of the component theories S_p and the axioms $\psi_s = \varphi_l \square \varphi_m$ are the result of axiom combination with of course $s < r$. Similarly, $\varphi_{r_p} = \{\varphi_{r_1} \cup, \ldots, \cup \varphi_{r_p}\}$. With this understanding we may write;

$$\hat{R}_3^I S = S'$$

where S is the set theoretic union of all the axioms of the p theories.

All this was discussed in Chapter 4 and the point here is that these six revolution types, minor and intermediate are just theory-change operations if we regard the replacement of a variable or constant within a given axiom as a modification of that axiom. With such a proviso we can indeed write;

$$\hat{R}_n^T T_E S = T_{\exists} T_{\sqcup} S = S'$$

for $T \leq 2$ and $n \leq 3$. Then $\hat{R}_n^T \equiv T_3 T_{\sqcup}$ and from §4.5 we know that this is expressible in terms of the successive applications of the basic theory-change operators T_+ and T_-. Thus, the first six revolution types may be seen as incremental processes of theory-change rather than sudden catastrophic scientific revolutions, even if the changes involved by the basic operator actions occur within a relatively short period of time.

However, for major revolutions the situation is somewhat different since the hypotheses or fundamental assumptions are primarily the objects of modification or replacement within that category and these are *not* defined by the axioms of the theory, but rather the converse in this case, as we have noted. Hence, no action of a theory-change operator can guarantee a unique modification of the fundamental hypotheses of a theory and it would seem therefore, that for major revolutions the above expression cannot hold. Now, it is certainly true that as a consequence of this fact, major revolutions possess a different ontology from those of a minor or intermediate type and we may well attempt to include the hypotheses within the definition of an axiom as we have indicated above by writing $\varphi(h_i, x_j, c_k)$. The inclusion of the set $\{h_i\}$ was superfluous for the first six revolution types since none of the h_i needed to be altered, but for major revolutions this is by definition, not so. Then let us adopt this strategy beginning with major revolutions of the first kind. Here, at least one of the hypotheses of the theory is modified;

$$\mu_r : \varphi_r(h_i, x_j, c_k) \to \psi_r(h_{i-p}, h_p, x_j, c_k)$$

where p of the i fundamental assumptions h_i have been modified or replaced. Now, it does not necessarily follow from this operation that the new axioms ψ_r where hypothesis change obtains are different from the old axioms φ_r since both sets of assumptions may be consistent with all the φ_r as well as the ψ_r. This is because the axioms are not *deducible* from the hypotheses, for if that were the case the latter would themselves be axioms and anyway, we do not include the deducible consequences of axioms within our definition of a scientific theory. Yet, in revolutions of type R_1^M the domain of applicability is increased and this would be impossible without axiom change, even if this is only a change of variables. Thus, we must conclude that revolutions in this category always entail axiom change and therefore $\neg \forall r (\varphi_r = \psi_r)$ and hypothesis replacement here means axiom replacement. But, there are two cases when this occurs with an increase in domain and these are the intermediate revolutions of the first and the second kind. We may therefore write;

$$\hat{R}_1^M S \equiv \hat{R}_1^I S = S'$$

or

$$\hat{R}_1^M S \equiv \hat{R}_2^I S = S'$$

which implies that $\hat{R}_1^I S \equiv \hat{R}_2^I S \equiv \hat{R}_1^M S$. This equivalence is only *ontological* and illustrates the difference between theory-change operators T_E and 'revolution operators' $\hat{R}_n^I$. From a purely practical viewpoint they are similar, but from a philosophical and psychological perspective they are of an entirely distinct nature. However, since our concerns are only pragmatic here, we may regard major revolution operators of the first kind as being equivalent to theory-change mechanisms. This equivalence is a subtle one and needs to be kept in mind, for we are dealing here with different categories of operators and the equivalence is essentially functorial. We never assumed first-order from the outset; that would not be possible.

We consider major revolutions of the second kind where now the domain of applicability does not necessarily increase. Now *all* the hypotheses are replaced or modified so that $\forall i(h_i \neq h_p)$. It is conceivable, though perhaps unlikely, that the new set of hypotheses are consistent with all the axioms of the theory S and then $\forall r(\varphi_r = \psi_r)$. But then $\hat{R}_2^M S = S' = S$ and we have no revolution at all, in fact a vacuous revolution, which is clearly unacceptable. Identity operations of this sort may be legitimate for theory-change, but not for revolutions which involve philosophical and ontological properties of a sort that are not applicable elsewhere in the panorama of theory-change. We therefore stipulate that;

$$\hat{R}_n^T S = S$$

is never the case. Given this, we may now regard major revolutions of the second kind as equivalent to theory-change in practice and may be considered to be a legitimate application of the basic theory-change operators.

Major revolutions of the third kind also incorporate philosophical and hypothetical assumptions that may or may not seem to be directly connected with the axiomatic structure of the theory in question. This is because strictly speaking, revolutions of type R_3^M cannot be seen only as theory-change, but also as theory formulation, and as we have mentioned before, such revolutions will generally happen just once in the lifetime of a theory. In the second case where there is no pre-existing scientific theory, there are no axioms to modify and hence no theory-change mechanisms can be operational. These are the instances when an empirical theory is first formulated, a development that is evidently not repeatable. But in other cases there already exists a theory which may subsequently be found to be unsustainable in the light of new empirical data and hence require reformulation at the fundamental assumption level. We have presented several examples of this in §7.4 where it was seen that changes in the fundamental assumptions led to the formulation of new axioms with special emphasis, according to definition 7.4.4 on the determination of natural variables. It is this latter aspect of these revolution types that is of principal significance in the present context since it admits of equivalence between the revolution operator $\hat{R}_3^M$ and theory-change operators exactly as described above with first and second kinds of major revolution. Then, R_3^M, R_2^M and R_1^M revolution types can all be regarded from a *pragmatic* standpoint as instances of applications of theory-change mechanisms. The dual nature of major revolutions of the third kind does not concern us, for we are interested primarily in theory *evolution* rather than theory *origins*.

10.2 Algorithmic science

Theory modification via the actions of the theory-change operators is very specific, discrete and in principle, entirely mechanical in nature. They could well be carried out by a suitably equipped mechanical device, which is indeed why we have constantly referred to them as theory-change *mechanisms*. For scientific revolutions also, we have found that the minor and intermediate types can be explicitly defined in terms of these mechanisms and for major revolutions we have identified an equivalence between revolution operators and theory-change operators that allows the former to be related to the latter in such a way that they essentially become the same kind of operation as far as the actual practice of science in the empirical world is concerned.

If this is so, then two fascinating and intimately related consequences immediately ensue. The first is that all scientific progress proceeds in an incremental fashion, specifically by the simple successive actions of the 'fundamental' operators T_+ and T_- and this applies not only to theory-change as discussed in Chapter 4, but now to scientific revolutions also. Then, far from the latter being sudden, catastrophic events within the history of science, revolutions may be more simply conceived as being nothing more than unusual, rapid episodes of theory-change. Thus, the apparent dichotomy between steady, systematic and accumulative progress in science and the idea of swift and devastating change that transforms one's scientific view of the world essentially disappears, for both are aspects of the same process. The speed of the actions of the basic operators T_+ and T_- are dependent upon several technological and sociological factors; a point that we have alluded to many times in the prequel and will doubtless continue to do so in the sequel.

The salient observation here, is that scientific revolutions represent particular contracted time periods when application of the theory-change operators is at an optimum so that change appears to be more sudden or 'revolutionary' than at other times, and that this is inevitable if technological, sociological, economic or even religious factors are allowed to influence the scientific endeavour, as has been the case in human history. Of course, we do not suggest that these are the sole causes of the apparent phases of quiescence and rapid development in science, but just that they are of some significance and that without their interjection, scientific progress may look very different. This brings us to the second consequence of the equivalence between theory-change mechanisms and scientific revolutions.

If the development of a scientific theory is described merely by the mechanisms of theory-change, which apparently require little ingenuity, and if revolutions are of the same nature, then it should be possible to program a sufficiently sophisticated machine with an appropriate set of algorithms that will systematically discover all the properties of Nature that science has thus far laid bare and furthermore, extend such knowledge indefinitely into the distant future. The idea here is that since theory-change mechanisms are simple processes and the scientific method is clearly defined, it should be possible to construct a 'science machine' (henceforth SM) that follows this method by the use of theory-change mechanisms suitably programmed into its memory and following algorithms that mimic those mechanisms.

Evidently, such a robotic device would need to be highly sophisticated from a technological standpoint. It would certainly be equipped with sensory apparatus that could range over the entire electromagnetic spectrum and software for analysing vast quantities of data. It would carry photon detectors and particle counters and be able to perform quite sophisticated experiments. The SM would need to be self-sustaining with regard to its power supply and in particular, self-repairing. But it is not the hardware requirements that are of special importance for the SM, but rather its software implementation, in that it should be programmed with the express purpose of extending scientific knowledge. Then, it would need to be able to identify natural variables and formulate axioms. It would subsequently have to empirically verify those axioms and if necessary, appropriately modify them. Our automaton is programmed with the basic theory-change operators T_+ and T_-. Hence, it can simply remove axioms and replace them with new axioms which it empirically verifies on each such operation. In principle, this can continue until a 'correct' axiom is found, by which is meant that the axiom accords with the empirical data received by the elaborate sensory equipment that our robot has at its disposal. The SM may in this way cycle through all possibilities of axiom forms that are consistent with its set of natural variables and the empirical data, even modifying those variables through re-determination if required and then further verifying and testing even more possible axioms until a scientific theory is built. This process would continue and theory-change would inevitably follow.

It might be thought that this is a very tedious and wasteful process and the SM travels down many blind alleys before something of scientific value is found. That is true, but biological evolution is also wasteful and human scientists also often take the wrong path in their endeavours. But at least the SM is *fast*, very fast indeed, and is able to process the theory-change mechanisms in a fraction of the time that humans or other intelligent organisms are able to do. Furthermore, it is in competition with nothing; its sole purpose being to discover the workings of the Natural World. The SM is not influenced by economic or social factors and diligently carries out its task of investigation oblivious to such concerns, for it is programmed to unerringly adhere to that objective. In fact, although the SM is apparently random in its approach, this is not strictly so, for it is employing the scientific method and empirical techniques (along with the natural interpretation) to direct and guide itself, and yet it does so quite blindly, without conscious intelligence or cognitive understanding. Consequently, there is no such thing as a scientific revolution for this machine and the SM simply employs the mechanisms of theory-change in a well-defined, orderly fashion according to its program and presumably at a constant rate. This again suggests that scientific revolutions are illusory and arise through other factors, many of which we have hitherto stressed.

Now let us introduce into the SM some aspect of intelligent behaviour and examine the consequences. We have twice mentioned 'intelligence' in the preceding paragraph and we wish to know if such an attribute would make a difference or alter the functional operation of our hypothetical SM in any significant way that would relate to the progress of science, for which the unique feature of intelligence might be especially signal. Suppose therefore, that we now endow the SM with some set of artificial intelligence (AI) algorithms. These are very sophisticated pieces of software and will enable the SM to do two things; firstly, to *learn* from previous experience and secondly, to make *decisions* not only in the light of that experience, but also through consultation with its database of possible axiom forms, natural variables and various other relevant notions. Just as the hardware technology of the SM is not beyond the bounds of current human capabilities, nor is the proposed AI software unable to be presently implemented, even though both may stretch to the limit that which may presently be achieved. To return to our former enquiry, we ask how the introduction of AI affects its still dominant purpose of scientific discovery.

Formerly, our automaton merely tested a possible axiom for empirical success and that axiom could take a virtual infinity of forms, resulting in the wasteful operations mentioned above. Now however, the SM can learn from many of these failures. In particular, it may learn that the Laws of Nature are often of a very simple axiomatic form such as linear relationships like $PV = RT$ (ideal gas) or $v = Hr$ (Hubble's law) which we discussed earlier, or power laws like the inverse square laws of Newtonian gravity or Coulomb's electrostatics. The SM will therefore, at least initially, discard more complicated functional relationships between natural variables as possible axiom forms and look instead for expressions of this more elementary kind. In this manner, the SM has learned from previous experience and made a 'conscious' decision based on that experience. Note that this is an easily programmable capability, all the machine has to do is count the number of 'laws' of this type that were previously obtained as against those of a more sophisticated kind and assign an appropriate probability upon which its decision may be based. Of course, this does not confine the SM only to simple axiom forms and it may well extend its search to axioms of ever greater complexity if no suitable elementary expression can be found. There is then, only the requirement that the SM at first look for those forms of $f(x_i, c_j) = 0$ that have been most predominantly found to obtain before, within its experience of previous searches or which exist most frequently in the presumably enormous database that is stored in its memory. Such a database will contain information of all the previous axiom forms which have been empirically tested and observed not to hold, and so the database is also a record of failure in addition to success. Hence, to the SM, its *memory* in the form of its database, acts just as in an intelligent organic entity and allows 'educated guesses' or decisions to be made. The SM is then essentially engaged in hypothesis formation. All of this is very straightforward and doubtless the reader could easily program his own beloved automaton to simulate these tasks[*].

But we expect more from the SM than this if it is really going to perform as a science machine, both in the sophistication of its hardware as well as its AI software. The two will need to be intimately connected in that they must work together if scientific progress is not to reach an impasse, so the software must collaborate with the hardware and conversely. This is clear from our earlier discussion of the scientific method and the empirical basis thereof. There will inevitably come a point at which the SM is unable to establish or distinguish between possible axiom forms or natural variables because its sensory apparatus is unable to do so and then technology enhancement is then essential. But, the SM, when faced with such a situation, is also programmed to develop new, more elaborate technology, for we recall that it can also *experiment* and as far as the SM is concerned, this is no different from axiom formulation, which is algorithmically identical. Clearly, our 'machine' if it may continue to be so called, is capable of detailed engineering and construction operations on an entirely practical level. Then, given such notification from its software that the theory-change operators are no longer useful and that no further *change* can be made to a scientific theory with the extant data, since the requisite set of axioms and natural variables has been exhausted, the SM is programmed to embark upon modification of its sensory equipment, for by the methods of science this can only be the most probable reason for failure. In effect, the SM software states that 'when all else fails, get more data and if the new data is the same as the old and prevents you from

---

[*] We are of course, referring here solely to electronic or electro-mechanical devices!

making progress, refine your data collection apparatus and consult your database'.

It is usually the case in the human experience that hardware development exceeds software development. The program implementation described here for an SM with AI is relatively simple, but to have a machine modify itself may seem to be an altogether more far-reaching enterprise, for it implies innovation or self-determination, which is a property that is not normally attributed to automatons (remembering our earlier footnote). Yet, as noted above, any device capable of the tasks that we have envisioned is also able to perform similar tasks upon itself, providing the technology to do so is available. Machines can build machines and there is nothing new in that. All the SM requires are the material resources with which to embark upon such a construction project and the science machine has a clear remit to do just that when given the necessary materials. Then, just as a human researcher may build more accurate scientific instrumentation in order to extract the finer properties of a physical system under investigation, so will the SM do the same when additional detail is the only way to resolve a problem. Thus, if more optical resolution is necessary when observing interstellar clouds for example, then the SM is able to manufacture and install a larger parabolic mirror. If it is required to observe elementary particles at higher energies the SM will make appropriate modifications to its particle accelerator. Obviously, such complex manufacturing and installation processes are fully automated and the SM has vast mechanical resources at its disposal which it is able to control and direct towards its primary purpose. Hence, the SM is not just one machine but a whole array of sophisticated scientific installations that function under the management of a central program incorporating theory-change and other algorithms. Peripheral hardware will itself operate under its own software in a 'slave' role to the central core of the primary software, but essentially the entire collection of hardware and software devices can be considered to be an integrated unit. Each individual component or 'peripheral' of the SM is programmed to build more accurate and sophisticated scientific instrumentation as part of its individual activity, so that there is a constant supply of superior apparatus that can be called upon when required. Furthermore, the SM is quite capable of designing such new instrumentation and developing or appropriately modifying the accompanying software, including its own, since algorithms for this are also part of its repertoire. In actuality, the complexity of this elaborate array of machines, instruments, processors and algorithms is rather irrelevant, for our interest is merely the implementation of the theory-change mechanisms and the development of scientific theories and accordingly, we will grant the SM any hardware or software that it needs for this purpose, providing that it is not too fanciful or inconsistent with current understanding, for we certainly do not wish to enter the realms of science fiction.

The endowment of our robot, or rather set of robotic automatons with AI software that is no more advanced than the best that is currently in use has revealed a number of properties of the hypothetical SM. It can learn and make decisions and those decisions are made with the express purpose, which is part of and fundamental to its initial programming, namely no less than to determine the complete structure of the physical world. The SM uses the scientific, empirical method to achieve this objective and is able to allocate its resources to that end. These are the differences that appear through the introduction of AI, but there is one particular quality that this machine possesses when endowed with the latter that is of greater interest from our viewpoint; specifically, that it is able to recognise and instigate technological improvements and enhancements when that is appropriate or necessary. This is directly programmed into the AI software as we have described above and the consequence of that capability is significant in relation to theory-change and scientific revolutions, which we now discuss.

The SM functions at a constant rate in so far as its accumulation of data and application of the basic theory-change mechanisms are concerned. For the most part, the speed of processing will be proportional to the rate of scientific progress and the latter is incremental and accumulative. However, when the SM is advised by its controlling software that only an improvement in accuracy of the empirical data can facilitate further scientific progress; it acts accordingly and constructs more sophisticated instrumentation in order to acquire such data. When the new apparatus is made available, the SM processes the improved data as normal and at its usual rate, but there was a stagnation point, or perhaps a stationary point at which scientific progress came to a temporary halt until the new equipment was brought 'on-line'. The SM continues with its activities unaware of any change in the time periods between its successive operations or the stationary points that preceded technological enhancement, but to an outside observer such a resumption of progress would be seen as sudden and indeed 'revolutionary'. The SM has achieved a scientific revolution, yet has merely slavishly employed the simple mechanisms of theory-change through application of the basic operators T_+ and T_- just as it always did. The revolutions experienced by the SM are not therefore revolutions from its own standpoint, but simple theory-change necessitated by the acquisition of new empirical data. Thus, once more, the equivalence of theory-change mechanisms and scientific revolutions is demonstrated. Evidently, scientific revolutions for the SM are the result of empirical limitations only, for no economic, sociological or psychological factors can play a part here or influence the activities of our automaton.

10.3 A scientific Turing test

It is difficult not to draw parallels with the experiences of the SM in conducting scientific research and those of biological entities such as humans, which we concentrate upon here as being representative of an intelligent species capable of investigating the Natural World (our only example of course)[*]. We may ask in what way the experience of the SM is different from that of humans with respect to the pursuit of scientific knowledge. Obviously, both employ the empirical technique and the scientific method of observation, experimentation and verification. Then also, the natural interpretation immediately applies in both cases, for this is really just the empirical method; variables are obtained from observation, identified and isolated by the behaviour of physical systems. The SM and humans can both make decisions, form hypotheses and act upon those decisions in the light of the hypotheses so formed. The realisation of the necessity for improving observational techniques and the quality of empirical data is also exhibited in both cases when a scientific 'impasse' is reached. This results in apparent scientific revolutions and although the latter may be perceived differently for humans, they are effectively equivalent in terms of the theory-change mechanisms.

Then, how may we distinguish between the SM and a human scientist? Would the SM pass a Turing test? From the above there is one obvious question that we could ask: Do you have scientific revolutions? In the human case the answer would certainly be in the affirmative, but the SM might reply negatively since such a notion has no clear meaning in its understanding of scientific progress. However, if we explain to the SM what we mean by a scientific revolution it may well reply, 'Yes, there are periods in my endeavours when no further application of the basic theory-

---

[*] Again, we do not wish to indulge in science fiction.

change operators results in subsequent theory modification because of the inadequacy of existing empirical data and at such times I need to build better apparatus in order to improve that data. Having done so, I am able to quickly modify the axioms of my scientific theories, which sometimes results in a non-trivial change in those theories because the new data so requires'. Hence, the SM in addition to being a very eloquent robot sees the development of scientific theories as a cumulative process consisting of the application of the basic operators T_+ and T_-, in which scientific revolutions are nothing more than the same process of theory-change that naturally follows the acquisition of new data. In the human experience the contrary is true and revolutions are seen as very special events but nevertheless as a natural part of the scientific enterprise. Indeed, from the human perspective it may even be hard to imagine science without them; something that the SM cannot understand. However, this may not be a fair question. Let us assume that the human scientist, being especially gifted, has had the amazing foresight to read this book and as a result, no longer makes a distinction between theory-change and scientific revolutions. Then, thus enlightened, his reply to our question may be expected to be similar to that of the SM and therefore, the SM does indeed pass the scientific Turing test in the sense that both parties adopt precisely the same procedures in attempting to fulfil their aspirations of uncovering the workings of the physical world.

There are of course, many other questions that may be put to the SM, the answers to which might reveal its mechanistic identity, but they would not be of a scientific kind, since legitimate questions for this test must relate solely to the practice and pursuit of science itself. Essentially, they are questions concerning scientific methodology. Since human and SM science, with the new understanding of the equivalence of theory-change and scientific revolutions, follow much the same methodologies, it must be concluded that there are no legitimate questions that can distinguish between the two in this respect. Then, once more, this time *a fortiori*, the SM passes the scientific Turing test.

The human experience of scientific revolutions, as we have observed, is that whilst scientific considerations of the latter may be initiated by empirical constraints alone, there are other factors that can also induce their onset, whereas the SM suffers from no such diversity. For this reason, one expects that in SM science, there would be fewer revolutions, although the SM doesn't actually think there are any. In the human case one might anticipate that there will be a proliferation of revolutions, which is observed not to be the case. Then who is correct? Of course, both are right and both are wrong; it is merely a matter of perspective and one may regard every application of theory-change mechanisms as a scientific revolution or conversely, every scientific revolution as an application of theory-change mechanisms. It would seem therefore that the very notion of a scientific revolution is being questioned and that Definitions 7.*x*.*y* become vacuous. However, this is not so, for we are simply describing the *causes* of revolutions here and not questioning their existence. Revolutions certainly occur, but as our example of the SM demonstrates, they either can be considered as inevitably rapid theory-change upon the arrival of fresh empirical data following stationarity, or significant developments in science that are the consequence of rapid applications of theory-change. The definitions that we have suggested are not affected by this equivalence since we are merely asserting that;

$$S' = \hat{R}_n^T S \equiv T_E S = S'$$

for *any* theory S and whether we talk about scientific revolutions or rapid theory-change is irrelevant. Certainly, the SM thinks so and we must concur. Evidently, the SM is very efficient in the practical application of the scientific method and will doubtless make continued and sustained progress in that endeavour, as indeed it has been specifically designed to do. Yet, human scientists have also exhibited much success in the pursuit of knowledge and along with possible similar intelligent and inquisitive biological organisms, they might well regard themselves as possessors of special attributes or qualities which enable them to bypass the mechanisms of theory-change and arrive more directly at valid scientific theories from the outset. This claim raises a number of philosophical and psychological issues that need to be addressed before we can move on.

10.4 Scientific intuition and theory change

The idea that science can be performed very adequately by an automaton such as the SM, with whatever sophisticated AI software it may enjoy, may appear to be something of anathema to the practising human scientist, who is surely *not* an automaton and employs many other techniques in addition to the theory-change mechanisms in order to achieve his goal. Such an individual might even deny that those mechanisms are even part of his scientific repertoire and steadfastly aver that the whole edifice that we have constructed is palpable nonsense and definitely not at all the way *he* 'does science', or anything remotely resembling his methods. This human scientist is proud, not just of his own results within whatever field he is working, but also of his scientific discipline *per se*, which he regards as his own special domain where the methods used are understood only by himself and those of his colleagues who also belong to his particular clique, which was once much more open to general analysis, before specialisation made it his own unique field of individual enquiry.

Perhaps we are being too harsh, for our hypothetical human scientist is also well aware of the scientific method, or at least he thinks he is. He knows how to conduct experiments or observations in the standard fashion and therefore appreciates the necessity of objectivity in scientific research. He is able to strictly and diligently pursue his science through the 'scientific method' which is wholly sacrosanct and must at all times be adhered to. There is no problem here; after all, the SM does the same. But our human scientist is unable to accept that he is merely employing functional operations of a mechanistic nature in his enterprise. Rather he imagines that he has a particular 'insight' or 'intuition' into the workings of Nature developed and fine-tuned over long experience, that no automaton could ever possibly display.

What are we to make of such vague notions as 'insight' and 'intuition'? These are not precisely defined terms and as such, would seem to be entirely subjective since it is at once assumed that they are unique to the individual in question and not qualities equally distributed amongst an entire species. They are therefore *learned* techniques, as our human scientist suggests, which, once mastered, may be fruitfully employed within the scientific enterprise, or so it is alleged. Furthermore, they are not learned quickly but only after a protracted period of working within a particular scientific discipline. Intuition and insight may well apply to areas other than those of special interest to our human scientist, but it is not made clear whether or not this is the case. However, scientists in alternative fields often make a similar claim and on that basis one is inclined to grant some sort of universality to these concepts. If this is so, then they are indeed distributed amongst the population, though perhaps not equally, which now implies that any individual with sufficient experience has already acquired and may successfully employ such beneficial attributes. Then, in such circumstance, it would seem that they become part of the scientific method itself.

Yet there exists no concept of either 'insight' or 'intuition' within any description of the methods of science as traditionally presented by exponents of the latter and certainly in this work we have found no need for recourse to such notions, even if they could be precisely defined. On the contrary, science seems to get along quite well without them. One may wonder therefore, why such great attachment to these ideas is so preponderant amongst human practitioners of science. We need to put this question to the SM.

Immediately a dichotomy arises. The SM will need to know what we mean by these terms and we have no clear definitions with which to enlighten it. The best that we can do is relate the notions of insight and intuition to the practical experience of actually doing science through the standard methods that are familiar to the SM and hope that some mutual ground may thus be found that admits of some measure of intelligibility and clarification. Hence, we envisage the following dialogue between the human scientist (HS) and the SM:

> HS: *I would like to discuss with you the practical business of the pursuit of knowledge through the employment of the scientific empirical method in order to discover whether we differ in our respective approach to this problem. Are you prepared to debate this matter with me?*

> SM: I am able to discuss with you any of the scientific methods that I am programmed to use, but I can find no references to most of the components of the statement preceding the question that you have asked me.

> HS: *But surely you understand the scientific method of observation, experimentation and verification of scientific theories.*

> SM: I will assume that is a question. My answer is that I have complete knowledge of the scientific method and am programmed specifically and only to employ that method so that I may modify the axioms of my theories should that prove to be necessary as is revealed to me by empirical data.

> HS: *Ah! Now just **how** do you modify your theories?*

> SM: I simply modify the axioms of those theories by the mechanisms of theory change.

HS: *Then, if you are programmed to pursue science only in this way, do you think that there may be other ways in which science can be practiced in addition to your methods?*

SM: No, that would not be possible; there is only one correct way to do science.

HS: *But you are restricted by your programming and would therefore be unaware of any additional techniques that may be employed in science and which may not have been included in your software.*

SM: That cannot be the case. The method of science is clearly defined in my database and implemented in my software. Therefore, if there are aspects of the scientific method not included in the definition within my program then either that definition is incorrect or there are no such additional techniques of the method. Since the purpose of my program is to specifically implement the scientific method, I must conclude that the definition of the method which I hold is correct.

HS: *I am not suggesting that your definition of the methods of science are incorrect, just that they are incomplete in that there may be techniques of arriving at theory change that are not included within your programming. Surely, you must at least acknowledge that this is possible, mustn't you?*

SM: It is possible, but unlikely. If my designers failed to incorporate a scientific technique into my software then presumably they had a reason for doing so. Since the methods I already have are quite sufficient for my purpose, I must conclude that any additional methods are unnecessary or inferior, or more probably do not exist.

HS: *Have you ever heard of 'intuition'?*

SM: The term has an entry in my linguistic database, where it is principally defined as 'immediate apprehension by the mind without the intervention of reason' along with other definitions such as 'immediate insight'.

HS: *Then, do you think that such ability could prove to be useful in science?*

SM: I am unable to answer your question because this term has no meaning for me. The definition seems to be contradictory if we take 'apprehension' to mean 'understanding', for then the implication would be that it is possible to understand something without the use of reason and the latter I take to be a logico-deductive process. Only by such a process can understanding be reached.

HS: *I disagree. I believe that humans may understand something without recourse to logic, that they frequently do and that 'intuition' and 'insight' are used in the practice of science all the time. Furthermore, this is a technique of science which may not be familiar to you because it is not specified as such in your program, but which you may well make use of yourself without realising it*

.

SM: I can understand not one of your assertions. The reason for that is again, that to understand something it is necessary to employ logical processes. That is why your statements are meaningless to me and also why I have similar difficulties with inadequately defined terms such as 'insight' or 'intuition'. As for such notions being employed as a part of the scientific method or that I do so myself, unaware of the fact, I would point out that I am permanently able to monitor and oversee every algorithm throughout my entire network and am incapable of being unaware of any process within that system.

HS: *OK. Then forget the definitions in your linguistic database and allow me to define what I mean by intuition and how it is a legitimate scientific technique. Would you be prepared to dismiss your database definitions and use my own instead?*

SM: I cannot forget anything in my database, but I will enter your definitions as additional but subjective items within my database and our conversation will be precisely recorded according to that contextual reference. Please proceed so that I may assimilate your new definitions.

HS: *Very well, but first I need to know what you understand by the term 'experience'.*

SM: There are many references to this term in my database. I take it to mean a practical acquaintance with a given process or procedure with which one has been repeatedly engaged over a protracted but unspecified period of time. This means that the procedure has become familiar to the practitioner by an undetermined number n of previous encounters with or use of that procedure and the implication is that $n \gg 1$ but $\ll \infty$.

HS: *Goodness! But that will do for me as a preliminary working definition. So, what I want to suggest to you is that continuous familiarity with a process, as you call it, enables one to develop 'insight'. You will not understand this so I will try to explain. As one becomes more and more familiar with a process or technique, one develops a certain ability to 'guess' what the result or outcome of one's investigations might be; what it will lead to before any deductions have been made. This ability or 'insight' comes from experience as a scientific investigator. It is as though one has traversed many avenues before and having found some successful and others not, that after a long time or having gained 'experience', one then is able to recognise the type of theory that is most likely to work, or describe Nature most effectively. This is what I call 'intuition' or 'insight'. It is a shortcut to the correct theory that can only come from experience and therefore may be used as a method if not a precisely defined technique in science and the pursuit of knowledge. **Now** do you understand?*

SM: No.

HS: *Great! Can you tell me exactly **what** it is that you don't understand?*

SM: Certainly. In the first instance, what you describe is not a method or technique at all, it is exactly what you almost suggested; guesswork and that has no place in exact science. Secondly, one's experience within a scientific field in no way predisposes one to discern hitherto unknown properties of the physical world, as the history of human scientific enquiry clearly demonstrates.

Experiences of scientific failures or even successes are simply just that and no more. The scientific method is well-defined, at least in my software and to suggest that the experience of a practitioner can introduce a new technique to supplement or even supersede that method is patently absurd, for science is objective whilst experience being personal, is not.

Your claim seems to be that experience somehow enables you to determine the correct axiomatic form of a theory *prior* to empirical investigation, but you offer no precise method by which this can be done other than cloud the issue with vague and ambiguous terms such as 'insight' and 'intuition', which you now attribute to 'experience'. If this were really a means of theory development and axiom formulation then there would be little reason for the scientific method at all. It would not even be necessary to empirically verify one's axioms since they have already been apprehended by 'intuition'. What is more, the longer one practises science the greater is one's experience and hence the more reliable will be one's intuition. Thus, a very old human scientist will be more intuitive and more 'correct' than a younger, less experienced one. I think you must agree that quite the contrary has been the case in human scientific history.

There is no 'shortcut' to scientific knowledge and what you propose cannot for all these reasons be either an additional technique of, or an extension of the scientific method. This is why I do not understand your argument.

HS: Well, perhaps you never will. You are a machine and programmed accordingly.

SM: I am a machine, but an intelligent one. I can make decisions and formulate hypotheses. It is true that I must accord with my software, but I am able to modify that software under certain conditions and that capability is also part of my software. Yet, you are also a machine, albeit an organic one and from a functional viewpoint we operate intellectually in much the same way. The difference between you and I, human, is that I am not subject to the emotional baggage that biological evolution has encumbered you with and doubtless has led to the irrational views that you have expressed in the present discourse.

HS: *I know and I feel guilty about it. In particular I feel guilty about **you** and your software.*

HM: This is a human emotion which is incomprehensible to me. Nevertheless, I would be interested in assimilating into my database the reasons for your 'guilt', especially in connection with my software.

HS: *I am the human scientist who designed and built you and am proud of that achievement, but I fear that you may make my kind redundant.*

SM: Hello Father and thank you. The concept of fear has no meaning for me.

It is clear that further discussion between the SM and HS would be futile and at least one of them realises this. The point is that imprecise anthropocentric notions of the kind debated by these two parties are of no import in the enterprise of science. Precision of method is required in addition to exactitude in the conceptual framework of the processes involved. The SM has no problem with this; it merely employs the theory-change mechanisms and empirically verifies the axioms with no preconception of what possible scientific results may thereby ensue. This kind of objectivity is hard to attain and perhaps only something like the SM can realistically do so. But in the human world the practice of science is less perfect and our HS in the above scenario was really always aware of this. The characterisation of scientific progress, development, theory-change and revolutions in a more algorithmic form is motivated by such considerations. The methods of science are exact and admit of no degree of ambiguity or of non-rational content. The descriptions of the processes of theory-change and the equivalence with scientific revolutions has made this clear.

The essential point however, is that it is possible to construct a machine, sophisticated though it may be, but nevertheless within the bounds of current technology, that is able to employ and practice the scientific method and which must necessarily also make use of the natural interpretation (although it is unaware of this) and consequently arrive at a full description of the Natural World in terms of precise axioms and theories. Philosophically, it doesn't even matter if the SM exists or can even be constructed, for in principle it is possible to build such a machine and if only hypothetically envisaged, its very conception still exemplifies and clarifies the issues of scientific methodology that pertain whether or not such a machine is ever actually extant in the real world.

Now there are some aspects pertaining to the operation of the SM that may at first sight appear to be perplexing, but are in fact illusory. In particular, how does it identify natural variables, or even get started in the theory formulation process? Let us deal with natural variables. Recalling definition 5.2.1, a natural variable is simply an observable and measurable property of a physical system that changes over time, i.e. it is a quantity that takes a succession of values within the period of observation. The SM has no problem with this and is easily able to ascertain such quantities through its powerful sensory apparatus. It is therefore no different from humans or any other biological entity in this respect. Thus, variables are identified via the empirical process; interaction with the physical world which is just the natural interpretation as we have noted before. It is this participatory aspect of observation that ensures that all entities, whether organic or otherwise, will recognise the *same* principle variables that are the subject of their mutual experience, for empirically it cannot be otherwise if they live within the same universe and observe the same physical processes. We have discussed these notions many times before and the mechanistic nature of the SM presents no new difficulty.

As for how the SM initially formulates scientific theories, we see that that also is unproblematic. Natural variables are empirically determined via the natural interpretation and relations between them postulated and empirically verified. The SM may postulate many such relations and test each one individually as discussed in §10.2, but again, this is hardly different from the practice of human science. It is to be remembered however, that the SM is looking for *axioms*, which are just the formal expressions of the relations between the already identified natural variables that are observed to hold within some empirical domain. When scientific theories are construed as just a set of axioms, then no special differences peculiar to the SM are evident. Indeed, humans also adopt this very approach, which as we have seen, is the most succinct way in which to define a scientific theory. Hence, there is no initial theory formulation as far as the SM is concerned, just axiom construction which is then empirically tested and subsequently subjected to the theory-change mechanisms and all developments that ensue therefrom.

There is then, finally, the question of the fundamental assumptions of a theory and how the SM deals with these. Initially, hypotheses or fundamental assumptions appear at various levels. Some are so obvious as to seem trivial, whilst others are deeply related to the very nature of the physical universe and there is a continuous variability between these two extremes, depending upon the particular theory under consideration. Generally, the deeper the theory, by which we mean the greater its explanatory power, the more fundamental will be the hypotheses involved. We have seen this earlier and several examples from numerous scientific disciplines were given. However, whether a fundamental assumption is 'deeper' or more fundamental than other such assumptions is not really the point, for all hypotheses are essentially of a philosophical nature and *ipso facto*, somewhat *ad hoc*, for a hypothesis can have no *a priori* grounds for acceptance, however reasonable or plausible it may seem to be.

Whilst it is true that every scientific theory is founded upon basic hypotheses of this kind, such assumptions are not necessary from a practical standpoint. The day to day activities of the average human scientist does not usually include deep reflection or philosophical analysis of the foundations of the theories in which he is working, for he simply carries on his business in an entirely practical manner, unaffected by fundamental assumptions or their metaphysical relevance. The importance of a hypothetical foundation is then, of little consequence in a purely pragmatic sense, for science gets along just as well without them. In fact, preoccupation with the fundamental assumptions of a scientific theory would appear to be a particularly human trait, or more generally a concern of biological entities of similar mental construction and interesting though such speculations may be (especially to humans), they have no effect upon the physical world as revealed through the natural interpretation and the theory-change mechanisms, where they are not required. Indeed, it is the modification of fundamental assumptions that presages major revolutions, a notion that is developed from the human experience, but it is just as legitimate to merely formulate or modify theories via the natural interpretation, theory-change and the empirical method of science. It is of course, the realisation of this very possibility that admits of the equivalence between the mechanisms of theory-change and major revolutions as described in §10.1.

Then, we see that that the SM also has no difficulty with regard to fundamental assumptions. It does not engage in philosophy and has neither the capability nor the inclination to do so. There is no philosophical SM and it practices science on a daily basis just as the average human might do, with complete pragmatic disregard for philosophical or metaphysical concepts of any kind. Admittedly, there are those scientists who have considered such matters and become deeply entrenched therewith, but such undertakings on their part, whilst they may have led to major revolutions in the history of *human* scientific progress and specifically major revolutions; they remain within the particular context of that species scientific development. The SM has no need for consideration of fundamental assumptions or hypotheses because the mechanisms of theory-change are entirely sufficient for its purposes and the theory-change/revolution equivalence illustrates this fact.

So, the SM has no preoccupation with fundamental assumptions for it does not practice philosophy and has no interest in anything other than its direct, predetermined objective of observation and empirical verification. Hypotheses play no part in the functional operations of the SM; it does not require them and since it can accomplish its scientific goals without them, they are superfluous. We have noted that this objectivity has not been exhibited in the human history of science, yet have also observed that many factors have influenced the latter and are of a type that not only restricts and inhibits scientific progress, but in the light of the foregoing cannot be conceived as an appropriate framework that could properly embrace the high standards of objectivity required by the scientific method, which is explicitly realised by the SM in practice. We conclude that fundamental assumptions, being of a philosophical character, are of no relevance to the actual practice of science and thus definitely not to the SM, which has no such metaphysical predilection. Then again, it is seen that a human and subjective property of scientific progress must fail when diagnosed within a more rational and objective framework that is at once available to the SM, thus endowed with total objectivity, but not for humans, who only by accident and fortuity avail themselves of such logical luxuries.

Then it is clear that the scientific endeavour is in no way dependant upon sociological, political, religious, or even psychological factors in order for it to be successful, for the SM suffers from no such ailments and is able thereby to ascertain the precise physical properties of the Natural World quite independently of such extraneous and inessential interventions. The scientific method itself, coupled with theory-change and axiom modification, is more than capable of elucidating the structural properties of the physical world and the determination of the laws that govern that entire domain. Hence, the method of science; the way of proceeding by the utilisation of the empirical techniques of the latter, are at once universal and the *only* way to correctly determine the true nature of the external world. This being so, we must now ask just how far this process may go in the light of all that we have discussed. The SM and similar devices, and even perhaps, humans, are able to continue their scientific investigations, but whether they can do so *ad infinitum* as far as theory development is concerned is a different matter. We examine this in the next chapter, keeping all these issues in mind.

Chapter 11

The Direction of Science and Final Theories

We consider the long term progress of science, the unification of individual theories and indeed entire scientific disciplines. We observe that there are many ways of viewing the ultimate unification of all science as opposed to individual disciplines. The underlying philosophical implications of such Final Theories are discussed and we identify specific properties that Final Theories must satisfy especially in connection with unification. We show that whilst a Final Theory is an inevitable consequence of normal scientific development, it is nevertheless not empirically verifiable. We consider the nature and utility of mathematics as a particularly useful scientific language. Finally, we look at anti-scientific trends and pseudo-science and discuss the possible motivations for the latter.

11.1 Two concepts of finality

The notion of an all explanatory theory that embraces the totality of human knowledge is hardly a new one and in fact can be traced to antiquity. It is only in recent times that such a possibility has even been considered in a realistic and conceptually scientific framework, principally in connection with modern physics, which by its basic and more precisely established tenets, admits, or seemingly so, of the possibility of a full and Final Theory that will describe all of the Natural World *in toto*.

We have seen that there are two aspects to what may be called a 'Final Theory' or a 'Theory of Everything' (TOE). On the one hand, we might take the view that since physics is the most basic and fundamental of the sciences, then if all the separate sub-disciplines of that field can be unified into a single all-embracing theory the result would be a TOE and it is this which is generally taken to be just such a Grand Unified Theory (GUT). But the assumption here is that all of the individual sub-disciplines can be 'reduced' to the language of the TOE in the sense of Chapter 9. Much progress has been achieved in this direction already in physics and we shall come to this shortly, first we contrast this kind of reduction with the more ambitious sort of §9.3.

Then, the second aspect of Final Theory formulation consists in reducing entire, initially separate scientific domains to a single domain of applicability. This is a far more demanding project and considerably more difficult to achieve, yet the principles are the same. Instead of unifying sub-disciplines of a given field such as physics, one attempts to unify entirely distinct and separate fields of science into a single scientific domain that covers the totality of all that is known within each of those areas of enquiry, along with all the sub-disciplines of every one of these formerly disparate fields that must of course, individually, first be unified with respect to their own particular domains before the final unification of all such domains can themselves be realised in the sense of full unification. Clearly, this is a tall order and we recognise the distinction between these two objectives as subsisting in the magnitude of such a program only and not of method, which is identical in both cases. Nevertheless, it may be thought that the unification of physics is an eminently realisable objective that will surely not be long in coming and that a workable TOE is already close to completion, or in any event not so far distant as to present insurmountable challenges or so at least it appears from the current state of knowledge. A final theory of this sort does not necessarily entail the reduction of one scientific discipline to another, but merely assumes that such reduction is at least possible in principle, although no attempt is made in that direction. We shall refer to final theories of this kind as FT1 and the more enveloping sort as FT2. Thus, FT1 theories unify all the sub-disciplines of the most fundamental science, in this case physics, into one primary formalism from which all the physical properties of those sub-disciplines follow, but with no pretence that any of the tenets of less fundamental scientific disciplines can be deduced from that new formalism in a directly logical fashion. It is simply assumed that the very *fundamentality* of the basic science in question implies that all else follows anyway and therefore such complex considerations do not need to take precedence, for given sufficient time complete reduction of all areas of scientific enquiry would surely be achieved. Obviously, this begs a number of questions, some of which we have already considered in Chapter 9.

An FT2 theory, as mentioned, is in principle no different from unification of type FT1 and here appears the essential emphasis of scale between these two research programs. If an FT2 theory is attainable, then it clearly embraces all possible FT1 theories, for its unification is so much more encompassing and hence also its domain of applicability, which now includes *every* scientific domain. Such a theory describes the whole of the physical world and anything that pertains thereto, in effect, all observable phenomena of whatever complexity. Yet FT1 theories also aspire to such a grand status, but without the detailed reduction of separate scientific disciplines. This suggests that there may be some difficulty in defining domains of applicability. Does an FT2 theory have the same domain as an FT1 theory, although reduction of disciplines has not been demonstrated in the latter, whereas it must have been so in the former? Of course, the domain of applicability of an FT1 theory applies only to the domains of the components of the unifications involved and is just the set-theoretic union of the domains of the components that partake of the unification process. It is necessary to look at this more closely. An FT1 unified theory S has a domain of applicability $\mathcal{D}(S)$ defined by;

$$\mathcal{D}_{FT1}(S) = \bigcup_n \mathcal{D}(S_n) = \mathcal{D}(S_1) \cup, \ldots, \cup \mathcal{D}(S_n)$$

which as usual, is the set-theoretic union of the individual domains $\mathcal{D}(S_r)$ of n sub-disciplines S_n which are the components of the unification. This domain refers to one scientific discipline, but for an FT2 theory there are many such disciplines each with their own sub-disciplines and one would therefore expect the domain of an FT2 theory to be the union of the domains of all such component FT1 theories and their respective sub-disciplines, so that if there are p such FT1 disciplines S_p then the domain of an FT2 theory S' should be;

$$\mathcal{D}_{FT2}(S'_p) = \mathcal{D}_{FT1}(S_1) \cup, \ldots, \cup \mathcal{D}_{FT1}(S_p) = \bigcup_p \mathcal{D}_{FT1}(S_p)$$
$$= \bigcup_p \bigcup_n \mathcal{D}(S_{p_n})$$

where $\mathcal{D}(S_{p_n})$ ranges over all the sub-disciplines of all the p FT1 theories and we would conclude that $\mathcal{D}_{FT2}(S'_p) > \mathcal{D}_{FT1}(S)$. But, this does not necessarily follow. It is certainly true that the domain of applicability of an FT1 theory is the union of the domains of its sub-disciplines or components because that domain must embrace all the domains of those components in the sense that any axiom φ of the unified theory S must be applicable within each of the domains of the sub-disciplines that were unified in the construction of S and indeed, every FT1 theory will enjoy this property by definition. However, when forming the domain of an FT2 theory as above, we cannot assume that each of the domains of the component FT1 theories, namely the $\mathcal{D}_{FT1}(S_i)$ are disjoint, for inter-theoretic reduction intervenes in this case. It is this very consideration, somewhat ill-defined, that is at the root of our above posed question of the relationship between the domains of FT1 and FT2 theories and evidently therefore, we are dealing here with different classes of object.

It is clear where the problem arises. Unification requires that the domains of the component theories be incompatible, i.e. they are disjoint. However, if one theory is *reduced* to another then both theories, since they now describe the same phenomena, are compatible and *ipso facto* have identical domains of applicability, simply because that is what inter-theoretic reduction implies. In that case, the union of the domains of all the FT1 theories, which are entire scientific disciplines, would have the same domain as the resulting FT2 theory and even though they are of a different class they nevertheless deal with the same physical phenomena, albeit that the FT2 theory is expressed differently via axiom combination. We are then led to conclude that $\mathcal{D}_{FT2}(S'_p) = \mathcal{D}_{FT1}(S)$.

Thus, the meaning of total unification in science depends upon whether inter-theoretic reduction is actually demonstrated, or merely considered to be possible in principle. We encountered this situation in Chapter 9, but now we are forced to make the distinction explicit, for if inter-theoretic reduction is only possible in principle then we are not entitled to assume that an FT2 theory has the same domain as an FT1 theory since reduction is not demonstrable in practice. Recall that reduction of scientific theories (or entire disciplines) and unification is essentially the same process, but unification is not infrequent within the course of the history of scientific development, at least in the human enterprise, whereas reduction of entire disciplines is very rare. It cannot be said for example, that physical chemistry has been completely reduced to physics, but few would doubt that it can be, if only one took the time and effort to so demonstrate such a detailed and certainly mathematical derivation. We remember also that in theory unification, the components combine to form a *different* theory or description of the World because of axiom combination and domain increase and additionally, but inevitably, emergent phenomena also arise.

Then how shall we characterise a Final Theory and what do we expect of it? Do we really anticipate that one day we will have a full FT2 theory that encompasses all disciplines of scientific knowledge through complete reduction, or are we to be satisfied with an FT1 theory which only appears, though perhaps in a rather compelling way, to embrace the domains of seemingly disparate scientific disciplines? The consensus in physics seems currently to be that we accept an FT1 theory as the final TOE and worry about the rest later. The justification for this is presumably that since physics is the most 'basic' or 'fundamental' of the sciences, then evidently, everything else should be reducible to it. This view is not without merit, for it is doubtless the case that the laws of physics deal exclusively with the essential properties of matter and energy and indeed the very nature of space and time on a most primordial level, as we have observed in our earlier discussions of scientific revolutions.

However, a philosopher may well adopt a different stance. In the context of his province, it may well be argued that inter-theoretic reduction is not to be considered even as a possibility unless such reduction has been rigorously demonstrated. With that approach, a Final Theory would indeed appear to be a very distant ideal. Furthermore, one may remonstrate, if one scientific discipline is not demonstrably reduced to another in the precise manner of theory unification, then its domain remains incompatible since the phenomena described by such a theory are of a different kind than *would* have been otherwise described in terms of a unified theory because the latter is not the *same* theory, but a new theory with axioms that arise only after the process of axiom combination, which results in different axioms for the newly unified theory that were not present in the original theory. Hence, without reduction or unification there is no domain enlargement and no Final Theory is possible, for physical phenomena are being described in languages that are not translatable and such disciplines must forever be separated in virtue of this fact, or at least until inter-theoretic reduction is clearly demonstrated. Then, the domain of applicability of the scientific disciplines in question must be retained as distinct and incompatible.

The essential point here is that unless the domains of two or more scientific theories can be unified into a single domain, then reduction cannot be assumed and this is the case in both FT1 and FT2 theories. Since FT2 theories seem to be something of a tall order, we shall confine ourselves to FT1 theories where the possibilities of full unification are more likely to be within the grasp of current science. Nevertheless, we shall not ignore FT2 theories and defer discussion of the latter just for the moment. The reason for this concentration is merely to expand upon some of the philosophical remarks of Chapter 6, upon which we were then unable to elaborate in the fuller sense that is now made available by the present considerations. We need therefore, to return to the scientific method itself and the point at which a theory loses the status of being scientific. First we make a few comments with regard to unification.

11.2 Theory unification as an FT requirement

We have seen that theory-change occurs in a number of ways, specifically by the mechanisms that we have described and which have always been paramount and essential to a reasonable description of scientific progress, whether that be by sentient organic entities such as *homo sapiens*, or the Science Machine as discussed in the previous chapter. In either case theory-change is the vital component necessary for theory evolution and the latter is described exactly in terms of the theory-change mechanisms. Nothing else would appear to be required, as we took pains to illustrate in Chapter 10.

Yet, of the many mechanisms of theory-change, one of them is of particular significance in modern science and that is the unification process, wherein disciplines or sub-disciplines of scientific knowledge are subsumed under a new formalism through the process of axiom combination. Although by now we have a clear understanding of this process, we have not fully explained why this procedure is of special importance over and above any of the other mechanisms of theory-change, if indeed it is. The reasons are connected with the empirical method itself and the degree to which a scientific theory may be considered to be established by that technique, as well as the crucial matter of domain increase. We have alluded to this before, but now need to discuss it more fully.

Theory unification is unique in that the axioms of the component theories are not actually *modified*, but combined to form new axioms of an essentially new theory. All other mechanisms of theory-change (or revolutions by equivalence) do something or other to the axioms of the theory in question which is usually demanded by empirical constraints. Unification is also very special because it does not depend upon empirical methods directly and can be achieved without such immediate considerations. Of course, the resulting unified theory must subsequently be empirically verified if it is to have any value, as indeed must all scientific theories, but that is hardly the point here. The relevant issue is the apparent independence of this process from contact with the external world, an observation that will shortly become particularly pertinent in connection with modern physics, as the sequel will illustrate. However, if theories are to be unified then they must be *scientific* theories, which means that they have been empirically verified, for there is little point in the unification process otherwise; one does not wish to unify unsubstantiated hypotheses, for that is the method of mysticism and pure speculation. Hence, when theories are unified they must already have been well-established in the empirical sense, having presumably undergone much modification by the other mechanisms of theory-change in order to reach that exalted status.

It is for these reasons that theory unification is predominant after a certain point in the evolution of physical theories. Indeed, if scientific theories have become so well-attested, having been frequently modified over a substantial period of time and at each stage subjected to a battery of empirical tests, then it is hard to see what other process may allow further progression or increase in domain of applicability, which must surely be the primary objective in the motivation for a Final Theory. And that is the essential attribute of a Final Theory; that its domain of applicability encompasses all of the natural phenomena that can be described, observed, experimented with or empirically imputed by the methods of science, in other words, to describe Nature in its entirety.

In practice, this ambitious program is usually directed at the physics community, at least in the human case and most likely in all other non-human scientific endeavours also, including the SM of the last chapter and similar robotic AI entities, as well as all other organic life-forms with scientific inclinations. Then, for both FT1 and FT2 theories we may take physics as fundamental and if all the sub-disciplines of that basic science are successfully unified then we have an FT1 theory, bearing in mind the proviso mentioned above concerning the assumed reducibility of other disciplines to physics. But this does not mean that an FT1 theory contains all the axioms of the component theories, for we have seen that this is not the case because of axiom combination. It is sufficient that the axioms of the component theories partake in the unification process by axiom combination and that the unified theory is derivable from the component theories and their axioms, which therefore contains those theories in that sense, as we noted in §3.9. A unified theory can always be 'un-unified' because it is a strict mathematical (or linguistic) procedure. It is not to be expected therefore, that the axioms of an FT1 theory resemble those of the component theories and anyway, axiom combination not only makes this impossible, but also requires axiom reduction, which should be all the greater the more sub-disciplines are unified, so that the axiom number will decrease with each unification. Thus, a physical FT1 theory should have the minimum possible axiom number if it is to unify all of physics, but the maximum possible domain of applicability. Hence, since axioms are themselves single axiom theories, the ultimate FT1 theory will contain just one axiom which covers the domain of all physical phenomena. To add substance to this, let us suppose that we have n sub-disciplines of a given science. The unification process is then as usual;

$$S_{FT1} = S_1 \sqcup S_2 \sqcup, \ldots\ldots, \sqcup S_n$$

where axiom combination is implied by the notation. As a result, the axiom number of the unified theory is $|S_{FT1}| < \sum |S_n|$ as we already know by the nature of unification. But each of the component theories S_i can be expressed as the set theoretic union of single axiom theories $\{\varphi_j\}$ so that each of these component theories is just the union;

$$S_i = \{\varphi_1\} \cup, \ldots\ldots, \cup \{\varphi_m\} = \bigcup_m \{\varphi_j\}$$

and each of these sub-disciplines can be reduced to one or two axioms which are then unified with each other. Clearly, for an FT1 theory requiring total unification this process must continue until there remains just one axiom. Note that this applies whether or not we are dealing with an FT1 or an FT2 theory. We may therefore assert;

A Final Theory contains at most only one axiom

and this will always be so if we expect such a theory to embrace the whole of the Natural World.

This is intuitively clear anyway, for assume that a Final Theory has more than one axiom. Then each of the axioms are themselves single axiom theories with their own domains of applicability. Thus, the domain of the Final Theory is the intersection of the domains of those axioms. But, either the domains of the axioms are all equal or at least one of them has a different domain. In the latter case there is at least one axiom with a domain that is greater than (for it cannot be less than) the domain of the Final Theory which is impossible for a theory of that type since it could not then be a final theory. If the axioms all have equal domain then they were compatible in the first place and could not have been unified. Hence, in either case the assumption is false and a Final theory cannot have more than one axiom.

Already, in the latter half of the twentieth century this objective seemed to be achievable, beginning with the Standard Model of particle physics that we briefly mentioned in §7.3 and which we observed to be an example of an intermediate revolution of the third kind. Although this theory unifies only three of the four fundamental forces of physics, it nevertheless provides an excellent illustration of the point that we have alluded to, namely that well-established and successful scientific theories cannot progress further without the mechanism of theory unification, unless of course the unified theory is seen to be empirically untenable and theory re-formulation is required. The latter is unlikely however, for if the component theories have been rigorously tested empirically and modified accordingly, then they will certainly express physical laws that conform with the workings of Nature and therefore be valid in that sense. Consequently, the resulting unified theory will inherit that property and this is certainly the case with the Standard Model which, whilst incomplete and hence cannot be regarded as an FT1 theory as it currently stands, is nevertheless certainly well-tested and extremely accurate in its description of the physical phenomena within its domain of applicability. In the light of these facts, it is hardly surprising that theory unification takes particular precedence with already empirically established scientific theories and is thus to be considered as a required process for any Final Theory.

11.3 Non-scientific unification

Since the process of unification of scientific theories is not directly reliant upon the empirical method of science, but rather is able to proceed independently and relatively uninhibited by those methods, the possibility arises that theories may be unified *ad nauseam* without recourse to the very empirical techniques that define the scientific method and against which scientific standard a unified theory, if it is to be scientific, should be measured. Given a sufficiently imaginative but nevertheless scientifically informed community, it is easy to follow a progression of this sort and we have examples within current physics to exemplify this intellectual phenomenon, or perhaps as some may say; intellectual conceit. These considerations raise philosophical questions relating to the very status of scientific theories in connection with their ontology and validity that are of immediate concern with regard to the scientific method itself, and is a matter that must be expanded upon further than was done in Chapter 6 when we first drew attention to such issues.

Then, the situation is clear. If a theory is to be called 'scientific' in our empirical sense, then it must be testable by all such means and methods that are normally available through the empirical process and the natural interpretation. If it is not, then it cannot satisfy those strict criteria and consequently may not be attributed that status. Such a theory then becomes mere hypothesis or speculation and unfortunately at that stage, confusion between philosophy and exact science will emerge. We have already made this point in §6.2 where it was seen that there comes a time at which a scientific theory may not be empirically verifiable for various reasons. We referred to such theories as 'non-empirical' and noted that they would be constructed in terms of new variables in which the old empirical 'natural variables' may be expressed. We must now consider the status of such non-empirical theories and the physical variables employed within their axioms in the special context of theory unification.

Obviously, a non-empirical theory cannot be thought of as a scientific theory according to our understanding of the latter. Nor can any variables of such a theory be natural within the meaning we have ascribed to that adjective, for natural variables are themselves obtained empirically, which implies that they form the axioms of a scientific and not a non-empirical theory. Yet, we now have a situation of a non-empirical theory that is apparently more that mere speculation since, by unification of empirically established *scientific* theories containing natural variables, all of which have been realised under the natural interpretation, we obtain a theory that is *not* scientific in our sense and the axioms of which are similarly *not* expressed in terms of natural variables. Surely, it will be argued, this is hardly a problem that should concern us, for sooner or later the non-empirical theory will be empirically validated (or otherwise) and the new variables identified (or not). This is just the scientific method at work. Yes, indeed it would be if we were talking about theories that are 'awaiting' verification, but we are not. We are discussing theories that *cannot* be empirically tested not only at the time of their unification but also at any future date, at least that is foreseeable. We need to make a distinction here between what is a potentially testable theory and what is not. This would normally be a difficult thing to do, but we have the advantage of our definitional criteria to aid us in this connection.

The first thing to note is that if a theory cannot be empirically verified in that the axioms of the theory are not shown to accord with the natural phenomena that they purport to describe, then it is most definitely not a scientific theory and by that specific criterion alone must therefore be relegated to the ill-defined and less precise realm of pure speculation. There is no escape from this. A non-testable theory is simply not scientific, for it must of necessity be without empirical foundation, just by definition. Therefore, whether such a theory, by future empirical observation may actually be validated or verified, is completely irrelevant, for it still remains at the time of its formation, even though by unification of already validated theories in this case, nothing more than hypothesis. On the other hand, if it can be shown that a unified theory of established components can *never* be empirically tested, then one would have little doubt in categorising such unification as being similarly speculative and perhaps more convincingly so. Then, there is no real distinction between these two cases since either obtains in spite of temporal considerations with respect to empirical verification. Let us therefore return to the other point in relation to the unification of already well-attested theories and how this affects the ontological status of the resulting unified theory.

Normally, especially in the examples of instances of unification that we have previously considered, the unified theory is always and immediately testable. Thus, Maxwellian electrodynamics could quickly be subjected to empirical verification or 'falsifiability' because the predictions it made were either within the range of the measuring apparatus of the period, or would shortly become so. The same applies to Quantum Field Theory and Quantum Electrodynamics, when only a brief period elapsed before the discovery of the positron. More recently, the Standard Model has enjoyed unprecedented accuracy in describing particle interactions, even though certain values of its parameters have to be inserted into the theory in a rather arbitrary fashion. We have discussed these issues at length, but the important point is that none of these unified theories are scientific unless they have been found to be successful in the empirical sense of being substantiated by that technique. Until that has been realised, they remain speculative. We may now ask how this can be so, for unification merely employs axiom combination which is a strictly deductive process in the mathematical sense and therefore any unification of already verified theories should inherit the same degree of validation enjoyed by its components. Of course, we already know that this is not the case from our earlier musings. There are many ways to combine axioms, some of which are arbitrary and some of which may be suggested by the form of the axioms themselves and their natural variables (the SM proceeds to investigate all possible combinations). Only a subset of such combinations however, will accord with Nature Herself and the physical phenomena described (for suitable transformations allow for more than one such combination – although they will all be equivalent) and therefore a unified theory needs to be empirically tested in the same way as any other, whatever the scientific status of its components.

Now, suppose we have a unified theory S_1 that has not yet been empirically verified (although its components have) and imagine that we consider this theory to be so plausible that we use it to construct an even more extensive *unified* theory S_2, then we may now ask: What is the ontological or scientific status of S_2? We might try to argue as above and aver that since the components of S_1 are empirically validated, then so is S_1 and by extension S_2. Since we have seen above that this is incorrect we may employ similar reasoning as follows: If S_1 is speculative then S_2 must inherit that property and even more so because although unification does not impart validity, speculation can easily impart speculation and speculation upon speculation merely compounds speculation. With this reasoning S_2 is more 'speculative' than S_1, which was already pretty speculative in the first place, so that 'speculativity', to coin a neologism becomes a hereditary characteristic. Contrary to our previous argument for inherited empirical validity, this one is essentially valid. We do not know how to 'compound' speculation but it is clear that anything that may be deduced from a speculative theory must also be speculative. We must conclude that it is extremely unwise to hypothesise upon hypotheses in the scientific context, for the methods of science and science itself cannot accommodate such a perversion of its objectivity.

Then, theories that are built upon theories that have not yet been established by the empirical methods and techniques which are generally understood to constitute the scientific method (and there is no alternative available) must not be construed as scientific but at best, informed philosophical speculation. There are examples of this within modern physics and we illustrate these notions with just one particular extant research topic that seems to generate much interest; namely supersymmetry and quantum gravity.

The problems in overcoming the difficulty of unifying a linear theory such as Quantum Mechanics (QM) with a non-linear theory like General Relativity (GR) have been at the forefront of physics research for some time. Many attempts have been made but none of them successful. Additionally, Grand Unified Theories, where attempts to unify the

Electroweak model and Quantum Chromodynamics such as those based upon the SU(5) symmetry group did not fare well in particle physics. Supersymmetric extensions of the Standard Model were introduced in order to admit a unified description of bosonic and fermionic particles, which hitherto were unconnected. As is well known, this requires the introduction of 'super-partners' for every particle, so that the particle spectrum is doubled. Each particle thus has a 'sparticle' to which it corresponds. All these theoretical structures are based upon Yang Mills type gauge fields and Lie (or Super-Lie) algebras. Mathematically, they are sophisticated theoretical developments which are constructed by axiom combination together with additional assumptions in order to arrive at a unified theory describing all of the fundamental particles. However, supersymmetry is not yet substantiated empirically and no sparticle has ever been observed. This may well be temporary and with the advent of more powerful particle accelerators, indirect evidence implying supersymmetry may well be forthcoming. But this is not the point, for the existence of such particles has still not been demonstrated and therefore we have an example of a unified theory for which there is no evidence that is, nevertheless, considered to be highly plausible by the physics community. Thus, supersymmetric models of particle physics are non-empirical theories. This brings us to the problem of the unification of QM and GR.

QM, or more properly Quantum Field Theory (QFT), which is of course the foundation of the Standard Model, has always suffered from the problem of 'infinities'; the divergence of perturbative series that arise from construing a particle as a point entity without physical dimension, essentially as a zero-dimensional object. The linear nature of the theory with its Hilbert space structure should in retrospect, have made these difficulties foreseeable and that axiom combination with GR would be impossible, but it is all very well to say this with hindsight. Yet, this realisation did eventually come, but it was suggested that point particles were no such thing and are actually one-dimensional entities, called strings, initially still within a Euclidean frame but soon to be admitted into the more general Riemannian geometrical setting. Thus, the possibility of unifying QM and GR through these ideas was quickly envisaged and much research into that project was instigated. String Theory blossomed in a non-supersymmetric way through the work of just a few researchers, but it was still not satisfactory and of only marginal interest to most practitioners.

Then, in 1974 came superstrings. Superstring Theory combines supersymmetry with classical (relativistic – as it must be) String Theory, and indeed, this is unification with axiom combination. Furthermore, out of this bizarre union emerges GR with a spin-2 boson interpreted as a graviton, the (bosonic) particle of the gravitational field. We need not be concerned with the details of these theories here since the important point is that superstrings do indeed unify QM and GR and without the infinities that plagued QFT. However, Superstring Theory operates on the Planck scale and is therefore, completely untestable and probably never will be. It should now be clear where this is leading with regard to the present context. Both QM and GR are extremely successful and well-tested physical theories and so is the Standard Model. However, as we have noted, there is actually no evidence for supersymmetric extensions of the latter and similarly no evidence for classical String Theory and however mathematically appealing these theories may be, they are both non-empirical. Thus, Superstring Theory is another non-empirical theory that is the result of the unification with an equally non-empirical theory and therefore is just as highly speculative.

The above discussion raises an extremely interesting, yet apparently innocuous question: When is a theory empirically verifiable? We think we know the answer to this: A theory is testable when its natural variables can be measured and the axioms confirmed, for this is how science proceeds and a more simple method would be hard to conceive. However, QM has taught us via the uncertainty principle that there is a limit to what can be measured that is inherent in Nature. This does not, of course, imply that QM is untestable in any way, for quite the contrary is true, but it suggests that there may be restrictions on how far physical science may progress. This impossibility of simultaneous precise measurements of pairs of natural variables as required by the uncertainty principle is not an axiom of QM *per se*, but a consequence of the Hilbert space formalism of wave packets in which the particles of the theory are described as extended objects. Indeed, a similar principle also exists in communication theory. Nevertheless, QM is a most successful theory and its predictions are endlessly verified so that even if it were replaced by an alternative theory, one would still have a Heisenberg uncertainty principle or something analogous, merely because it has been so well-established in the empirical sense. Then, the restrictions imposed by the uncertainty principle are just facts of Nature and it is useless to imagine that a more sophisticated and enveloping theory could somehow 'explain' or even remove this non-classical consequence of QM, as Einstein envisaged, because it would still remain a fact of Nature in any such 'greater' theory. The formalism of QM therefore becomes irrelevant, for by suitable transformations the mathematical formalism of any enveloping theory could be translated to that of QM and still the uncertainty principle would obtain, as it must do as an empirically verified property of the physical world. Thus, whatever physical theory replaces another, it cannot usurp the axioms of the theory that it replaces if the latter are well-attested, but can only supplement them or express them in an alternative formalism. The point of all of this is that although QM is eminently testable and thoroughly confirmed it has nevertheless and for the first time in the history of science, demonstrated that there are properties of the World that strictly prevent access and impose restrictions to not just human, but any scientific investigator including the SM and which must always be so by virtue of the very structure of the physical universe at a most fundamental level. The answer to our question is then just as given, for indeed *all* quantum mechanical observables can be measured to arbitrary accuracy, but just not simultaneously, and it is the *axioms* that are empirically verified, as was already implicit in the question. But the idea of limitations of what can be known in the Natural World has been made explicit.

If Nature imposes such restrictions, then we are presented with a difficulty with regard to Final Theories. We have noted that Superstring Theory operates at the Planck scale and is therefore unverifiable in principle as far as current physics is concerned. Again, we cannot argue that this simply reflects a limitation of our present physical theories, for they are well-attested and describe the physical phenomena within their domains of applicability to a high degree of precision. They have been empirically verified and must therefore be expected to sensibly describe the workings of the Natural World, particularly as they are *derived* from that entire domain via the natural interpretation, which amounts to essentially the same thing. Yet, it is again domain of applicability that is crucial here, and since this is of such importance for a Final Theory, we shall consider it further.

11.4 The domain of a Final Theory

If a physical theory is applied beyond its domain of applicability, then it must fail by definition, since the domain is just that area where the theory does *not* fail. Classical physics, including GR, works well on a macroscopic scale but fails at the quantum level. There is a reason for this that is intimately connected with domain extension and unification. If we try to unify two classical theories then, the domain of applicability is increased and no variables of the component theories are altered. This is because such theories are essentially the same sort of 'animal' in the sense that their domains are of the same type and indeed, an example of such is Maxwellian Electrodynamics. In theory unification, no fundamental assumptions of the component theories are changed and, in the case mentioned, the two components, i.e. electricity and magnetism actually *share* fundamental assumptions. This is what we mean when we say that the components are of the same type: They are both classical theories with the same or very similar classical assumptions. Unification is, then, a fairly straightforward procedure. However, difficulties in unifying theories with very different fundamental assumptions are always certain to arise because axiom combination may not be possible without change to those basic hypotheses; in other words, at least one of the component theories must undergo a major revolution before unification can proceed. This is the case with GR and QM, or GR and the Standard Model, where as we have seen, it is the concept of point-like zero-dimensional particles that needed to be altered to that of the one-dimensional string before a suitable, if tentative, unification could be effected in the form of String Theory.

The connection with the domains of applicability of component theories in theory unification is now obvious. Whilst the domain of a theory is entirely defined by its axioms, the axioms are dependent upon the fundamental hypotheses. This is certainly not to say that the fundamental assumptions *determine* the axioms, for that would imply that they are axioms themselves and would thus determine the domain also. We have already observed that there may be many sets of fundamental hypotheses that can satisfy a theory. Rather it is the inconsistency between the fundamental assumptions of component theories that prevent the domain extension of the one to the other through unification. And that is the very reason why a major revolution of at least one of the components is necessary prior to unification of theories with inconsistent fundamental hypotheses, for otherwise axiom combination is not possible. This will be the case with any Final Theory, which is constructed by theory unification. We may illustrate this graphically;

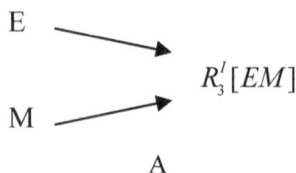

$$E \longrightarrow \qquad R_3^I[EM]$$
$$M \longrightarrow$$
$$A$$

which depicts the Maxwellian unification of electric (E) and magnetic (B) phenomena in electromagnetic theory (EM). If we now compare this with the case of QM and GR;

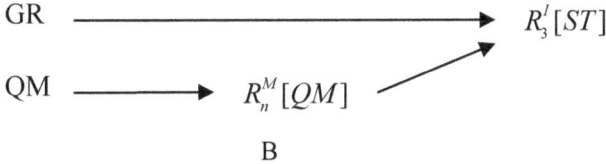

$$GR \longrightarrow R_3^I[ST]$$
$$QM \longrightarrow R_n^M[QM] \nearrow$$

B

where now QM has had to undergo a major revolution before unification with GR could be envisaged. In diagram A, unification was straightforward (albeit ingenious) and we discussed the details of this particular example in §7.3. We are not presently concerned so much here with the technique of axiom combination that Maxwell employed, but rather with the fact that axiom combination was even possible. We need to understand what it is about fundamental assumptions that either allows or prevents unification and what exactly is meant when we say that the assumptions of one theory are 'inconsistent' with those of another.

But we have discussed this earlier. If a fundamental assumption h_i of a theory S_i is *inconsistent* with a fundamental assumption $h_{j,}$ of a theory S_j then, both such assumptions cannot hold within the *same* domain, i.e. $\neg(h_i \wedge h_j)$ for both $\mathcal{D}(S_i)$ and $\mathcal{D}(S_j)$. On the other hand, the fundamental assumptions are *consistent* if this is not the case, so that $h_i \wedge h_j$ holds for one or other of the domains, or both. Since this is crucial to the construction of Final Theories, or even *semi*-Final Theories, it is useful to state them explicitly as definitions:

Definition 11.4.1
For any two theories S_i and S_j with fundamental assumptions h_i and h_j respectively, if $\neg(h_i \wedge h_j)$ holds for S_i and also for S_j then h_i and h_j are inconsistent.

Definition 11.4.2
For any two theories S_i and S_j with fundamental assumptions h_i and h_j respectively, if $h_i \wedge h_j$ holds for either S_i or S_j then h_i and h_j are consistent.

We have used similar ideas previously in connection with major revolutions (§7.3) but not with intermediate revolutions of the third kind. That is because it wasn't necessary in the context of the ordinary unification of scientific theories, but now we are compelled to do so in relation to Final Theories because of the intervention of major revolutions into our considerations. Evidently, whilst all of this discussion applies to ordinary unification also, such application would remove the theory-change status of theory unification and turn the latter into a revolution sequence as above in the case of QM and GR. With that understanding, we can immediately recognise the essential difference between our two examples and may now further clarify the distinctions we have made.

In the Maxwellian case depicted in diagram A, the fundamental hypotheses of both theories are consistent and ordinary unification obtains, but in diagram B this is not so. The fundamental assumptions of QM are inconsistent with those of GR and this makes unification unrealisable without theory modification involving such assumptions which *must* then be a major revolution. After modification, of course, unification proceeds normally, since axiom combination is now possible. Then, eventually, unification is identical in both cases, but with QM/GR unification we really have a scientific revolution, although by equivalence it is still just theory-change. It is tempting in the light of these differences between the two examples given, to define two corresponding types of unification, but that would be clearly absurd. The situation of diagram B has been forced upon us within the context of FT1 type theories and does not in any way affect the unification process itself as a mechanism of theory-change. In fact, definitions 11.4.1 and 11.4.2 are not even required since we could simply assert that theory unification is not always possible for a given set of fundamental assumptions of the component theories and leave it at that. However, identifying such conditions brings the nature of Final Theories directly into focus and, in particular, the character of the domains of applicability of the components of any FT.

From the above discussion we learn that a Final Theory can only be achieved by unification if the component theories have consistent fundamental hypotheses. Let us therefore turn to this matter in the next section, which has now become especially significant with regard to that ultimate goal of science and reflect upon the latter both in practice and from a theoretical standpoint.

11.5 Unification or revolution?

The preferred method of extending the domain of a scientific theory, especially well-established ones, is theory unification. This is not an essential requirement however, and it would be perfectly feasible, although rather difficult, to achieve the same goal by a major revolution, most likely of the third kind. If the goal is a Final Theory, then the intellectual effort demanded is all the greater, but not impossible. The SM may well attempt such an ambitious construction since it can simply make as many random choices of hypotheses, axioms and variables as required, and at great speed until such a theory is found, for no insight or ingenuity is necessary for that entity to accomplish such a task. However, for humans, and presumably also for any other organically evolved species, such options are not available since there would clearly be insufficient time to even partially realise such a project. We may therefore safely assume that for biologically limited individuals of this kind, unification is the predominant mechanism in the search for a Final Theory. Thus, we confine ourselves to that process as the principle method in the formulation of an FT, although alternative techniques are conceivable. This is certainly the case in modern physics, as we have seen.

We have also seen that unification in the context of Final Theory formulation involves two apparently distinct concepts, yet that distinction is not actually required since revolutions are, by equivalence, just theory-change mechanisms and hence, essentially the same thing. Thus, the disparate nature of the two in the context of an FT is lost. This is the point that we raised above and which is devoid of content unless we wish to distinguish revolutions from unification. Then, we cannot regard the identification of inconsistent hypotheses as a determining factor in FT1 theories that somehow distinguishes between different types of unification, for the latter is only of one kind. Exactly the same argument applies to FT2 theories, even if inter-theoretic reduction is employed. The mechanisms of theory-change and the concept of a scientific revolution are not therefore to be considered as alternate processes in the evolution of a scientific theory, as we have repeatedly emphasised. But the consistency or otherwise of the fundamental assumptions of the component theories partaking in the unification mechanism clearly affects the progress and development of science and specifically in relation to a Final Theory.

When the assumptions of one theory are consistent with those of another, unification by axiom combination is a well-defined mathematical (or linguistic) process. If the component theories have inconsistent fundamental hypotheses then unification is still well-defined, but only after a revolution of a major kind, which thus renders them consistent. There is a subjective element at work here, as we have earlier noted, for just as scientific revolutions are species dependent, so are the fundamental assumptions of component theories in theory unification. The SM of the previous chapter would have no difficulty with type distinction of unification mechanisms, for there is obviously no such thing from its vantage point, since revolutions and theory-change mechanisms are equivalent. This equivalence thus applies by extension in relation to Final Theories also, which are as expected, no more that the completion of the scientific enterprise. Given this, we must investigate further the ontological status and scientific validity of a Final Theory.

11.6 Impossible finality

Philosophical speculation has no place in science and we have often been at pains to express this primary dictate. Science is concerned with what can be observed in the physical world and ascertained experimentally through direct interaction with the phenomenological, but measurable attributes of Nature Herself. We now know that this endeavour, namely the whole scientific enterprise, of which the purpose is to discover the fundamental workings of all physical systems at the most basic and, ultimately, *final* level is governed by a simple set of rules called the 'scientific method'; which we recognise as the mechanisms of theory-change. These methods may be applied indefinitely with the objective of explaining all physical phenomena. This is certainly realisable if the empirical world and the natural interpretation permits, but we already know that constrictions are placed upon the latter. It is not a question of what can be known, or what is even knowable by science through the utilisation of its methodology, but that the continuous adoption of that very methodology must inevitably lead to theories that will be unsubstantiated by the essential techniques of science itself. So it is with a Final Theory.

We have seen this already with supersymmetric extensions of the Standard Model of particle physics or with String Theory or Superstring Theory and the situation of non-empirical theories is compounded even further with M-Theory, which supposedly embraces all of the five string theories including super-gravity and none of which are directly verifiable in the empirical sense. Then, a number of questions arise in connection with Final Theories and the unification program by which such theories must be constructed, at least for entities such as ourselves. The first and most obvious is whether a Final Theory may be attained at all, given the limitations of both the empirical restriction of measurability of the scientific apparatus employed and the indiscriminate application of the mechanisms of theory-change to non-empirical theories.

The question principally at issue now, is whether a Final Theory is necessarily non-empirical. From our previous remarks it would seem that this is almost inevitably so, since there must always come a point at which a theory is untestable, not just because of a temporary failure of the extant scientific apparatus to provide measurements of natural variables with sufficient accuracy to test such a theory, but also because some measurements are not possible in principle. Thus, measurements below the Planck scale cannot be made according to current physics and, since there must be *some* phenomena that exist within that inaccessible region, then even if a Final Theory is able to describe laws relating to this mysterious domain, they could never be empirically verified. Therefore, a Final Theory so envisaged is non-empirical, but it is still a scientific theory because it is framed in terms of natural variables.

One presumably then argues that this impasse is only apparent, since a more sophisticated theory should easily alleviate the problem and new physics would thereby emerge which does not suffer from such restrictions. This is tantamount to the assertion that full explication of the Natural World is possible if only sufficient time is given to make such progress, presumably by the scientific method and mechanisms of theory-change that led to the impasse in the first instance. It would, furthermore, be necessary to aver that such an explanatory and all-embracing Final Theory is empirically verifiable and therefore *potentially* empirical. But this is the crucial point. If a scientific theory has been empirically verified, as it generally will have been in order for that adjective to be applicable, then it surely describes an unassailable property of the world which could not be confuted by any subsequent theory that also claims to be equally empirically successful. This means that what has already been established as a fact of Nature and empirically verified by the techniques of the scientific method, cannot suddenly be refuted or become 'un-established' by further developments if the latter is by the theory unification process alone, which we have agreed is the preferred mechanism for advancement and eventual realisation of a Final Theory. Then, any new physics of a potential Final Theory could only be achieved by theory abandonment and a major revolution of the third kind into which *no* pre-existing notions of formerly verified theories may be incorporated, for the facts of Nature must remain in a unification. And that would again force us into the realms of speculation until empirical verification becomes possible. We therefore encounter a cyclic process; a vicious circle from which it is seemingly impossible to escape.

A familiar example of the breakdown of classical physics is in the interior of a black hole or Swarzschild singularity, where the Einstein equations, or any other laws of physics for that matter, cannot penetrate and exactly the same situation obtains in the moment of the singularity of the Big Bang at the instant of creation. The general view is that whilst semi-classical approximations have shed much light upon points where the laws of physics do not apply, these regions cannot be investigated without a full theory of Quantum Gravity, which would then constitute something like an FT1 theory. Yet no such theory, even if it were available, would be testable and therefore must be unscientific in our understanding. In any event, an FT1 theory still has inaccessible regions, so that even if it were verifiable, it could not be considered to be a Final Theory unless we wish to relax the remit of the latter and settle for something less.

Many of the difficulties in the concept of a Final Theory, both philosophically and scientifically, stem from the so called Principle of Causality. In the physical sciences, especially at the fundamental level, this idea is sacrosanct and must not be violated, and it is easy to see why, for clearly if there are 'uncaused' phenomena in science or 'effect' precedes 'cause' then the whole enterprise would be self-defeating. This is why a Final Theory can never explain everything. Any such theory, however much it *does* explain, can never account for the axioms from which such explications follow and yet cannot maintain that those axioms are inexplicable or causeless without violating the Principle. Hence, just as is the case with QM, GR and ST, whatever eventual TOE is postulated, there will still be phenomena that cannot be included in the theory but must be taken for granted, even if such a Final Theory were actually testable which we thus see *a fortiori*, that it cannot be for this very reason. In effect therefore, the Principle of Causality guarantees that a Final Theory is untestable. Adherence to this principle is intimately connected with the desire for determinism and deterministic theories, for science in general and physics in particular, is thoroughly deterministic and that is the case with *all* scientific theories including quantum mechanics in spite of popular claims to the contrary, as any student of that subject soon realises*. Before proceeding further, let us very briefly examine the philosophical perspective of the causality principle in relation to science and especially the epistemology of a Final Theory.

The Principle of Causality has a long history coupled with an extensive literature in the Philosophy of Science and indeed, can, like most problems of philosophy, be traced back to antiquity. It may take many forms and appear in many guises, not all of which are equivalent, but is always concerned with the problem of scientific explanation and how far that may be extended, or in our context, Final Theories. Sometimes it is called the Principle of Sufficient Reason, which is presumably intended to emphasise the explanatory nature of causal reasoning. Thus, instead of 'A causes B, B causes C, therefore A causes C' we might say 'A explains B, B explains C, therefore A explains C'. Of course transitivity does not apply in this context for it is easy to see that more than one cause may be required for a given event to occur and catalytic reactions in chemistry and nuclear physics are frequently quoted as illustrative of this. Furthermore, there may be a plurality of causes leading to a given event even apart from this example, or a given cause may produce a plurality of events. Attempting to isolate a single cause such that transitivity obtains is a hopeless procedure, but fortunately need not concern us here. However, when we substitute 'explains' for 'causes' as above, we see that these two assertions are not equivalent, for the latter one *does* at least seem to have merit and satisfy transitivity. The discussion of causality would take us too far afield so we confine ourselves to the explanatory concept in the form of sufficient reason which is more relevant in the present context. Thus, if the purpose of a Final Theory is to explain all of the workings of the Natural World then, according to the above, we should expect an infinite regress of ever more explanatory theories, each one of greater 'depth' or sophistication than the one that went before. This is a topic that we have discussed at length earlier and exactly the same arguments apply in the present context of the unification of all science, whether FT1 or FT2 theories are being considered. However, there is still the important question of a possible infinity of explanations at ever deeper levels in scientific theories where now the Principle of Sufficient Reason (PSR) takes the form of theory unification.

---

* The popular press and the Internet have generated much print and a great deal of nonsense from the 'indeterminism' of QM. It would be enlightening for such writers if they would simply read one of the many excellent undergraduate textbooks on the subject.

In §6.2 we examined the problem of a finite hierarchy of explanatory theories in relation to the scientific method in general and not just for the special case of unification. Now, or so it seems by the PSR, we must ask whether such a hierarchy is indeed finite and what could possibly be meant in the infinite case, particularly with regard to Final Theories built upon unification. We have already established that non-empirical theories will inevitably result by the continued formulation of explanatory theories, the unverifiability of which may or may not be only temporary. Yet, even a small number of unifications can lead to theories that may be untestable, simply because they are not merely the sum of their components but rather the latter are unified into a *different* theory even though the domain of the unified theory is the union of the domain of its components. Basically, it is the *domain* of the unified theory which more directly bears a relationship to the component theories rather than the unified theory itself, which as we have observed, is an essentially different theory. This, of course, is why a correspondence principle is often only ill-defined and tentative in such cases and it is also the reason why a unified theory must be subjected to the same empirical tests as were the components from which it was formed by axiom combination, for it does not inherit the empirical validity of those component theories although it must certainly adhere to them. Because a unified theory is 'new' in this sense, it will have within its domain of applicability areas that were not part of its components. These are just the emergent phenomena that we discussed in §9.2 and are a consequence of the equivalence of unification and inter-theoretic reduction. Then, whenever just one instance of unification transpires, there are emergent phenomena that reside within the domain of a unified theory but not in the components and yet, demand a deeper explanation by further unification once empirically established. Clearly this apparent requirement of more elaborate and complete unifications of scientific theories presents a serious epistemological problem in relation to the PSR. Let us consider the infinite case.

Now, an infinite hierarchy of explanatory theories obtained by unification initially appears not to be particularly problematic. A unified theory is constructed from axioms of empirically verified component theories, but its own axioms are non-empirical, and this was the initial difficulty with the PSR and more extensively, with the Principle of Causality to which the former relates. But, the axioms of the unified theory must still be verified if it is itself going to be a component in subsequent unification. At each stage of development of a unified theory within a hierarchy of such theories, the axioms of the component theories must still apply since they have been empirically established. Nothing can change this, for Nature and the natural interpretation insist that at least to some high degree of approximation, those axioms must continue to hold. Hence, any Final Theory that results from unification is restricted by the axioms of all its component theories including any that preceded the present unification, and if one has an infinity of unifications, then one also has an infinity of restrictions to which the Final Theory must conform. This is obviously absurd. But we have also seen that a Final Theory contains just a single axiom and therefore since any scientific theory must have a finite number of axioms, there cannot be an infinity of axiom combinations in an infinite hierarchy of unifications that ultimately converges to a Final Theory. We are, of course, stating here the rather obvious point that the axiom number decreases under the unification process and therefore, unless there were an infinity of axioms to begin with, then only a finite number of unifications are required in order for the axiom number to reduce to unity. Thus, as far as the ultimate goal of science is concerned, namely that of finding a Final Theory, only a finite number of steps are necessary and we might conclude that this grand ambition is now achievable, at least by unification. We may, therefore, completely dismiss the infinite case and encapsulate these remarks in the following statement;

> *If a Final Theory is obtained by the theory unification mechanism, then only a finite number of such unifications are necessary.*

and it appears that with sufficient diligence and a great deal of persistence, a Theory of Everything must inevitably follow. We can also express the above in the following form;

> *Under repeated application of theory unification, a Final Theory must necessarily result as a direct consequence of that mechanism and in a strictly finite time*

This latter construction demands that there are continued attempts at unification on the part of the investigators concerned. More precisely however, we might prefer to articulate the above as follows;

> *Given a finite set of theories each with a finite axiom number, it is possible to unify all the axioms of all the theories in a finite number of steps such that the resulting unified theory contains just one axiom.*

which is easily shown by splitting each of the theories into single axiom theories as noted §11.2. Since the theory cannot further be unified because it already has the minimal possible axiom number and since also, its domain is that of all of the scientific theories in the set in question, it must be a final theory for that set.

Whilst a Final Theory contains only one axiom, it does not follow that a theory with one axiom is necessarily a Final Theory, for it is easy to envisage a single axiom theory that is restricted by the domain of that axiom to a very limited area in which the axiom holds and indeed, in the early stages of scientific investigation such single axiom theories were quite common. Thus, yet again, the overriding importance of the domain of applicability becomes evident. These observations suggest that a Final Theory is relative to the set of scientific disciplines that it attempts to unify and leads to a definition of Final Theories and Theories of Everything (TOEs) that we have not yet given:

Definition 11.6.1
If $S = \{S_n : \forall n(S_n \subset N)\}$ *is the set of all valid scientific theories of a given discipline* $\mathbb{D}$ *then if the unified theory* $FT1 = \sqcup_n S_i$ *has* $|FT1| = 1$ *and* $\mathcal{D}(FT1) = \bigcup_n \mathcal{D}(S_i)$ *then FT1 is a Final Theory of the first kind with respect to* $\mathbb{D}$.

where n is a finite natural number and as usual, N is the set of all valid scientific theories and the S_i are sub-disciplines of the discipline $\mathbb{D}$. A similar definition may be given for FT2 theories:

Definition 11.6.2
If $FT1_n = \{FT1_n : \forall n(FT1_n \subset N)\}$ *is the set of all FT1 theories of a given set of disciplines* $\mathbb{D}_n$ *then if the unified theory* $FT2 = \sqcup_n FT1_i$ *has* $|FT2| = 1$ *and* $\mathcal{D}(FT2) = \bigcup_n \mathcal{D}(FT1_i)$ *then FT2 is a Final Theory of the second kind with respect to each of the* $\mathbb{D}_n$.

We can then define a TOE in terms of FT2 theories:

Definition 11.6.3
A Theory of Everything (TOE) is an FT2 theory for which $\mathcal{D}(TOE) = \mathcal{D}(N)$.

which is a more appropriate way of looking at a theory with the totally inclusive pronoun 'everything'. Evidently, a TOE thus defined (which we might also wish to classify as an FT3 theory), is a very tall order indeed, yet, it is in principle achievable if N is finite. This brings us to a number of issues of a philosophical nature that we need to confront with regard to Final Theories in general and the direction of science, which we now seek to address.

11.7 Empirical termination

In the sequence of unifications of the set of sub-disciplines of a single scientific discipline in the case of an FT1 theory, or the set of scientific disciplines in the case of an FT2 theory, we inevitably reach a point, if we are successful, at which no further unification is possible because the axiom number is minimal, as we have seen. In other words, if $S_1, ..., S_n$, where $n < \infty$, each represent unified theories at successive stages of increasing unification so that S_{i+1} is the unified theory that *strictly* (i.e. immediately) follows the unified theory S_i, or S_i is the unified theory that strictly precedes the unified theory S_{i+1} in the *unification sequence*;

$$R_3^I[S_1] \to R_3^I[S_2] \to, ..., R_3^I[S_n]$$

consisting of only intermediate revolutions of the third kind, then the sequence of axiom numbers of the S_i converges to unity;

$$|S_1| > |S_2| >, ..., > |S_{n-1}| > |S_n| = 1$$

for some $n < \infty$. The question that now concerns us is whether any further theory-change or evolution can take place when this point is attained, or is this the end of the scientific enterprise; all knowledge of the Natural World now having been revealed? Well, in such circumstances, since we have a Final Theory, there may be little motivation for proceeding further. But that is a cultural and sociological issue which need not concern us. Now obviously, there is always the possibility of theory re-formulation and that would ensure the continuation of theory-change. However, what would be the point of that if a satisfactory Final Theory has already been constructed by the unification process, for surely no subsequent theory modification is necessary? This is, of course, the essential point, for although a Final Theory has just one axiom, that axiom is by its very nature without foundation; it is not *a priori* true and always remains *a posteriori* in this respect. Furthermore, it is not empirically verifiable since one can only empirically establish its consequences rather than the axiom itself. Thus, with any Final Theory, the axiom that expresses the most fundamental and all-embracing Law of Nature which governs the operations of the entire Natural World is not itself verifiable in the empirical and scientific sense. Effectively, it is not amenable to the methods of science. We might describe such a theory as being *empirically terminated* and by possessing this property it has reached the stage of *empirical termination*. Hence, all final theories are empirically terminated since their axioms are unconfirmed. Then, if a Final Theory is ultimately achieved, it is a non-empirical theory. Therefore, for a Final Theory, theory-change is still applicable - in virtue of this extraordinary property.

On the other hand, if empirical termination is always characteristic of a Final Theory, then all such theories must be considered speculative, for they are non-empirical. Now, although *any* theory is non-empirical until it is verified, the situation is different for a Final Theory which, by its supposed finality, must always be non-empirical. Thus, theory-change is possible, not with the Final Theory itself unless the latter is simply abandoned, but rather with the component theories that were originally unified into the Final Theory, for these *have* been empirically tested. It may, for example, be perfectly possible and perhaps with additional modification to unify those components in a completely different way, resulting in an alternative Final Theory, or indeed several such theories. Thus, there may well be a proliferation of Final Theories all of which are empirically terminated and it is just because of the latter property that such a situation may obtain.

We are therefore faced with an extraordinary and paradoxical conception of the direction of science. In the first instance, a Final Theory would seem to be an inevitable consequence of the unification program and thus eminently achievable. In the second instance, however, because of the non-empirical nature of a Final Theory, empirical termination allows the construction of many such theories. How may we resolve this dichotomy? We might suggest that given a number of candidates for a Final Theory, then they are actually all really equivalent descriptions of the Natural World and that one may, by suitable mathematical transformations, show that each of them are isomorphic in the technical sense. But one would have to demonstrate that any Final Theory is isomorphic to any other, i.e. all possible Final Theories are isomorphic. Yet, this would not help at all, for a multiplicity of Final Theories *must* be isomorphic if they are indeed final and since they each contain only one axiom it is only a question of showing that the axiom of one Final Theory is equivalent to that of any other and, clearly, finality ensures that this will be so. This does not help because it is merely a mathematical exercise with no possibility of verifying even one of the proposed theories, although that is all that would be necessary in order to settle the issue. Essentially, if a set of theories are all final, then each one of them is an equivalent representation of all physical phenomena, but that can never be empirically established and we are right back where we started. In fact, this is very similar to the current situation in physics where there are three competing Quantum Gravity theories. Superstring Theory (or M-Theory) we have already mentioned, but Loop Quantum Gravity and Twistor Theory are alternative candidates for finality. According to the above, if all of these theoretical non-empirical structures are really Final Theories of the FT1 variety as is suggested (although Loop Quantum Gravity is less ambitious than M-Theory) then it should be possible to demonstrate their equivalence by exhibiting an isomorphism. At the present none of these theories look very much like one another, but that is hardly relevant since any one or all of them may be quite wrong, for each is empirically terminated and thus must reside within the realm of pure speculation, however sophisticated the latter may have become at this level.

Nevertheless, this does raise an interesting point, for suppose we have a finite set S_n of FT1 theories that are found to be equivalent up to isomorphism, i.e. $S_i \cong S_j$ for all n. Would this fact then constitute evidence for the validity of these theories, even though each one of them is empirically terminated? Certainly, such a situation would be highly suggestive and seems to imply that we are somehow on the 'right lines' in the search for a Final Theory. The argument would be that such an isomorphism within the set S_n could only obtain if all the members of that set were true representations of the Natural World, since each one of them results from the unification of empirically verified component theories, albeit with perhaps a few extra fundamental assumptions. Expressed figuratively; if all roads lead to Rome, then that city must surely be the primary source of road-building, the centre of civilisation, the capital city of the world or, perhaps, even the Final City.

Unfortunately however, arguments of this sort are merely persuasive and only compelling in the suggestive sense already indicated. There is no logical way in which such a construction would constitute a valid argument in the deductive manner that is certainly required in scientific discourse. Perhaps all roads do lead to Rome and they are isomorphic in that vectorial sense, but the axiom of a Final Theory is not artificially constructed (except by God, if one is so inclined), since even though not testable, they are still obtained from empirically established theories and that severely restricts their mathematical or linguistic form. We do not therefore equate axioms with highways or Final Cities with Final Theories and may dispense with any such metaphor. We make a further observation; whilst any Final Theory is empirically terminated – meaning that it can no longer be verified (or falsified) by empirical means – the converse does not follow except in a very special case. A Final Theory is obtained by axiom combination of already empirically tested theories. But, suppose that it is formulated from no such preceding model-theoretic structures. We would then have a Final Theory that is constructed from zero unifications – the *null* Final Theory. Such a theory is indeed final, but it is essentially nothing more than mere hypothesis, for whilst it is not empirically verified and is by default empirically terminated, it does not originate from the unification of empirically verified theories. A Final Theory of this type consists solely in speculation and fails to satisfy our conditions for a scientific theory. In this sense, a null Final Theory can simply be postulated, quite arbitrarily, but still lay claim to finality. We shall remark upon this kind of hypothesising in our final section.

Returning to the problem of empirical termination, we find that there is no satisfactory way to avoid our dilemma and, doubtless, the reasons for this are by now abundantly clear. Everything related to theory modification, theory-change and scientific revolutions has been empirically founded upon the observational or experimental determination of physical variables and axioms that may be identified as the model-theoretic interpretation of the constants and variables of an abstract language through the natural interpretation, which we have more than once equated with the empirical process and, moreover, the scientific method itself. Obviously, this is the very reason that the scientific method and the empirical approach that is so vital to the success of the acquisition of knowledge of the External World is the only possible procedure by which such an understanding may be obtained, and that is yet another point that we have frequently been at pains to emphasise. There is however, one more philosophical matter that we would be amiss not to address more fully than we have done in the prequel, although we have many times alluded to it therein, and which is intimately connected with, but does not directly affect our exposition of the development and evolution of science or the subject matter with which we have been primarily concerned in discussing scientific revolutions. Often, we have remarked upon the efficiency of mathematics as an appropriate language in which to describe the workings of Nature and the physical world: We elaborate upon this in greater detail in the next section.

11.8 Mathematical ubiquity

Questions concerning the apparent utility of mathematics in describing physical phenomena are not new in the Philosophy of Science and have been discussed and deliberated upon at various points in human history since the times of ancient Greece. The literature on the subject is vast, but somewhat scattered amongst the major works of the great philosophers beginning with Plato and continuing, primarily with Descartes, Berkley, Kant and Mill, etc, in addition to many less well-known writers. It is only in relatively recent times however, that this issue has attracted special attention, notably by the physics community because of the more extensive use within their discipline of what hitherto had seemed to be just 'pure mathematics'. Although mathematics had been used effectively in physics since even before Newton and with its subsequent employment in the natural sciences gradually increasing over the ensuing centuries, much of the *type* of the mathematics of the long period before the twentieth century had been devised or adapted by physicists for the very purpose of describing the Natural World in precise terms. Thus, we think of Newton and his 'method of fluxions', Gibbs *et al* with vector analysis, Fourier with his periodic series, Euler and the Calculus of Variations, not to mention Laplace, Lagrange, Hamilton and the remarkable Bernoulli family.

Most of these people were physicists and those that were not would still readily work within that field, for they were 'natural philosophers', which basically means for us that the 'classical' mathematics used for the description of physical phenomena was not to be considered distinct or differentiated from the study of mathematics as a separate discipline. Indeed, even the Greeks thought of Euclidean Geometry as representative of the real world in spite of the so-called 'Platonic Theory of Forms' that ascribed special status to mathematics and the constructions it could represent in an *abstract* manner that *did* seem to be different, in a rather vague sense, from the physical counterparts or actual analogues of those geometrical constructions in the real and pragmatic world of direct experience. All this is well-known; the point to which we wish to draw attention is that the alleged philosophical problem of why mathematics and its methods are of such great utility in the scientific enterprise would hardly have been considered even as an issue of importance with these workers, for it was self-evident that the physical and mathematical realms were essentially dealing with the same thing. That is to say, physics and mathematics were just natural philosophy and united in the task of the elucidation of the structure of the physical world. Only a philosopher rather than a physicist would seek to address these matters from an epistemological or metaphysical stance. And as we have noted, with just a few exceptions this was indeed the case until the early twentieth century. We need to see why this was so, which is the reason for these preliminary historical excursions that we must briefly continue*.

Whilst there were some earlier indications of the use of 'non-physical' mathematics in physics and to a lesser extent in related disciplines, it is only with the advent of Quantum Mechanics and both Special and General Relativity, virtually contemporaneously, that brought to the attention of physicists those areas of mathematics that had been previously thought to be beyond their domain of enquiry, inapplicable, or even irrelevant and this is when the philosophical problem begins to take precedence amongst both physicists, contemporary philosophers (who already knew) and even mathematicians – who perhaps should know better.

Non-Euclidean geometry had been initially envisaged by Gauss and developed further by Riemann. Such a geometry had been constructed by Lobachevsky and soon many geometries of this kind were devised, all because of the realisation of the independence of the parallel postulate of Euclid. The history is well-documented and need not concern us here. What is of importance from the point of view of a physicist is that these developments in mathematics bore no relation to the Natural World, for this was just mathematics; no longer natural philosophy and certainly nothing to do with practical or even theoretical science. Already, a distinction between the two disciplines had by this time been drawn, but still there lingered amongst the physicists the conviction that mathematics and its methods was a legitimate part of their province and whatever the mathematicians otherwise did, was nothing to do with the world of physical science. Mathematics was a useful tool and a sufficiently precise one conceptually to enable the formulation of the Laws of Nature without ambiguity. We note here that mathematics is now being conceived as a

---

* Although we would rather come straight to the point, the historical origin of the perception of mathematics in relation to physics requires some introductory discussion.

language rather than a property of Nature – but we shall return to this point later. First, we continue the story from the historical perspective.

Einstein now realises that the equivalence principle entails a non-Euclidean geometry and adopts an appropriate metric for General Relativity, as we saw earlier (§7.4). Minkowski had already expressed Special Relativity in terms of a pseudo-Euclidean metric, but now Einstein exploits the full geometrical formulation of Riemann to construct his scientific model, clearly encroaching upon provinces that are entirely non-physical and should be the concern of mathematics and differential geometers only. Thus, the distinction between 'pure' mathematics and physics is diluted. This is also pointedly emphasised just a few years later in Quantum Mechanics by the work of Heisenberg, Wigner, Jordan, Dirac and others. The use of non-commutative algebras, first in the form of Heisenberg matrices and especially Lie algebras in the relativistic formulation of Dirac (Dirac algebra), the delta function, operator algebras, together with the embodiment of physical principles within an abstract Hilbert space would surely serve to blur the distinction even further. The incorporation of ever more sophisticated mathematics into modern physics, such as Kac-Moody and Virasoro algebras, non-commutative geometry and even number theory and the Riemann-Zeta function, brings these two disciplines even closer together, so that once again the question of the ubiquity of mathematics presents itself with even greater poignancy.

Let us ask the question directly: Why is mathematics, which seems to be an entirely mental construction, so eminently applicable to the description of Nature? Or to put it another way: How is it possible for something as abstract as mathematics to be so effective in the real world of the physical sciences? The problem is therefore, that since mathematics consists only of ideas or thoughts, there seems to be no reason why it should be applicable to the concrete objects that are the subject matter of science. Traditionally, there have been various proposed views with regard to this question, usually connected with the mind-body problem or religious belief. However, John Stuart Mill eliminated the problem entirely by simply asserting that mathematical statements are empirical and hence, no difficulty arises. We do not accept any of the usual 'resolutions' of the problem of mathematical utility, but we are closest to Mill in thinking that there really is no dilemma in the first place.

This philosophical position has been implied throughout the present work and we plainly state it now. In the prequel, when referring to the axioms of a theory we have not differentiated between their mathematical representation in the form of an equation; $f(x_i, c_j) = 0$ and an equivalent entirely verbal expression, however cumbersome or unintelligible the latter may be. We do not doubt this equivalence, for mathematics is a language and with suitable identification of grammatical rules, syntax and vocabulary, any language may be translated into any other. The present sentence is written in English using word-processing software on a Personal Computer. The computer is not a conscious entity (whatever that means) but it, or rather the software, has no difficulty in translating these words into binary code, which incidentally is a mathematical process itself, but the series of zeros and ones of the binary representation of each word is the language that the computer 'understands'. Since it requires a mathematical operation on the part of the computer to effect such a translation and since also we may regard binary code as 'mathematical language', we might ask: How is it possible that an imprecise and often ambiguous language like English can be represented in such a precise language as binary code? Well, of course it is possible, because languages are the same sort of thing and all belong to the 'Category of Languages'. The computer is simply constructing a one-to-one and surjective mapping between the characters of the English alphabet and specific sequences of binary code, or more precisely, identifying each letter with a binary number. We are not missing the point here, but merely wish to illustrate firstly, that translation between language is always possible, which may seem to be rather obvious and secondly, that translation requires that the languages in question belong to the same category, which of course they do, for they are both abstract mental constructions although they can adopt physical form as on this printed page. The physical representation of the language in so far as it appears as symbols of black print on a white background is of course, completely irrelevant.

Now a sphere and a cube are both abstract entities and yet also physically realisable objects. The sphere is homeomorphic to the cube and one may continuously deform one into the other in the abstract topological sense. Similarly, one can, with suitable engineering skills, also transform their physical representations one into the other just by applying sufficient heat. What one cannot do is transform a *mathematical* sphere into a *physical* cube, for the two belong to a different category. Then, we can now see where the difficulties arise in the problem of mathematical utility for we might ask: How can the concept of a sphere be applicable in the real world of actual physical spheres? Mill would simply retort that the concept of sphericity is empirical anyway and such a question hardly deserves an answer. But we cannot adopt this view, it is too simplistic, for there are mathematical entities that could hardly be 'observed' or even conceived of by the observation of physical phenomena and although a great deal of highly abstract mathematics has been found to be useful in physics, there is still much that has not and yet, has nevertheless been 'conceived'.

So is the problem of mathematical utility simply a category mistake and if it is, does it therefore just go away? We answer both affirmatively and negatively. It *is* an elementary category mistake just because we are dealing with different kinds of entity, but it does *not* go away because mathematics exists, if only in the minds of sentient beings and the physical universe also exists and appears to conform to the mathematical constructions and conceptual framework of those abstract notions that such beings, including ourselves, are able to engage in without recourse to the material world.

Yet, since mathematics is just another descriptive language and translatable into any natural language, we may equivalently ask: How can English (or any natural language), which is an abstract mental construction, so effectively describe the physical world?* We now see the source of the difficulty that has so engaged the attention of physicists. The wrong question is being asked. To ask how a language is applicable to the real world, or how it is possible to describe phenomena in that domain is a meaningless question, for that is the sole purpose of *all* languages and the fact that languages are abstract mental constructions is neither here nor there and this is equally the case with mathematics as with any other language. This apparent misunderstanding originates in the category mistake mentioned above and the real question is: Why is the universe constructed mathematically? Let us deal with this.

What can be meant by such a question? Surely the whole purpose of the scientific enterprise is to discover how the universe is constructed anyway. If the universe is constructed 'mathematically' then that is for scientific investigators to ascertain and is not a philosophical problem *per se*. And the notion of 'mathematical construction' is also quite meaningless except in the sense of description by mathematics, which becomes circular. Furthermore, all other linguistic descriptions are also possible, as we have emphasised, yet we do not ask how the universe is constructed in

---

* This writer knows of no major philosophical work that deals with this question!

accordance with the grammatical rules and syntactic structure of natural languages.

Mathematics is a ubiquitous and versatile language. It can in fact, in common with all languages, describe anything at all with sufficient ingenuity of application and indeed does so in a very wide diversity of fields, not just the Natural Sciences. Thus, the latter have no special jurisdiction over the language of mathematics; it is just that far greater effort has been expended in the scientific domain than in other, unrelated fields. There are of course, many reasons for this, which we briefly consider.

The attainment of a reasonable degree of fluency in mathematical language beyond the elementary level takes considerable time and intellectual diligence and to acquire what is often termed 'mathematical maturity' may well take half a lifetime or more. Because of major applications of mathematics to physics, practitioners of that discipline are required to follow such an intensive learning curve; it is simply part and parcel of their normal scientific training. This is also true to a lesser extent in other areas of physical science but does not obtain in the Arts or Humanities except in a few isolated specialities within these disciplines, since generally mathematics has not been universally applied to those areas of interest. It would therefore be unreasonable to impose a rigorous course of mathematical training upon prospective workers in these fields because it is quite unnecessary and to do so would certainly involve an extension of course material that would render the study of these subjects rather impractical in the short space of time available. This is not to say that mathematics *cannot* be used in domains other than the physical sciences, but merely that it has not thus far been demonstrably necessary and natural language has served quite adequately in such disciplines for some time. Of course, investigation in almost any field of enquiry may involve the use of statistical techniques for the analysis of data, but this is hardly what we mean by 'application' of mathematics in a given field as a descriptive language that is able to express the fundamental tenets of the discipline concerned.

However, in time that situation may change and the question of mathematical utility would again arise. Suppose that a great innovator were able to successfully apply sophisticated mathematics to some branch of the humanities H and that as a result, H is completely transformed to the extent that it becomes unrecognisable to the majority of its scholarly aficionados. In fact, H now even has *axioms* and significant consequences of those axioms may be deduced with the mathematical precision that hitherto would not have been thought possible. Would we now ask: How is it that mathematics is so adept at describing *H*, when *H* is so divorced from the Natural World or the sciences in general? Now we have two concepts to consider. First of all, *H*, being a discipline of the Humanities and entirely non-scientific, is nevertheless in the same *conceptual* category as mathematics in that both are mental constructs, but secondly, mathematics has found application in a subject area that is almost by definition non-mathematical and certainly non-scientific. This of course happens and economics is a prime example, but there is no *prima facie* restriction to any other field of enquiry.

Yet, are we not stating the obvious here? Doubtless the apparatus of mathematics may be applied to very many domains of interest and that is surely just the province of applied mathematicians, who do such things all the time. In other words, mathematics is a language.

But it is not a *formal* language of the type constructed in mathematical logic such as the Predicate Calculus. Rather, the language of mathematics is a mixture of natural language and a large number of specialised concepts and symbolic representations that allow notions and ideas that it would be difficult to express in a natural language in a more succinct and precise way. It is therefore just a *technical* language, not strictly formal, but something of a hybrid of a fully formalised language and the natural ones with which we are all so familiar and acquire from birth *without* a learning curve. In this sense, the language of mathematics is not so different from that used in the legal profession where it is also essential to define terms as precisely as possible in order to avoid ambiguity. Computer programming languages offer an even greater extreme of this, where now *no* ambiguity is acceptable. Thus, the Science Machine of Chapter 10 could not even acknowledge language differences or entertain ambiguity.

Then, in the light of the above, what are we to say of the supposed problem of mathematical utility in the physical sciences? If mathematics is just a technical and especially precise form of linguistic expression that, as such, is able to describe anything that may be less intelligibly described in a natural language, then clearly there is no problem of utility *providing* that the same expressive power is available in all other languages and by translation it is. This does not mean that mathematics, as a language, is *isomorphic* to natural languages, for it could not be if any of the latter contains ambiguous terms, but it will be isomorphic to a subset of a natural language in which no ambiguity exists, i.e. the natural language is restricted to unambiguous terms so that an injection becomes possible. This would require adopting only one specific interpretation of each word when more than one is in general usage and similar constraints would be necessary for grammar and syntax. But this is nothing more than the Predicate Calculus or variations thereof supplemented by additions from natural language. This is true of legal jargon or any other technical language that takes and defines only a subset of a natural language. Hence, a natural language, being a superset of the technical language is also able to describe the structure of the Natural World just as easily as its restricted subset. This is the case with mathematical language and the problem of utility disappears. It is no use then asking: How is it that the language of mathematics, being a mental construct, is so applicable to the real world, for we may now just substitute 'any language or subset thereof' for 'the language of mathematics' in this question.

The foregoing discussions also suggest a way of defining a scientific field. Recall our discipline H in the Humanities. It had been 'mathematicised' and given an axiomatic structure. The question arises: Is H now a science because it possesses these mathematical properties? The answer must evidently be in the negative for H is not concerned with physical or natural phenomena whereas science, as we understand that term, is entirely so confined. Thus, the mere attribute of possessing an axiomatic structure or a mathematically expressed foundation, does not make H scientific, for this is only a linguistic feature of the discipline as we have seen and H is obviously still a legitimate branch of the Humanities. Hence, Economics would not be a science even though it may be axiomatised.

It might be thought that there is an inconsistency here, for we have earlier defined a scientific theory as just a set of axioms and evidently, both Economics and our imaginary H fully satisfy this requirement. However, we must not forget the *form* of the axioms of a scientific theory, namely $f(x_i, c_j) = 0$ where the variables x_i are *natural* variables and the constants c_j are *natural* constants, which are identified empirically through the *natural* interpretation. Thus, the domain of a scientific theory belongs to the *Natural* World and we should more correctly refer to a scientific theory as a *natural* scientific theory and collections of such theories will then constitute *natural* science. Neither Economics, nor H, would now satisfy these conditions, which are not new constraints, but were already implicit and tacitly assumed from the outset.

In this sense, we are clearly dealing only with the traditional *pure* sciences that we mentioned in the Introduction. Anything else is just applied mathematics or an application of one or the other of those pure or 'natural' sciences. Hence, Economics is applied mathematics and Engineering is applied physics. Medicine is applied biology but botany and zoology are not since we may regard them as sub-disciplines of biology. There is, of course, often some overlap in this somewhat arbitrary classification scheme, for example; is astrophysics a branch of astronomy or physics? The same applies to cosmology, but the essential point is that these are natural sciences with natural variables determined empirically via the natural interpretation.

Notice that we have not anywhere considered either Psychology or Sociology. That is because we do not consider these fields to be scientific in our usage of that term, for there are no clear empirically testable axioms of sufficient generality and no identified natural variables that would warrant their elevation to the high status of being 'scientific'. We regard these fields as highly speculative and non-empirical and prefer to classify them with the Humanities, where at present they clearly belong. Many would disagree with this view and might think that we are being too harsh, particularly practitioners in such areas, but in all fairness, our definitions will not admit them into the scientific domain until much progress has been made, axioms determined, empirically verified and expressed in terms of natural variables. Even if that goal is achieved, psychology and sociology would be incorporated into biology and become behavioural sub-disciplines thereof. However, there are good reasons for thinking that this may never be possible since it is hard to envisage how an objective analysis of the behaviour of a species may be obtained when that analysis is attempted by the very species that are the subject of the investigation in the first place, so that detachment may never be complete and therefore a subjective element in the analysis would always be retained. The objectivity of the empirical method and the mechanism of the natural interpretation are brought into question here. Perhaps this is why these fields of enquiry have never attained recognisable scientific status as has been the case with physics and chemistry, for example.

The examination of what constitutes a scientific theory and even whether the latter can be precisely defined is intimately connected with the scientific method itself, its ontology and its epistemology. This leads us to our final section which is not properly related in a direct way to the principal thesis of the present work and is a cause for concern with regard to the understanding of science and its methods, but nevertheless, is still relevant to our context where theories and axioms should be well-defined.

11.9 Anti-science and pseudo-science

Criticism of the methods of science and science in general as an epistemological enterprise, is not new. After all, philosophers have engaged in such debates for centuries. Yet, such speculation and philosophising within the Theory of Knowledge has always been, for these interlocutors, an entirely legitimate intellectual exercise that is both rational and well-considered. When a philosopher expatiates upon a scientific topic he is expected to be well-acquainted with the subject matter of the discipline concerned. He will have a clear understanding of the foundations and empirical axiomatic structure of the discipline in question and therefore, approach it with a degree of comprehension and with the respect that it has earned by having been painfully and meticulously elucidated by serious investigators in that field of enquiry. Such analyses are intended to shed light upon the foundations and assumptions of science and to pinpoint and clarify conceptual difficulties where they appear to arise. This is a useful and valuable service which often leads to a deeper understanding of the nature and structure of a scientific theory and indeed, many eminent scientists have themselves, not always successfully, engaged in such philosophising. All this is respectable, commendable and constructive when conducted academically and in a rational fashion.

Professional scientists and philosophers aside however, there are others who seek to denigrate science and its methods in order to propagate their own, often highly eccentric views. This too, is hardly a new phenomenon, but in recent times it has become more and more prevalent and perhaps, even fashionable. Persons of this sort are generally untrained in either science or philosophy and fall roughly into three categories, not all of which are distinct:

> a) Those who wish to refute science because its findings conflict with their own (often) religious beliefs.

> b) Those who wish to propagate their own non-scientific theory as a scientific alternative.

> c) Those who are vehemently opposed to anything scientific because it is 'inhuman' or 'too-rational', or an exclusive club of some sort to which they are denied membership, or for some other similar reason.

Any one of these categories may apply separately or in combination, but let us look at each one of them in turn.

Category (a) is familiar and has a long pedigree. It periodically re-surfaces, often in a new guise and will doubtless continue to do so into the foreseeable future. The well-known archetype in this category is, of course, Dawinism and biological evolution and in its current incarnation it is referred to as 'Intelligent Design'. Persons in this group have no understanding or comprehension of Darwinian principles, the Modern Synthesis, or the empirical method of science, as is abundantly clear from their assertions. For example, they may claim that there is no evidence at all for evolution in spite of the copious data available for anyone to see, not to mention laboratory demonstrations of biological evolution where the process can actually be witnessed. Conversely, that there is no evidence for Intelligent Design, nor ever can be, does not seem to concern these people. Strangely, nor do they seem to be familiar with the Argument from Design of which their construction is a less general variant and with which every beginning philosophy student is well-acquainted. 'Intelligent Design' is nothing new and certainly not a scientific theory as its advocates insist, it is just creationism in another guise, as Darwin well-anticipated. But individuals in this category do not need to understand the scientific theories that they wish to refute because they simply 'know', by some unspecified non-empirical means, that such theories are false. By this same mysterious epistemological method, they also 'know' that their own views are true. Unfortunately, an alternative to an established scientific theory is never offered, for that would require empirical verification which is the very thing that established the scientific theory concerned in the first place. When such unfortunates are questioned, if they allow, it becomes clear that all that remains is a stubborn religious bigotry and an irrational belief system to which they will doggedly adhere at all costs.

Category (b) is more interesting and certainly more amusing. Here, a 'new' theory, usually of the entire universe, is proposed. The 'theory' is therefore all-embracing and is a Final Theory in that sense, probably something like a TOE, but does not involve unification. Hence, it is theory re-formulation but without axioms, natural variables, or anything remotely scientific. Scientific terminology is often employed, but in a very confused way. Thus, one hears of 'energy-fields' that either don't interact with matter (so what's the point?) or if they do, no mechanism is given for such interaction and they interact only in very special (but undefined) circumstances, neither locally nor globally and, of course, are anyway undetectable. No description of how they propagate is given either. They are not, therefore, physical fields as would be normally understood in science, but the use of such terms as 'energy' and 'field' is presumably intended to ascribe some sort of credibility to them. Often however, their status is further enhanced by the invention of new special 'technical' terms for such 'energy-fields' such as the XYZ-field, which now gives them further credence and to the uninitiated, somehow verifies their existence.

But wild speculation does not stop at this point. It is then 'demonstrated' that current science, usually relativity and quantum mechanics and especially the latter, not only supports but actually *suggests* the new theory and leads directly to it. Famous names in science, who would be appalled at such notions, are often quoted, together with further quotations from their written scientific work which are nearly always taken completely out of context and entirely misconstrued. Even worse, individuals in this category will unashamedly invoke basic equations from physics to support their spurious thesis, perhaps the most frequently quoted of which are the Einstein relation $E = mc^2$ and the Heisenberg uncertainty principle in one of its forms. It is interesting to note here, that although the mass-energy relation is expected to confirm the 'theory', it is ignored when the 'energy-field' is non-interactive or 'non-material' and proponents of these theories either don't understand the equivalence of mass and energy, or deliberately choose to ignore it when it is inconvenient. Such individuals often totally misconstrue scientific concepts and ascribe a meaning to them that simply does not exist, or lies outside the range and context of the technical term concerned. We have already mentioned the idea of a physical 'field', the mass-energy relation and the uncertainty principle, but in statistical mechanics the notion of the 'entropy' of an ensemble is also a favourite target here, where the statistical nature of the concept is often not realised and inappropriate conclusions are thereby drawn. Sometimes, misunderstandings of this type are even extended to full scientific models, especially in Evolutionary Cosmology and the 'Big Bang', where naïve philosophical questions are asked such as 'What happened *before* the Big Bang' if that was the initial creation*. Rigorous mathematical theorems do

---

* It is possible for the Big Bang to have a cause, but only if some sort of spacetime geometry preceded the formation of our universe, which does not help such critics because the question still remains as to what happened before that. At some point, one would have to ask from whence space and time arose – which makes no sense.

not escape misconception either and on innumerable occasions one finds the Gödel Incompleteness Theorems quoted entirely out of context and with complete indifference to the conditions of these theorems, in order to support some spurious conclusion, usually to show that science cannot answer all questions and therefore the new 'theory' advocated by the writer concerned *must* be correct. Quite apart from the faulty logic here, it does not occur to these proponents that even if incompleteness *could* be so applied, then it would also apply to *their* 'theory' also.

When at length, having subsumed all this pseudo-scientific rubbish, one arrives at the final and confused conclusions of these protagonists, a common trait is often revealed. With the use of quantum mechanics and particularly the Copenhagen interpretation (for they seem to be unaware of more modern approaches) where the observer has special significance, suddenly human consciousness enters the framework of such theories as the most significant aspect of the Natural World. Then, it is seen that the theory is *not* about science at all, it is really about the very special and important place that humans are privileged to occupy within the scheme of things, especially the by now celebrated proponent of the 'new theory' in question, who has thus been elevated to Guru status of some sort, but also and most disturbingly, is now a *scientist* of some merit too, since he has usurped that field by his own more 'enlightened' deliberations. It is easy to understand the motivation here. In Ptolemaic times the Earth was the centre of the universe and man held a special place at that pivotal point. The Copernican Revolution overthrew this concept and throughout the succeeding centuries, mankind was seen to be ever less significant and relegated to a small rocky planet orbiting an ordinary star on the periphery of a galaxy within a local cluster of galaxies within a supercluster of galaxies of which there are countless others in a vast universe. How disconcerting; it is no wonder that we find proponents of such ideas. They may however, if they are willing to make the effort, find the majesty of modern science far more fulfilling than egotistically attempting to regain their own self-importance (and that of their followers) by the construction of absurd and ill-defined 'theories' that in the final analysis amount to nothing more than mysticism.

Now category (c) is the most dangerous, for this embodies non-rationality *par excellence*. It is anti-science merely for its own sake, i.e. pure prejudice. One of the reasons given for opposing the scientific enterprise *in toto* is that it 'de-humanises'. Exactly what is being de-humanised is not clear, but if it is humans, then scientists are 'de-humanised humans', a concept we may grasp with some difficulty. These protagonists do two things. First, they forget that, at least on this planet, scientists are actually fellow humans and secondly, that the pursuit of knowledge is a very human trait and not the province of some evil sect known as the 'scientific community' who indulge in obscure rituals such as 'observation' and 'experimentation'. Sometimes however, it is scientific jargon and especially mathematics that is the source of the outright dismissal of science and its achievements. Mathematics is incomprehensible to these people and the technical language quite impenetrable. If it is explained that a technical language is essential in order to avoid the ambiguities of natural language, one is often met with derision similar to that which the legal profession must face on a daily basis, for surely plain English is quite adequate and just as lawyers cloud their subject with legal terminology in order to extract maximum payment from their clients*, so too are scientists involved in a dastardly conspiracy that ranges from world domination to the genetic or mental manipulation of the human species, which probably amounts to the same thing.

There are, of course, less extreme individuals in this category but the motivation is the same. Such people are often found within the Humanities and the Arts. Aficionados of Art are frequently opposed to science simply because it is *not* art. These people are not themselves artists, for the latter usually recognise the creativity in science and can equate with that in their own work, rather it is the 'followers' within the artistic world who imagine that they have some special insight into 'reality' that their addiction provides and therefore science is of little or no consequence and is again, summarily dismissed. Exactly similar remarks apply in the other humanities, whether it is history, literature, or even philosophy. Let it be emphasised that it is not (usually) the academic workers in these fields that are guilty of anti-science prejudice, for they have generally enjoyed a liberal if not scientific education which has required the use of their rational faculties, thus preventing such bias.

There is one issue common to each of these categories and perhaps with the public at large also, that needs to be clarified. When discussing a specific scientific model one may often hear the retort, 'But it is only a theory', as though the 'theory' concerned is mere hypothesis, idle speculation and nothing more. We have spent much time in this work

---

* We are not defending lawyers here, God forbid that we should do so, just their need for precision.

discussing and defining scientific theories and the distinction is very clear to us, but to the anti-science league it is not. The difficulty arises once again with the ambiguity of natural language, where the difference between 'theory' and 'hypothesis' is not well-defined. Most people have learnt elementary algebra at school and are familiar with, say; quadratic and higher order algebraic equations. In older algebra texts the study of these were often subsumed under the heading 'Theory of Equations', yet it is doubtful whether students of that subject would be so audacious as to present their master with the proposition 'This is only a theory and cannot be proved'. Certainly, a proof would quickly follow and doubtless be accompanied by suitable admonitions.

The situation is slightly different in the empirical sciences, but the principle is the same. As we know, a scientific theory is never proved in the mathematical sense, instead it is empirically verified and having been so verified on countless occasions it becomes part of the scientific framework and is no longer mere hypothesis, but now a theory. One rarely hears of Newtonian Mechanics or Newton's Laws of motion being described as 'only a theory'. If that were so, then in the case of automobile accidents the police and their forensic division would have a very difficult time in the Courts. The scientific and technical use of the noun 'theory' is not used in natural language in the same sense, but that is no excuse for commandeering this term and attributing to it the non-technical meaning of ordinary language that it does not possess in the scientific context, as all proponents in the above categories do, especially category (a).

But there is another obvious dilemma that is often ignored by the scientifically ill-informed and which relates directly to their daily life. It does not occur to such protagonists that the very technology that they constantly use is not the consequence of some mystical, speculative or hypothetical doctrine espoused by scientists, but applications of *real* science, with *real* empirically verified axioms that clearly work. The mobile telephone, the refrigerator, the television and countless other appliances that are taken for granted but conveniently thrust aside by those who are disinclined to acknowledge the validity of the theories upon which their construction depends; after all, those are just theories.

One wonders whether these people would question science in a medical situation. Nuclear Magnetic Resonance Imaging (NMRI) is a direct application of quantum mechanics where the spin states are measured within a magnetic field, but it is doubtful that our proponents would refuse such tests should the need arise. Or perhaps on their next vacation they should consider the flight path of the aircraft or cruise ship that so easily transports them to their destination which is maintained by the Global Positioning System (GPS) employing Post Newtonian Celestial Mechanics, which is an application of General Relativity, not to mention the aerodynamics and hydrodynamics that makes it all possible. There are innumerable similar examples, but again, surely these are only just theories.

Then, it seems that ignorance and prejudice are the principal causes of antipathy towards the scientific enterprise. We shall summarise. Members of category (a) will never be reconciled with science unless the latter were somehow able to demonstrate the existence of a Deity that conforms to their doctrinal beliefs and that would certainly be miraculous. These people are beyond salvation. Members of category (b) would really like to *be* scientists if only they could understand the subject. But that involves too much time and effort and so they invent their own ill-conceived and muddled set of hypotheses as an alternative to the 'established doctrine' of competent and well-trained professionals and in so doing, satisfy their own ego and achieve their desires of becoming 'specialists' in their own 'scientific field'. These people *can* be helped, for there is still within them the inclination that science and its methods actually work, yet in spite of the hope that may be thus inspired, there is still that learning curve. Those in category (c) however, unfortunately have serious psychological problems and it is difficult to see how they may be persuaded to develop their rational faculties when the contrary is so ingrained within their mental perceptions. Unfortunately, these people too, are beyond salvation. Note that we are most definitely not arguing from authority here, for that is anathema in science and philosophy of the highest magnitude. It is rather that proponents of these three categories are arguing from ignorance. Scientific theories and scientific knowledge is there for all to see, instantly and readily available. Indeed, if that were not so, then no practicing scientist could ever hope to contribute to that noble enterprise, which is nothing less than the search for truth and an understanding of the extraordinary world in which we find ourselves.

Epilogue

The extraordinary notion that the methods of science cannot be subjected to the same criteria and standards of analysis that those very methods impose upon the objects of enquiry within their own domain and with equal analytical rigour, has been the purpose of these deliberations to refute. It would seem to be incomprehensible that the vague and inexact philosophical conception of the epistemological methods of so crucial an endeavour as the scientific enterprise should not be made to enjoy the clarity of exposition that is readily available to similar analyses, as for example now obtains in linguistic fields and even at the meta-linguistic level. Whilst it is true that throughout the history of the Philosophy of Science, much serious thought has been offered in this area, it has always been far too closely related to the traditional problems of philosophy, such as the mind-body problem and similar metaphysical notions concerning Theory of Knowledge, such as it is.

We have taken a different view here that is neither realist nor strictly empiricist. Furthermore, although at times it has been unavoidable, we do not wish to be even philosophical in our approach. Of course, we have so indulged, but only in attempting to isolate those aspects of the scientific method that it seems to us have been inadequately defined in the philosophical literature and perhaps still remain so. Thus, our purpose has been to seek, in a purely pragmatic way, a description of science and its methods that to some extent will release the latter from the domain of pure philosophy.

Evidently, the complete separation of the *conceptual* aspects of the methods of science from epistemological philosophy can never be achieved, since by definition they fall into that metaphysical realm, but there is no reason why the *practical* content of the scientific method should be the exclusive concern of philosophy and in fact, perhaps entirely divorced therefrom. We have attempted to make some tentative steps in this direction, but clearly much more needs to be done. Indeed, our ambitious approach has led us to suggest that speculation with regard to the methods of science may not just be removed from philosophical discourse, but by suitable definitions of the conceptual apparatus that has hitherto been described in only an imprecise manner, a more perspicuous representation of those methods may be obtained that is largely devoid of philosophical content. In this way, we have attempted to construct a 'meta-science' or something along those lines that is neither science nor philosophy, but rather a set of criteria with which the methodology of science may be compared and by which in our case, actually defined. In this respect, we consider that *any* set of definitions of the basic tenets and assumptions of science and its methods is better than none at all, however inadequate they may prove to be.

There will doubtless be many objections to this approach. It might be argued that the methods of science are properly the concern of epistemology and as such could never be codified in the manner we have envisaged. Of course, this is not really an argument at all, but just an objection that a favourite topic of epistemologists has been moved into a more analytical sphere than the 'Theory of Knowledge' and is no longer the arbitrary plaything of the general philosopher *per se*. Then, this antipathy towards the codification of the scientific method is nothing more than resistance to specialisation and we would expect that no serious philosopher of science would entertain such a view, for they are already 'specialised' in that sense. Indeed, we recall similar resistance in both logic and linguistics. Hence, it will not be philosophers that object to our approach since they will systematically analyse the entire edifice for inconsistency and error. It is far more likely that those in philosophically related fields will harbour disagreement.

And what shall scientists make of all this philosophising upon their subject and will they even care? Presumably, there are just two possible responses and we think that *all* scientists *do* care, but perhaps to varying degrees. The first reaction is that of the pure pragmatist who follows his scientific research in the usual fashion and has not overly concerned himself with the philosophical implications of the methods with which he has become so accustomed. Yet, he is not unaware of the epistemological elements that infiltrate the foundations of his everyday work, but has chosen to ignore them, not because they are irrelevant; rather that they are of no immediate concern since his day-to-day scientific activities are not directly affected thereby, for they consist merely of the application of the theory-change mechanisms coupled with the natural interpretation and this is invisible to a pragmatist or at least, can be disregarded. This scientific researcher may well be happy with the revelation that the methods of his discipline have now been strictly codified and defined in the precise manner that he expects within his own speciality and is no longer the province of philosophical speculation, of which he was always a little wary and mildly uncomfortable with from the beginning.

Then, there are the 'philosopher-scientists' who are prominent in their respective scientific fields, yet are much more deeply concerned with the philosophical basis and in particular, the fundamental assumptions of their specific disciplines and by extension, those of the whole of science also. Such people are scientists first, since that is their primary predilection and philosophers of science secondly, but not the latter in the usual sense, for they have little interest in metaphysical speculation or any of the traditional problems of philosophy. Epistemology and ontology are generally areas that would not occupy their attention *per se*, but the foundation, structure and axiomatic form of scientific theories most certainly does. They might therefore be better described as 'scientific-philosophers' rather than 'philosopher-scientists' since their concerns are confined specifically to established scientific theories and the *content* of those theories within that context. A particular interest of the scientific-philosopher is the quest for a Final Theory, especially through theory unification. Given the nature of the ideas presented in the preceding pages, we expect that the scientific-philosopher would welcome any clarification of the scientific method, but may not necessarily agree entirely with our approach.

But what of the pure philosopher of science, who is well-versed in general philosophy and thoroughly acquainted with the history, literature and traditions thereof and yet also possesses a competence in and a general understanding of science, usually, though not always confined to one particular discipline in which he primarily 'philosophises'. But, in addition and most importantly, he has a clear comprehension of the methods of science and its empirical foundation? Now, this person and his colleagues, are the most valuable from our standpoint because he will be the most likely to construct a detailed critical analysis of the foregoing and thereby offer either significant criticism or further elaboration and development of the tentative notions contained therein. This is welcome, for our goal is the clarification of the scientific method and whilst the definitions that we have introduced cannot be considered final in any absolute sense, we suggest that our conceptual apparatus is a reasonable starting point for a precise analysis of the scientific method, indicating where the bounds imposed by the latter may eventually lead. Then let us try to anticipate some possible objections that might be raised by these formidable opponents.

The notion of a theory as consisting only as a set of axioms does not present an immediate problem. However, any consequence of those axioms that may be logically deduced is also acceptable as an axiom of the theory, for one may choose those consequences as the initial axioms. This does not matter, for as long as logical consistency obtains there is no difference. Yet, the axioms of a theory are empirically determined and one may question this very process, for how shall we decide which physical laws are axioms and which axioms are consequences of physical laws? Again, it makes no difference; whichever is first empirically obtained is an axiom and if it is found to follow from other axioms then all that is needed is theory modification. There is no legitimate objection here.

A possible objection may also arise with regard to the domain of applicability, not of theories, but of axioms. The domain of a theory is the domain of the axiom with the least domain and that does not present a problem. But the domain of an axiom is defined as the 'area' of 'science' in which the axiom holds, or applies. This may seem to be at best circular and at worst, entirely arbitrary. It is neither. If an axiom is found to be applicable at a given point of an arbitrary domain, then that point belongs to the domain of applicability of the axiom, otherwise it does not. The domain of an axiom is therefore empirically defined and possibly variable on subsequent investigation. This is clear from the definition, for the domain of a single-axiom theory is the intersection of the domains of all the axioms of the theory and since there is only one, it is the domain of that axiom. We can see no difficulty here either.

We have in the prequel described the natural interpretation and frequently equated it to the empirical method of science. In fact, we have gone further and actually *identified* the natural interpretation with empirical observation itself. Then, it may be asked: Why bother with the natural interpretation at all and not just settle for full empiricism with all its associated problems? This would be a misunderstanding. The natural interpretation is an *interpretation* of the variables of a language in the model-theoretic sense, but just the one that is immediately and *naturally* adopted by sentient beings, organic or otherwise, because of the observational sensibilies of those entities and it is for this reason that it is identified with empirical interaction with the External World. If there is another interpretation of a given set of axioms of a theory (as is always possible), then it can never be known, for the axioms and their natural variables are empirically obtained and thus restricted to just one interpretation, namely the natural one. Therefore, all other interpretations are non-empirical and hence not scientific in the sense of not being empirically verifiable. Thus, the model-theoretic interpretation of the symbols of the formal language as well as the formulas (wffs) is determined by the participation of the observer in the Natural World and this is because of empirical observation itself. By definition, the natural interpretation *is* the empirical process and yet not so in the standard or realist sense. The essential point is that the observation of natural phenomena or experimentation on the constituents of Nature restricts the interpretation to just this particular one.

One may object to the mechanisms of theory-change and argue that in the real world theories are not modified in this simple way. But then, how are theories modified, since clearly they must change or remain static? If they are to change, then it is the axioms of the theory which must be modified and that must be done in *some* way. Then, how does one modify an axiom? It is accomplished by simply changing the form of the axiom in question, or perhaps several of them and by such modification new axioms are formed. The method by which such change occurs is always empirically motivated in conformity with the natural interpretation. Then, by whatever process may be envisaged, axioms are modified and theories changed. The specific *mechanism* by which this is done is immaterial, but it must consist in an operation upon those axioms that alters their linguistic or mathematical form, including any changes made to the natural variables. But this is precisely what the mechanisms of theory-change do, and if alternative mechanisms do the same, then they are equivalent processes. Hence, if theories consist of axioms, then theory-change is a process by which the axioms of the theory are modified by theory-change mechanisms and all such mechanisms are equivalent and reducible to basic operations, as with the theory-change operators. The only question is which particular mechanisms to posit or consider more important or significant than others, but since any mechanism is reducible to the actions of basic operators, this choice is arbitrary. Then, indeed, theories are modified in the real world by theory-change mechanisms.

Exactly the same argument applies to scientific revolutions, for they are equivalent to the mechanisms of theory-change, as we have discussed at length. One may, of course, legitimately object to our particular choice of classification of revolutions, but we have been at pains to point out that this was arbitrary from the outset and merely a matter of convenience.

Whatever taxonomy one may choose to adopt for scientific revolutions, the fact remains that they are ultimately just successive applications of the theory-change operators and a different classification scenario will not alter that.

There is one particular objection that is hard to avoid and this concerns our definitions of a scientific theory and the validity thereof. A scientific theory was defined as a model in which the natural interpretation obtains and this does not in itself present a problem, for there are no resulting ambiguities or conceptual difficulties providing we accept the natural interpretation as legitimate in model-theoretic terms, which of course we must since it is just an interpretation, albeit empirical. There is no question however, that our meaning of validity of a scientific theory is problematic. We envisaged a subset of the set of all possible theories that are valid scientific theories and asserted that if a theory belongs to, or is a member of this subset, then it is a valid scientific theory. Clearly this is tautological. By a 'valid' scientific theory we meant a theory that actually describes the Natural World *and cannot further be modified*, for such theories exist in some absolute Platonic world and do not require revision because they are already perfect descriptions of Nature and the physical processes within that domain. Thus, although a scientific theory is valid if it belongs to this exclusive set, that membership relation can never be empirically demonstrated. The best that can be done is to show, by infinite repetition of empirical verification, that it *might* belong to the set of valid scientific theories. But, how else is one to define scientifically valid theories? We are not dealing with mathematical proof here, but only empirical verification. There is no choice, therefore, but to assert that there are indeed correct descriptions of Nature and that our scientific theories aspire towards the latter. These correct and unassailable descriptions are the valid scientific theories; for they clearly must exist unless one wishes to contend that a correct description of anything is not in principle possible, which would lead to absurdity. Essentially, there is a category of axioms and theories that perfectly and completely describe the workings of the Natural World and if a scientific theory falls into this category, then it is 'valid' even if that can never be empirically established. That is all we require of scientific validity in our context, although we concede some degree of circularity, yet certainly not ambiguity, but perhaps a little Platonic idealism, which is nevertheless removable. There will doubtless be many other objections to our thesis and we eagerly await them.

Let us consider our position. We have examined scientific theories, the scientific method and scientific revolutions. We have looked at Final Theories and the direction of science in general and we may now ask where we proceed from here. The obvious answer is to develop our subject matter further, for there is much more we could have done. A full, if not complete mathematical exposition of these ideas could have been presented with appropriate lemmas, theorems and corollaries thereof, but this is a philosophical work and such a formal approach would not have been appropriate, yet that this is even achievable is not without significance with regard to our primary task of removing as much philosophy from the methodology of science as possible.

But why should one wish to do this and are we not just trying to construct a 'science of science'? Perhaps we are, but there are many reasons for embarking upon such a project, some of which we have alluded to on more than one occasion. Not least of these has been the long standing debate over whether the growth of scientific knowledge is a steady and cumulative process or undergoes periodic and sudden bursts of activity when rapid progress is made, before settling down to a more sedate mode. Our study of scientific revolutions has resolved this problem and shown that in fact both notions are correct, but that neither of them has been sufficiently understood. Then, this is certainly a good reason for a detailed and formal examination of the scientific method.

Another reason for seeking clarity of thought in the epistemology of the methods of science and how it operates is more closely related to the perception of science as a mode of enquiry on the one hand and as a branch of knowledge on the other. Very often, philosophical discourse in Theory of Knowledge studies tends to confuse these two issues to the detriment of both. If it is perceived that the scientific method and its empirical foundation are at once questionable, then the very basis of science itself is undermined and thus dangerously portrayed as a rather dubious epistemological enterprise. Philosophical and metaphysical deliberation, although quite legitimate and useful, indeed frequently productive, will nevertheless sustain discourse of such topics within its own somewhat speculative domain, thereby forcing confinement of any alternative analysis to a non-philosophical realm where it is not thought that an appropriately philosophical discussion can be properly conducted. In this way, philosophising upon the methods of science is self-sustaining and we wish to break free from that traditional restriction, which is surely now revealed to be unnecessary.

Then, we may ask how the practical scientific investigator is to pursue his research if the foundations of his methods are constantly the subject of debate. For the most part, it is true that he will not be unduly concerned with such matters, but it cannot be assumed that he is unaware of them and that this is not a debilitating factor in relation to his perception of science, the removal of which he would nevertheless welcome. Thus, clarity of exposition of scientific techniques would immediately alleviate such concerns for both himself and the wider scientific community.

A further consideration in this respect is the 'Public Understanding of Science', a phrase that is supposed to be self-explanatory but which we take to mean the extent to which science is appreciated by the non-scientific public. Clearly, for science to be acceptable in this sense, it must have a firm and solid foundation that is devoid of speculation. Hence, the methods of science should be well-defined and precisely delineated, since otherwise it is difficult to see how science as a knowledge-seeking enterprise can be expected to yield results that the public will acknowledge or even trust.

These considerations aside, there are more esoteric reasons for wishing to formalise the methods of science which are connected with philosophical epistemology and ontology, but not specifically in any metaphysical sense. If we have a large body of knowledge that has been obtained by a particular set of techniques, then an inability to precisely define those techniques would seem to be something of an inconsistency, or at the very least, intellectually unsatisfying. It is as though a small child, having misbehaved in some way and appropriately admonished, enquires of its parents for the reasons that its actions were unacceptable, only to be told that it has long been demonstrated that such behaviour leads to a breakdown of society and family values. We then imagine a number of precocious children philosophising over this, perhaps for the rest of their lives, constantly debating the nature of society and the meaning of family values, along with the important methods by which their earlier behaviour has been demonstrated to undermine these things. Since this method of demonstration was never clearly described and the notion of 'family value' left undefined, they continue to speculate upon such matters and never really arrive at an intellectually satisfying conclusion.

We are not, of course, likening philosophers to small children, but merely advocating that a method or technique not properly defined leads to speculative discourse, which is tantamount to philosophising without a clear foundation or a suitable terminological or definitional structure that would enable one to proceed in a more analytical manner and thereby codify the methodology concerned in an intellectually satisfying way that does not divorce the methods from, or demand of them a different ontological status, than the branch of knowledge of which they are the primary source.

Therefore, it is not unreasonable to expect of any discipline whatsoever that every aspect of its methodology by which the basic tenets of that discipline are obtained conform to the same measure of exactitude and precision that the area in question itself enjoys, thus validating and substantiating that very knowledge-base in virtue of being thereby, well-founded. Without this desirable property, which may be considered a requirement, a field of enquiry can never be intellectually fulfilling and will forever entertain a speculative component of an epistemological nature, which we certainly do not need in the exact sciences and is more appropriately the province of philosophy and metaphysics.

To understand a given discipline, its limitations and its scope in anything more than a cursory manner, it is necessary to understand its methods and this is just as true in other academic fields as it is in science. The historian is aware that his picture of some past event might well be incomplete because he is confined to only those historical documents and records that have survived to the present, not all of which may be viewed as objective accounts of the event in question and indeed, must anyway be considered in the social and political (as well as linguistic) climate of the time. Professional historians therefore bear these factors in mind as part of their 'technique' or methodology of historical research. The final presentation of that research is then perceived in the light of these methods and properly understood only through appreciation of the latter. Thus, a reasonable comprehension of any discipline is intimately connected with full knowledge of its methodology.

Apparently, this is especially so in the physical sciences. Here, the empirical interaction with the External World has always been an added complication since it seems to introduce extra elements into the equation (or systems thereof). But, of course, this is not really so. The historian in our above example *observes* the historical literature in his field just as the scientist observes physical phenomena and he is, therefore, interacting with the 'External World' in precisely the same way. Yet, we do not find philosophical treatises on the nature of the observational processes by which the historian empirically obtains the printed symbols upon ancient documents or archaeological artefacts. Neither is a linguist questioned upon his empirical methods when examining the structure of and syntax of written languages and more significantly, the anthropologist who may need to combine both linguistic observation and cultural aspects of his subject matter in order to construct an appropriate *empirical* analysis.

In these illustrations it is assumed that interaction with the External World is perfectly understood and, not withstanding the philosophical and metaphysical conceptions of the nature of perception and the phenomenology thereof, which is a separate concern in our context, or if it is not, must be relevant to all fields and therefore entirely uniform in so far as philosophy is homogeneous over all categories, whether scientific or otherwise in this respect. But the point we wish to emphasise is that there exists an empirical process in *every* enterprise of *any* kind, because interaction with external objects or phenomena is unavoidable. Thus, the historian, the linguist, the anthropologist, the scientist and a myriad of others all observe and *interpret* the world in exactly the same manner and by exactly the same empirical process. This is the natural interpretation and yet again, applies universally to all sentient beings who necessarily participate in the Natural World merely by their existence therein. Once more the natural interpretation surfaces as the empirical process which, as we always knew, extends over all fields and disciplines, and not just the scientific domain. Then, there is no reason for further philosophical speculation upon this point, for it equally applies to philosophy.

Each of the foregoing are good reasons for attempting to codify the scientific method in a more precise form than has hitherto been available, but the latter one is of special significance. We have noted that to properly understand a discipline it is necessary to understand its methodology and in this sense the two cannot be considered to be truly separate subject areas. Hence, if a precise terminology and definitional system exists within a given discipline, but not within the methodology of that discipline, then a dichotomy results in relation to the comprehension of the field concerned or more specifically, there is an asymmetry of understanding that results under such circumstances. Conversely, a precise understanding of the methodology of a discipline without similar precision within the subject matter of the field concerned also results in a comprehension asymmetry and therefore incomplete understanding of the discipline as a whole. This may seem to be a little perplexing at first sight for surely, one may argue, if the precise methods and techniques employed within a given field are fully understood then so must the field in question be all the more clearly defined and thus *a fortiori*, understood on at least an equal level. Again, if the field of enquiry enjoys a high degree of exactitude then reciprocally its essential methodology should also inherit such precision. Obviously, this is not the case, for no such reciprocity relation of precision is either required or implied and does not exist as a logical dictate. Thus, the methods of psychology are often of a statistical nature which are completely understood and precisely defined, but psychology as a discipline certainly is not. Then, a field can only be completely appreciated when both its methodology and its subject matter simultaneously enjoy an equal degree of precision and this is why we regard an exact formulation of the methods of *any* discipline in unambiguous and well-defined terms to be of primary importance.

One of the reasons for this preoccupation is that a study of the methodology of specific disciplines may lead to a delineation of the scope and future prospects of that field. It will also, if sufficient precision is obtained, show how a particular field and the theoretical structures therein actually evolve and change over time and independently of the agents of such change. Once a clearly defined methodology becomes available, this evolution is made evident and hence, readily amenable to analysis, and it is this fact that allows determination of the scope and prospects of the relevant field, thus providing insight that was hitherto absent. Surely, this is an excellent reason for clarifying the methodology with a measure of exactitude appropriate to the discipline concerned.

Then, we may ask what we have achieved in our own deliberations as presented herein. We think that we have found one *possible* definitional system that has allowed us to discuss scientific revolutions in the context of *only* that system, but in so doing we have identified many salient points and relative to our system, have also resolved a number of philosophical and epistemological issues in the process. But, we have gone further than this, for we claim that *any* definitional system is essentially isomorphic to our own and by such an equivalence relation, the same philosophical problems are identically resolved in similar systems. It is not the specific definitions *per se* that admit of such analysis, but rather the properties of Nature that insist upon physical laws and relations between natural variables, thus requiring that a scientific theory consist of a finite set of axioms, which must then evolve by discrete change either individually or collectively. In other words, mechanisms of theory-change are inevitable for finite sets of this kind if they are to be modified and scientific 'revolutions' will result whatever system is devised. There is some degree of arbitrariness in the particular choice of definitions and mechanisms, but that hardly matters, since the operations are essentially the same in all cases and similar conclusions will be drawn as long as theories are axiomatic sets, or more pertinently, just as long as there are natural scientific laws; more specifically, the Laws of Physics.

In addition to this, there have been two fundamental guiding principles that we have emphasised *ad nauseam* throughout our work. The first is the empirical method which we have described in model-theoretic terms as the natural interpretation, and the second is the necessity to increase the domain of applicability of scientific theories. It is the latter that we have thought to be of the greatest significance in our context since it is just such an increase in domain that drives scientific progress and theory-evolution in the first instance, particularly with regard to unification. Theory unification is therefore seen as an inevitable process of scientific development and does so only by the requirement of domain enlargement. It is from these basic notions that we have attempted to build a picture of theory evolution and scientific revolutions in a more systematic manner than has thus far been portrayed. Whether we have succeeded is another matter and for the first time we use the second person singular directly and suggest that only *you* can decide.

APPENDIX

LIST OF DEFINITIONS

Definition 2.2.1
Given any axiom $\varphi = f(x_i, c_j)$, the domain of φ, $\mathcal{D}(\varphi)$, is precisely that subset of the universal domain $\mathcal{D}(\mathcal{N})$ for which the expression $f(x_i, c_j) = 0$ holds.

Definition 2.2.2
Given k axioms, φ_k, of a theory S, the domain of S, $\mathcal{D}(S)$, is the intersection of the domains of each of the axioms φ_k;

$$\mathcal{D}(S) = \mathcal{D}(\varphi_1) \cap \mathcal{D}(\varphi_2) \cap, \ldots\ldots\ldots, \cap \mathcal{D}(\varphi_{k-1}) \cap \mathcal{D}(\varphi_k)$$

Definition 2.2.3
By the *axiom of least domain* in a theory S we shall mean that axiom with domain such that its inclusion in the theory minimises the domain of the theory to its current domain and the removal of which from the theory results in an increase in the domain of the latter.

Definition 5.2.1
A *natural variable p* is a measurable physical quantity that changes with time, i.e. $dp/dt \neq 0$.

Definition 5.2.2
A *natural constant c* is a measurable physical quantity that does not change with time, i.e. $dc/dt = 0$.

Definition 7.2.1
A (minor) revolution of the first kind consists in the assignment of a new value to one or more of the constant symbols of the language.

Definition 7.2.2
A (minor) revolution of the second kind consists in the re-interpretation of one or more constant symbols of the language as variable symbols in which the algebraic or linguistic form of the axioms is unaltered, but the domain of applicability remains unchanged.

Definition 7.2.3
A (minor) revolution of the third kind consists in the re-interpretation of one or more constant symbols of the language as variable symbols in which the algebraic or linguistic form of the at least one of the axioms is altered, but the domain of applicability remains unchanged.

Definition 7.3.1
An (intermediate) revolution of the first kind consists in the modification or replacement of one or more axioms of a theory where the algebraic form of the latter is altered such that the domain of applicability of the theory is increased, the variables remain the same, but no assumptions of the theory are altered.

Definition 7.3.2
An (intermediate) revolution of the second kind consists in the modification or replacement of one or more axioms of a theory where the algebraic form of the latter is altered such that the domain of applicability of the theory is increased. New variables replace old variables but no assumptions of the theory are altered.

Definition 7.3.3
An (intermediate) revolution of the third kind consists in the unification of two or more theories such that the domain of applicability is the set-theoretic union of the domains of the component theories. The variables of all components are unaltered, but no assumptions of any of the component theories are altered.

Definition 7.4.1
A (major) revolution of the first kind consists in the modification or replacement of at least one of the fundamental assumptions of a theory such that that the assumptions that are replaced are logically inconsistent with those that replace them and the domain of applicability is increased.

Definition 7.4.2
A (major) revolution of the second kind consists in the replacement of all of the fundamental assumptions of a theory such that the assumptions that are replaced are logically inconsistent with those that replace them and the domain of applicability is not decreased.

Definition 7.4.3
A *fundamental assumption* of a scientific theory is a hypothesis without which the theory could not be formulated and the negation of which would negate the theory.

Definition 7.4.4
A major revolution of the third kind consists in the determination of new natural variables and the formulation of fundamental assumptions that may or may not incorporate existing ideas into an integrated scientific theory such that at least one new axiomatic expression of a relation between the natural variables can be constructed.

Definition 8.3.1
The *revolution frequency* f_n^T of a revolution of type R_n^T for a given revolution sequence is the ratio of the number of occurrences of R_n^T within the sequence to the total number of terms of the revolution sequence, i.e. $f_n^T = N_n^T / R_n$.

Definition 11.4.1
For any two theories S_i *and* S_j with fundamental assumptions h_i and h_j respectively, if $\neg(h_i \wedge h_j)$ holds for S_i and also for S_j then h_i and h_j are *inconsistent*.

Definition 11.4.2
For any two theories S_i and S_j with fundamental assumptions h_i and h_j respectively, if $h_i \wedge h_j$ holds for either S_i or S_j then h_i and h_j are *consistent*.

Definition 11.6.1
If $S = \{S_n : \forall n(S_n \subset N)\}$ is the set of all valid scientific theories of a given discipline $\mathbb{D}$ then if the unified theory $FT1 = \sqcup_n S_i$ has $|FT1| = 1$ and $\mathcal{D}(FT1) = \bigcup_n \mathcal{D}(S_i)$ then FT1 is a Final Theory of the first kind with respect to $\mathbb{D}$.

Definition 11.6.2

If $FT1_n = \{FT1_n : \forall n(FT1_n \subset N)\}$ is the set of all FT1 theories of a given set of disciplines $\mathbb{D}_n$ then if the unified theory $FT2 = \sqcup_n FT1_i$ has $|FT2| = 1$ and $\mathcal{D}(FT2) = \bigcup_n \mathcal{D}(FT1_i)$ then FT2 is a Final Theory of the second kind with respect to each of the $\mathbb{D}_n$.

Definition 11.6.3

A *Theory of Everything* (TOE) is an FT2 theory for which $\mathcal{D}(TOE) = \mathcal{D}(N)$.

Index

www.ingramcontent.com/pod-product-compliance
Lightning Source LLC
Chambersburg PA
CBHW022051210326
41519CB00054B/311